ADVANCED CARPENTRY & JOINERY

metric edition

By the same author

CRAFT TECHNOLOGY FOR CARPENTERS AND JOINERS

BUILDING GEOMETRY AND DRAWING

Advanced Carpentry & Joinery

metric edition

FRANK HILTON, A.I.B.I.C.C., A.R.S.H.

Senior Lecturer Building Department Bolton Technical College

With Drawings by the Author

LONGMAN

LONGMAN GROUP LIMITED
London

Associated companies, branches and representatives throughout the world

First published 1963
Second edition (Metric) 1969
Second impression 1970
Third impression 1973

cased ISBN 0 582 42516 6
paper ISBN 0 582 42515 8

Set in Monophoto Times New Roman and made and printed by offset in Hong Kong by Sheck Wah Tong Printing Press

Preface to First Edition

In 1946 Sir Winston Churchill warned the nation of the danger of falling behind the rest of the world in technical education.

Since the famous Fulton Woods speech, the Government has issued two White Papers on technical education, the first in 1956 and the second in 1961, the latter being entitled *Better Opportunities in Technical Education.* The Government's purpose in submitting this paper is stated in its concluding sentences:

> Our natural resources are limited, and in an increasingly competitive world our future as an industrial nation will depend upon our success in developing the native skills of our young people. The system of technical education will only be able to make its full contribution if it is geared to the rapidly changing requirements of modern industry. It must also provide every student with the best opportunities for making the most of his ability and aptitudes.
>
> A heavy responsibility rests upon all who are concerned with technical education and training, whether in technical college or industry. The Government believe that the reorganization proposed should help greatly in vigorous development of technical education for the technicians, craftsmen, and operatives of the future, whose skill, knowledge and adaptability are as much wanted by the nation as a whole as the development of their capacities and interests is essential to their own satisfaction as individuals.

Detailed proposals in the White Paper included: Craft courses will be modified, both by broadening them and by providing suitable terminating points for the competent craftsman on the one hand and the exceptionally gifted craftsman on the other.

To this end the City and Guilds of London Institute's revised schemes in the building crafts represent a new approach to the educational requirements of building craft apprentices and of those journeymen who aim to become craft or general foremen. A series of courses and examinations is provided which will be within the capacity of and of direct interest and relevance to young men at their various stages of development, and which will prepare them for the positions they are likely to occupy as each stage of the training has been completed.

The schemes include a 'Craft Certificate Course' designed for the average craft apprentice, and also cater for the potential first-class craftsman, the craft foreman and the general foreman. The needs of the latter are met by progression through the Advanced Craft Certificate course to a Full Technological Certificate course with extension course, and then to General Foremanship studies or to the advanced years of the Higher National Certificate in building.

An endeavour has been made to provide in this book a fairly complete course of instruction in Advanced Carpentry and Joinery, to meet the requirements of the above courses for the student who can achieve a high and comprehensive standard of proficiency and understanding in his craft, and has given evidence of capacity to profit from further study.

A special feature has been made of the illustrations used to convey as much information as possible, facing in most cases the relevant text which has been kept to a minimum.

Every effort has been made to cover the syllabuses of the City and Guilds and the Regional Unions and students preparing for the revised examinations of the Institute of Builders and Certified Carpenters should also find this work useful.

That there are many existing books on this subject is a well-recognized fact; that some of them have excellent matter, and others have good illustrations is also duly acknowledged, but from a wide experience in building and many years in teaching and examining, the author has been forced to the conclusion that something more in the way of text-books is needed in the light of new syllabuses, new techniques and modern practice.

Bolton 1962 *Frank Hilton*

Preface to Metric Edition

Following a report advocating the early adoption by the United Kingdom of both decimal currency and a metric system of weights and measures, the President of the Board of Trade announced in 1965 that:

> The Government consider it desirable that British industries on a broadening front should adopt metric units sector by sector, until that system can become in time the primary system of weights and measures for the country as a whole.

As a result of this decision the British Standards Institute has prepared a programme for the change to the metric system in the Construction Industry. The programme provides that drawings and documents will begin to be produced in metric terms in 1969 and that in 1970 builders will change to construction based on those drawings and documents. A similar changeover in syllabuses, courses of instruction and examinations also requires to be programmed to include greater emphasis in the use of metric units.

The techniques described in this book have not altered substantially in the short time that has elapsed since publication of the first edition. Hence, apart from minor amendments and the inclusion of the new Building Regulations affecting stairs, the preparation of this new edition concerns the conversion from Imperial to metric terms.

Metric values below one metre (m) are expressed in millimetres (mm). The full stop is used throughout this work as the decimal marker. It is used on illustrations to separate metres from millimetres. Thus two metres one hundred millimetres is written 2.100, the figure could also be read as two thousand one hundred millimetres.

In conclusion, the author desires to express his gratitude for the friendly and appreciative reception accorded to his labours by his critics and the public, and trusts this new metric edition will prove as useful as its predecessors.

Bolton 1968 *Frank Hilton*

Acknowledgements

Sincere thanks are tendered to the following firms and organizations for their courtesy and help in supplying valuable technical information, and for their permission to include in this book, extracts from publications and reproduce photographs:

Acrow Ltd., Robert Adams Ltd., A. H. Anderson Ltd., Austins Ltd., S. N. Bridges & Co. Ltd., Black & Decker Ltd., British Equipment Co. Ltd., Centec Machine Tools Ltd., Commercial Secretary, High Commissioner for Canada, P. G. Henderson Ltd., William Kay (Bolton) Ltd., Linden Doors Ltd., Plywood Manufacturers Association of British Columbia, Stanley Works (Gt. Britain) Ltd., Timber Research and Development Association, Tomo Trading Co. Ltd., William Newman Ltd., Westland Engineers Ltd.

F. H.

Contents

List of Illustrations

Chapter One: Carpentry

Site Hoardings

When alterations to existing buildings are to be carried out and where new work is to be developed involving site excavations, temporary hoardings are required to be erected to protect, not only the site and building materials, but also the general public to ensure their safety at all times.

Hoardings erected on large contract sites, which may be in position for some considerable time, are constructed to serve not only as protection whilst work is in progress, but used also as a means of advertising. Usually special provision is made for displaying posters and painted signs on framed signboards.

For smaller types of work, such as shop alterations, hoardings are much simpler in construction. Sheet 1.1 shows such a hoarding where sleepers may be used as uprights, bolted to stumps buried 610 mm in the ground, well tamped or concreted, in pre-prepared holes. A normal height is 2.100 m, with rails 100 mm by 50 mm in thickness, to which close boarding, or alternatively corrugated sheeting is nailed. The posts should be plumbed and the top edge of the boarding cut to a line to ensure a neat finish or finished with a capping mould as shown. Access for materials and personnel is provided by a lift on and off gate, or a hinged door, provided with hasp and staple and padlock. For double gates to admit vehicles, the posts should be of extra size to carry the weight of the gates and should be set deeper in the ground. A head piece is sometimes used to tie the two posts, providing there is sufficient headroom for vehicles to pass below.

Sheet 1.2 shows a section through a hoarding similar in construction, 2.400 m high, using 150 mm by 75 mm posts, the covering being sheets of 2.400 by 1.200 m exterior grade plywood, nailed to the rails.

Sheet 1.3 shows a temporary or permanent hoarding, 3.600 m high, with timber posts bolted to concrete stumps concreted in the ground. A hoarding of this height is subject to considerable wind pressures and must be suitably fixed and braced to withstand this. Sheet aluminium is used for the covering, providing a good finished surface to the hoarding.

A similar hoarding is shown in Sheet 1.4, incorporating 1.800 m high signboards. An alternative ground fixing is shown, where mild steel channel sections are concreted in the ground and the posts bolted in the channel through the web. Corrugated sheeting is used on the lower portion of the hoarding.

When part of a street pavement is to be taken up with a hoarding during building alterations, a type which may be employed is shown in Sheet 1.5. Where steel scaffolding is to be used in the work, this may be used as a fixing for the hoarding, although bracing back to the building can often be done. A timber baulk serves as a sole piece for the scaffold and a fixing for the base of the hoarding. Hook bolts fixed round the tubular steel scaffold and bolted through the 100 mm by 50 mm posts, fix the hoarding to the scaffold.

With a pavement hoarding of this kind, fanguards or splayed boardings are often fixed to the top of the hoarding, similar to those in gantry construction. The purpose of the fanguard is to give further protection to pedestrians from falling debris. Bearers are nailed to the sides of the posts and the boards fixed to them. Further stiffening of the guard is obtained by fixing bracing in the form of raking struts at 2.400–3.000 m intervals.

Pavement hoardings are required to have red warning lights placed at each end and along the hoarding during darkness, to indicate a narrowing of the pavement and obstruction. They are often colour washed or painted.

Temporary Site Buildings

The number and size of temporary buildings will be decided by the size of the contract, the number of men employed and the size of the site. A medium-sized contract requiring supervisory staff on the site would expect to have offices varying in size from 3.600 m by 2.400 m to 7.200 m by 3.600 m for clerk of works, agent or general foreman, engineer and quantity surveyor. A site office, stores huts, possibly a mess hut and workmen's lavatories are other site buildings required to be erected.

Generally these are of sectional construction in interchangeable units or panels of a convenient width for easy handling and quick erection, timber being the usual material employed.

The siting of offices, stores, etc. is most important. They should be erected in positions where they are to remain throughout the job, preferably near to the front of the site, so that they are easily accessible from the road for unloading stores, etc.

On confined sites, the huts may have to be erected on the site of the actual building or placed on a gantry over the pavement or on a steel scaffolding platform.

Offices should be well lighted by windows and electricity, positioned with

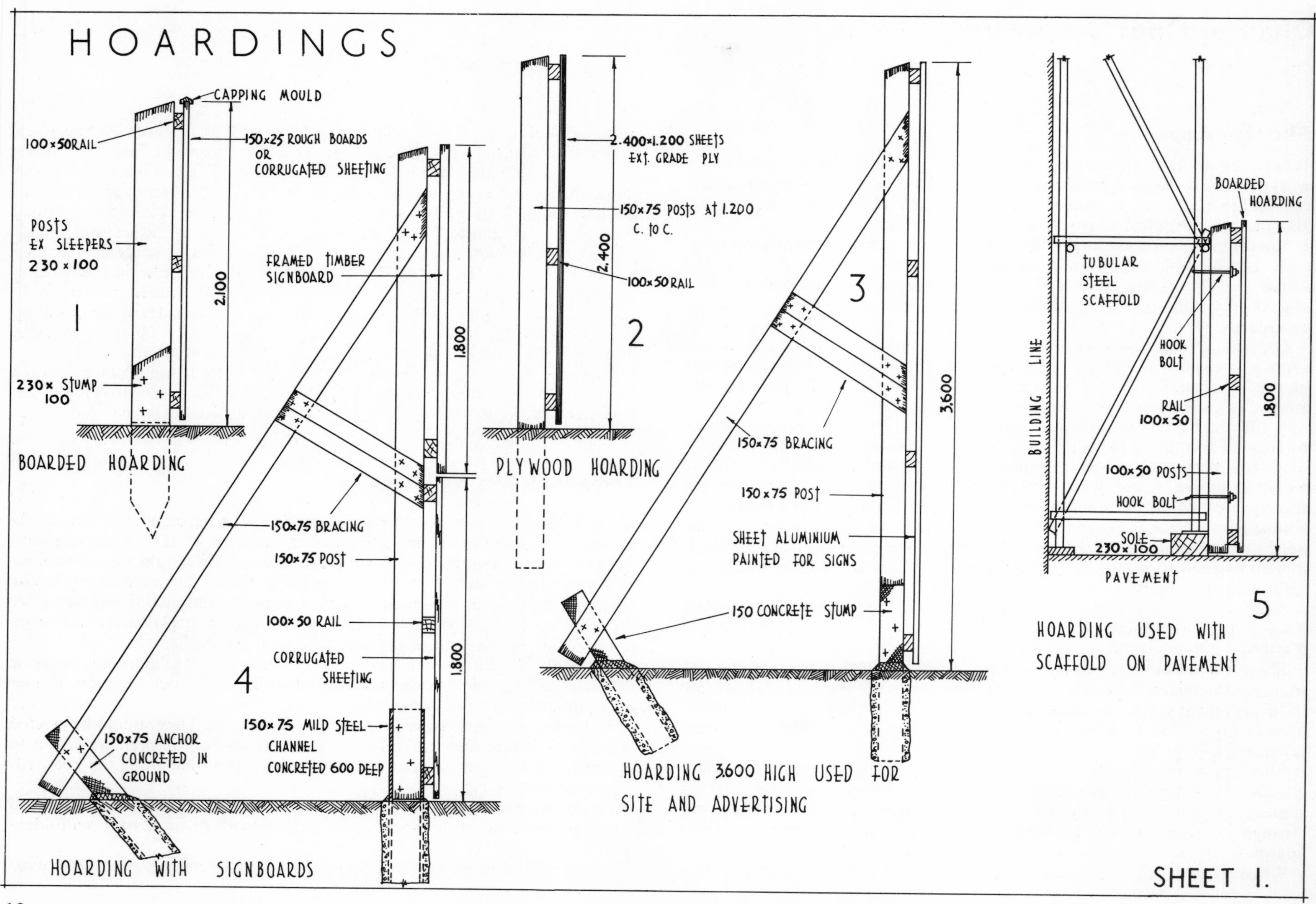
HOARDINGS
CAPPING MOULD
100×50 RAIL
150×25 ROUGH BOARDS
OR
CORRUGATED SHEETING
POSTS
EX SLEEPERS
230 × 100
2.100
1
230× STUMP
100
BOARDED HOARDING
2.400×1.200 SHEETS
EXT. GRADE PLY
150×75 POSTS AT 1.200
C. TO C.
2.400
100×50 RAIL
2
PLYWOOD HOARDING
FRAMED TIMBER
SIGNBOARD
1.800
150×75 BRACING
150×75 POST
100×50 RAIL
CORRUGATED
SHEETING
4
1.800
150×75 MILD STEEL
CHANNEL
CONCRETED 600 DEEP
150×75 ANCHOR
CONCRETED IN
GROUND
HOARDING WITH SIGNBOARDS
3
3.600
150×75 BRACING
150×75 POST
SHEET ALUMINIUM
PAINTED FOR SIGNS
150 CONCRETE STUMP
HOARDING 3.600 HIGH USED FOR
SITE AND ADVERTISING
BOARDED
HOARDING
TUBULAR
STEEL
SCAFFOLD
HOOK
BOLT
RAIL
100×50
1.800
BUILDING LINE
100×50 POSTS
HOOK BOLT
SOLE
230×100
PAVEMENT
5
HOARDING USED WITH
SCAFFOLD ON PAVEMENT
SHEET 1.

the windows overlooking the site. A typical example of a clerk of works office is illustrated on p. 14. This is equipped with a desk provided with drawers for plans, a cupboard and wide level-topped desk. A wash bowl and provision for some form of heating are other requirements.

Sheet 2.1 shows the arrangement of the framing of the panel sections. These are 1.200 m wide, as are the floor and roof sections, so that the lengthening or shortening is easily done, when the bolt-hole positions are bored to a standard rod. For an office of this size, the type of roof shown in Sheet 2.3 is suitable. This is formed using 230 mm by 38 mm roof bearers tapered from the centre to 100 mm at the eaves. 75 mm by 50 mm blocking pieces are nailed between the roof bearers to provide for the fixing of the roof sections by bolts to the wall frames. A soffit and fascia boards are fixed to the bearers to complete the eaves finish shown in Sheet 2.6.

The detail at the floor is shown in Sheet 2.7 where sleepers are used as a base for the sectional floor. Where vertical boarding to the hut is used, this is allowed to cover the boards of the floor to ensure a watertight job.

Horizontal weather-boards are often used on site buildings, and where this is the case, the boards are set in from the edge of the vertical studs and a fillet nailed in after erection.

The jointing at the corner between a side and end is detailed in Sheet 2.4 with a detail of the arrangement at the doorway shown in Sheet 2.5. The framed door is set in flush with the inside of the office with planted door stops and threshold as shown. An inside lining to the office has not been included in the drawing; where this is used, it provides insulation and a better appearance over bare framing. It also provides a good flat surface for pinning up progress charts and plans, etc.

For offices of larger spans than the one illustrated, roofs of steeper pitch in sections are generally used. These may have light roof trusses introduced in them to support the purlins and boards, or boarded partitions introduced instead. The covering is usually a good quality felt stripped with narrow battens with a capping at the ridge.

Sheet 2.8 shows a sketch of an alternative design for a sectional site office. The single pitched roof is in two sections and overhangs the front forming a canopy. The cladding is of horizontal red cedar weather-boarding. The flush door opens out with opening lights above.

Timbering for Excavations

This work is rarely carried out by carpenters, but is so intimately connected with building that its superintendence will sooner or later fall within the duties of those to whom this book appeals. In continuing earlier work in timbering for trenches, it is proposed to cover the timbering for deep trenches, basements and shafts.

The excavation of earth for this type of work may be taken out by hand or by mechanical shovel.

Larger buildings having lower floors and basements, such as stores and factories, require foundations which are much deeper than those for normal building work. The walls have to be strong enough to carry the weight of the superstructure and resist the lateral earth pressures. The timbering for deep trenches in work of this kind is assembled in situ and inserted stage by stage as the excavating proceeds.

In trenches above 1.800 m deep, or in loose soil, the sides of the trench should be close boarded, Sheet 3.1. It is important to prevent the escape of earth between the boards; this is liable to take place after heavy rains and any movement behind the timbering could result in the collapse of the supports, with danger to men working in the trench. For this reason timbering must be inspected regularly by a competent person and action taken immediately there are any signs of movement.

The poling boards are held in position by walings 230 mm by 75 mm in section. Struts are placed between the walings at 1.800 m intervals, so as not unduly to impede the excavation work. Trenches that have to remain open for any length of time should be more heavily timbered than those to be refilled quickly. Puncheons are used with heavy walings in deep trenches placed vertically between them for extra support. A continuous plank spiked to the walings at intervals prevents any collapse, should there be any loosening of the struts and wedges. Stagings for the passing up of earth have been omitted from the drawing.

Sheet 3.2 shows the timbering for a large trench or wide excavation. The timbers are somewhat heavier than those shown in the previous example. 38 mm poling boards or runners 3.000 m long and shod with iron or pointed are used. As much of the earth is taken out as is possible without the sides of the excavation falling in. Runners are then inserted, waled and strutted. The struts, which are 230 mm by 230 mm, are heavy to handle and require either lip blocks or props to support them. They are placed in position and wedges driven at the ends. Puncheons support the walings with struts propped where necessary.

When the ground has been excavated to a depth of the runners, a second system of runners is placed slightly in advance of the former, and the ground excavated as before. Earth removal is done by using buckets lifted by crane and deposited on the site or into vehicles and tipped.

Sheet 3.3 shows the timbering to a basement excavation. Poling boards supported by walings are held in position by inclined shores. The feet of the latter abut against a sole piece. Sheet 3.4 shows the timbering to a shaft. These are seldom over 1.800 m square and are required for piers, foundations, etc.

The earth is excavated to a depth of 1.200 m or 1.500 m and the sides of the

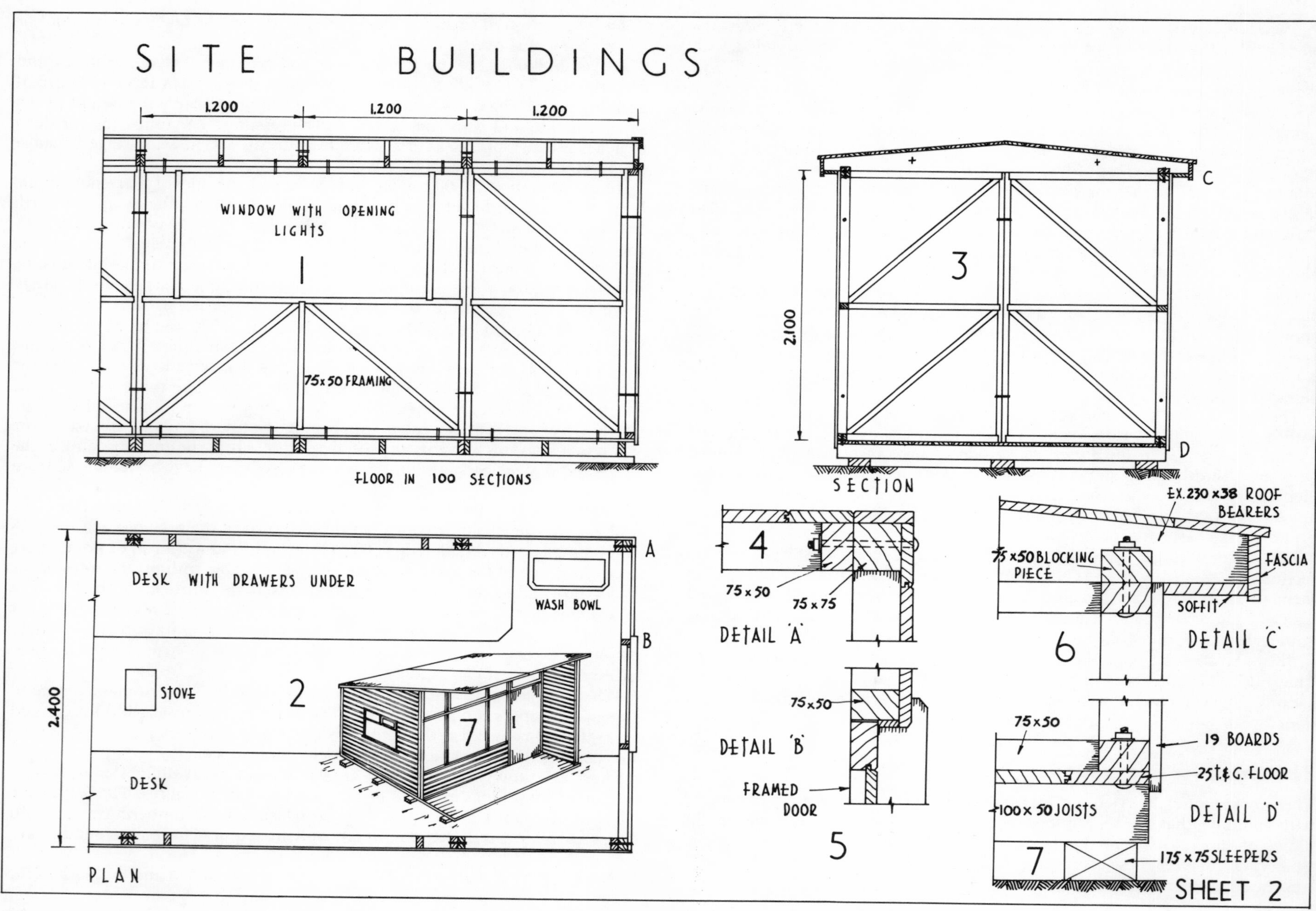
SITE BUILDINGS
1.200
1.200
1.200
WINDOW WITH OPENING LIGHTS
1
75x50 FRAMING
FLOOR IN 100 SECTIONS
3
2.100
C
D
SECTION
DESK WITH DRAWERS UNDER
WASH BOWL
A
B
STOVE
2
7
2.400
DESK
PLAN
4
75 x 50
75 x 75
DETAIL 'A'
75x50
DETAIL 'B'
FRAMED DOOR
5
EX.230 x38 ROOF BEARERS
75x50 BLOCKING PIECE
FASCIA
SOFFIT
6
DETAIL 'C'
75x50
19 BOARDS
25T.&G. FLOOR
100x 50 JOISTS
DETAIL 'D'
7
175 x 75 SLEEPERS
SHEET 2

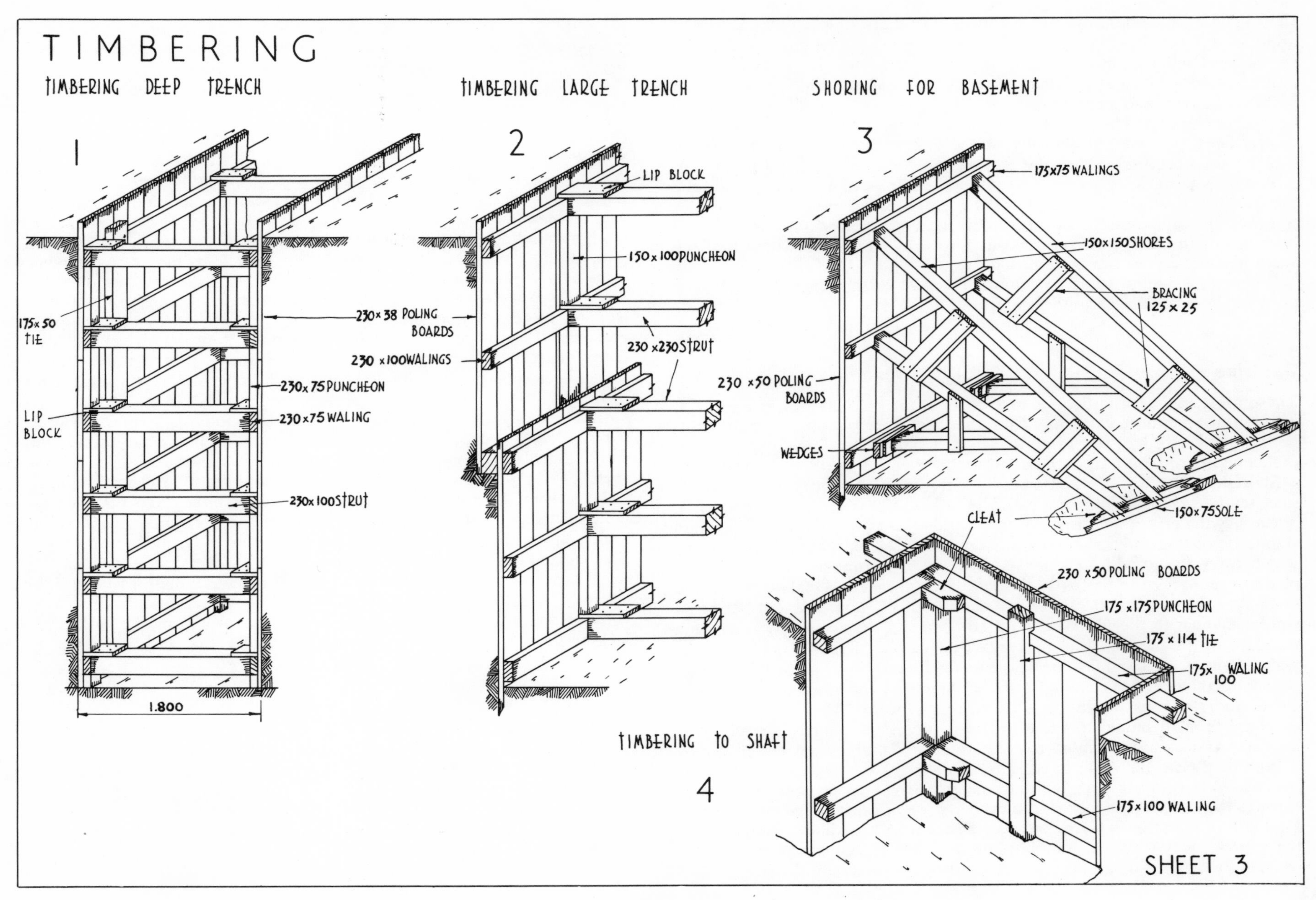
TIMBERING
TIMBERING DEEP TRENCH
TIMBERING LARGE TRENCH
SHORING FOR BASEMENT
1
2
3
175x 50 TIE
230×38 POLING BOARDS
230x75 PUNCHEON
LIP BLOCK
230x75 WALING
230x100 STRUT
1.800
LIP BLOCK
150 x 100 PUNCHEON
230 x 230 STRUT
230 x 100 WALINGS
175x75 WALINGS
150x150 SHORES
BRACING 125 x 25
230 x 50 POLING BOARDS
WEDGES
150x75 SOLE
CLEAT
230 x 50 POLING BOARDS
175 x 175 PUNCHEON
175 x 114 TIE
175x 100 WALING
175x100 WALING
TIMBERING TO SHAFT
4
SHEET 3

shaft lined with 230 mm by 50 mm boards. These are supported and strutted apart by 175 mm by 50 mm walings placed opposite each other and at the same level. On the opposite sides further walings are driven in position to form a frame. Cleats nailed at the corners secure the members. Further excavation is then taken out and a second system of poling boards placed in. Frames of walings are separated by puncheons in the corners of the shaft and a tie at the centre. The top waling projects beyond the shaft excavation on each side and forms a good bearing on the solid ground.

Where buildings are to be erected on the banks of rivers or near tidal waters, it is necessary to adopt other methods than those of timbers or trenches. In such cases, the site may be enclosed by interlocking steel sheeting piles driven through the river deposits into firm ground below, so that the earth inside the piles can be removed.

Where it is not possible to remove the soft earth, timber piles having pointed shoes are driven until firm ground is reached.

Gantries

These are temporary structures erected in front of buildings in course of erection or alteration, extending from the face of the building over the footpath to the kerb edge. The structure acts as a staging for unloading and storing materials, and may accommodate a scaffold erected as the work proceeds.

Sheet 4.1 shows a part elevation and section of such a gantry with a tubular steel scaffold erected from the platform. As will be seen from the section, Sheet 4.2, the erection consists of two parallel rows of 230 mm by 230 mm standards resting upon runners and supporting 230 mm by 230 mm heads, which carry the joisted and planked decking. The runners are first laid in position spanning the footpath, the standards placed at 3.000 m centres are dogged to the runners, the heads are placed on the standards and fixed in a similar manner to the sill. Where a joint occurs in the head, it is made on a standard with a short piece of timber or bearing plate fixed between the head and the top of the standard, the whole being dogged together. Struts 150 mm square are then cut and fixed between the standards, the lower ends resting on cleats and the tops butting against straining beams. 175 mm by 50 mm diagonal bracing bolted to the standards stiffens the gantry laterally. The platform or decking is supported on 230 mm by 75 mm joists with 230 mm by 75 mm deals laid flat edge to edge. Scaffold planks are often used for this purpose in two layers with building paper between to prevent dust falling through the joints. A fanguard is required to prevent workmen or debris of any kind falling from the platform to the street below. A hoarding is fixed beneath the gantry, away from the building line, to allow workmen to work, and is provided with a lift on and off gate or door to give access to the site. The reduced footpath may be planked for pedestrians and to prevent any persons stepping into the roadway, a 100 mm by 75 mm handrail in forked cleats is used. A fender or baulk 300 mm by 300 mm is placed against the curb or against the outside frame, to prevent any damage to the gantry from vehicles in the roadway.

Scaffolds

Although scaffolding cannot be called work of the carpenter, timber scaffolding is rapidly being replaced by tubular steel scaffolds. Economy, light weight, ease and speed of erection are some of the advantages of the latter. On large modern buildings, scaffolding forms a most important part of the work and is normally carried out by specialist firms. Scaffolds are of two kinds: (1) bricklayer's; (2) mason's.

Sheet 4.2 shows a mason's scaffold consisting of putlogs or transoms, ledgers, standards and bracing. Decking, guard boards and guard rails complete the scaffold. Base plates, joint pins, transom, swivel and double couplers are used at positions shown in the drawing. This scaffold is called an independent scaffold and should be tied back to the building through window openings.

In a bricklayer's scaffold, only the row of standards farthest away from the building is used with putlogs flattened at the end, secured into the brickwork joints, thus doing away with one row of standards. Sheet 4.3 shows sketches of the types of couplers and joint pin used in steel scaffolding.

To prevent lateral movement and to give a more even distribution of material loads on the scaffold, diagonal bracing is required. Scaffolds over long frontages must have diagonal bracing to prevent swaying by high winds. This is fixed at the angles and at the centre by swivel couplers.

All types of scaffolding should conform to the regulations of the Home Office. The safety of the workmen must be ensured at all times so that working widths of platforms, permissible loads, provision of guard rails and kicking boards, are specifications which must be adhered to.

Shoring

Shoring is the term given to the temporary framing providing support to structures which show signs of instability, or which are liable to disturbance through alterations, or the removal of some of their supports for repairs.

The form of shore applied to a building which shows some sign of movement or bulging due to settlement is known as a raking shore.

Shores are known as systems, single when only one raker is used, double when two rakers are used and so on.

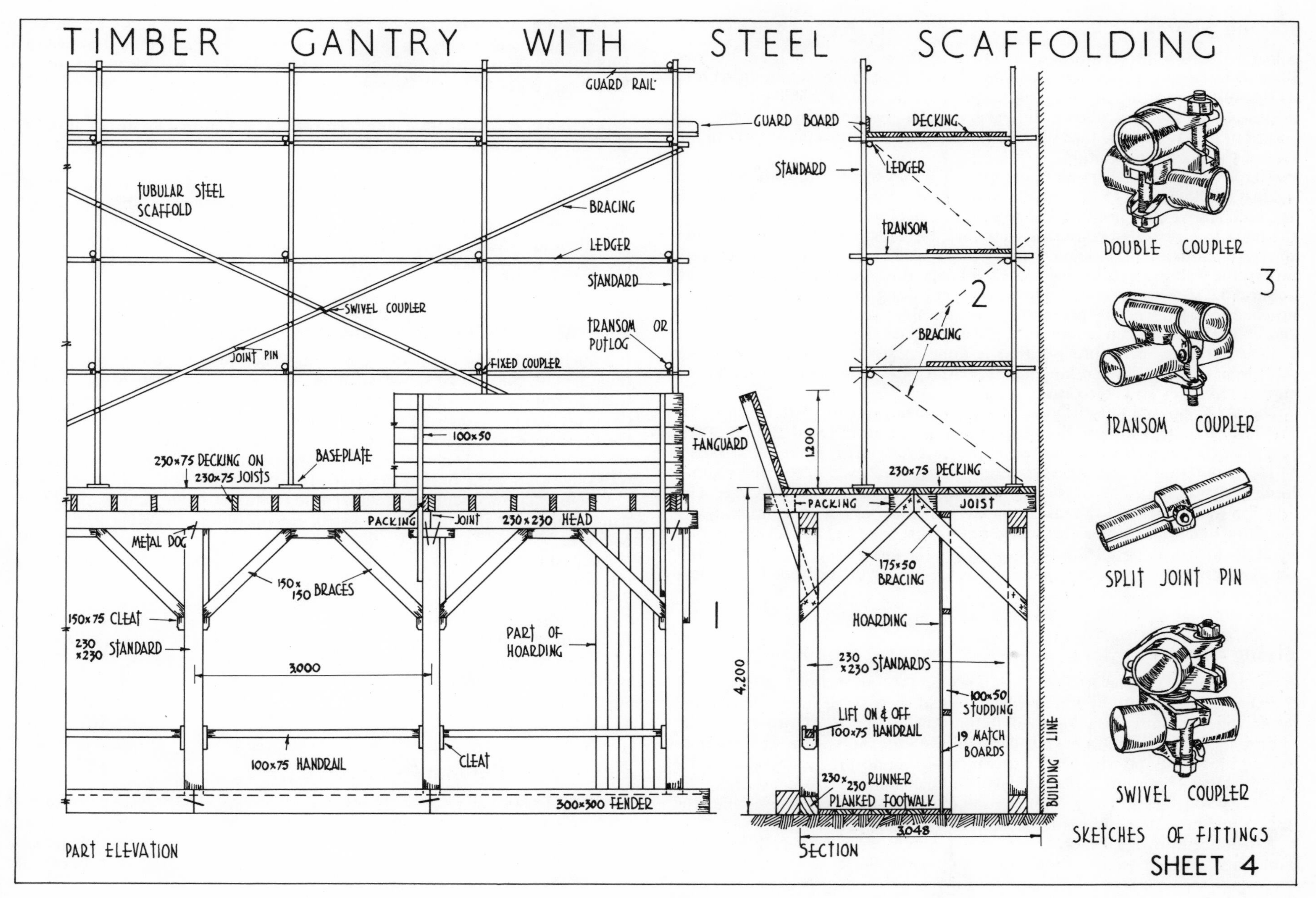
TIMBER GANTRY WITH STEEL SCAFFOLDING
GUARD RAIL
GUARD BOARD
TUBULAR STEEL SCAFFOLD
BRACING
LEDGER
STANDARD
SWIVEL COUPLER
TRANSOM OR PUTLOG
JOINT PIN
FIXED COUPLER
100×50
FANGUARD
230×75 DECKING ON 230×75 JOISTS
BASEPLATE
PACKING
JOINT
230×230 HEAD
METAL DOG
150×150 BRACES
150×75 CLEAT
230×230 STANDARD
PART OF HOARDING
3,000
100×75 HANDRAIL
CLEAT
300×300 FENDER
PART ELEVATION
1
DECKING
STANDARD
LEDGER
TRANSOM
2
BRACING
1200
230×75 DECKING
PACKING
JOIST
175×50 BRACING
HOARDING
230×230 STANDARDS
4,200
100×50 STUDDING
LIFT ON & OFF 100×75 HANDRAIL
19 MATCH BOARDS
230×230 RUNNER
PLANKED FOOTWALK
BUILDING LINE
3,048
SECTION
DOUBLE COUPLER
3
TRANSOM COUPLER
SPLIT JOINT PIN
SWIVEL COUPLER
SKETCHES OF FITTINGS
SHEET 4

Raking Shores

Sheet 5.1 shows the elevation of a system of three rakers. As these timbers have to resist compressive stresses they should be as nearly square in section as possible. The group of shores are received at the foot by a sole piece which prevents slipping and allows the rakers to be tightened by levering into position. Cleats and distance pieces are spiked into the sole piece to keep the feet of the rakers in position.

The heads of the rakers abut against a 230 mm by 50 mm wall plate and against a 100 mm by 75 mm needle which passes through the wall piece into the wall. Above each needle a 100 mm by 75 mm cleat is housed and spiked to the plate to give added resistance to the needle. To stiffen the rakers making the system into a unit, bracing is used being nailed as shown across the rakers and to the wall piece at approximately 90 degrees to the outer raker.

Sheet 5.2 shows the detail at the head of the raking shore which is housed or notched round the needle preventing any lateral movement at the head of the shore. This can be seen clearly in the sketch, Sheet 5.3.

Levering the shores into position is done by using a crowbar in a slot mortise at the foot of the raking member. Metal dogs are then driven on each side of the raker and sole piece.

Sheet 5.4 shows a sketch at the foot of a raker and the detail of a rider. For lofty buildings an additional member is often introduced to reduce the length of the outer raker which would be necessary. This is termed a rider.

The inclination of the rakers depends on the area available at the ground level, but normally any angle between 45 and 60 degrees is suitable. The position for the head of the rakers is determined by taking the centre lines of the floor and wall to give the centre line of the raker as at 'A', Sheet 5.5. Here the joists run parallel to the wall. When the joists are carried by the wall the centre line of the raker is taken from the corner of the floor joist, Sheet 5.6.

Flying Shores

Sheet 5.7 shows the elevation of the shores used to support two parallel walls 7.800 m apart in lofty buildings where a road or passage runs between and where one or both show signs of failure. They consist of two horizontal shores, termed flying shores, cut between the two walls with wall plates to spread the thrust and with needles and cleats to support the timbers. They are stiffened by inclined struts placed above and below the horizontal shores at an angle of 45 degrees and by 150 mm by 150 mm posts between. These stiffen the shores and provide more points of support to the walls.

When one horizontal shore is used the system is called a single flying shore, and when two are used as shown, a double flying shore.

The method of dealing with the struts, needles, cleats and wall plates is clearly shown and similar to that described for raking shores, but notice that with the lower struts and shore the positions of needles and cleats is reversed. Straining pieces are used for the struts to bear against the wedges used for tightening and easing the system.

A detail of the horizontal shore with supports and wedges at 'D' is shown in Sheet 5.8, while a sketch of the arrangement of members at 'C' is shown in Sheet 5.9.

The spacing of both raking and flying shores will depend on certain site conditions, such as where windows and doorways occur, but they are usually spaced about 3.000 m apart. The hoisting into position of these shores, especially long rakers, requires special care. Mobile cranes, where these can be employed, are the answer, otherwise ropes and ladders have to be used.

Shoring for Buildings of Unequal Height

It sometimes happens that two buildings which require shoring are not the same height. Sheet 6.10 shows such an example where a single flying shore with two raking shores has been used. The positions for placing the needles in the walls and for mortising the wall plates will be found in the same way as the previous examples. When possible, the horizontal shore should be fixed against cross walls. The method of dealing with the struts, needles, cleats and wall plates is similar to that described for raking shores.

Where window openings occur and any signs of movement in the brickwork are visible, it is advisable to strut the openings to prevent any further collapse.

Internal floors and the roof may also require support when strutting of the interior from floor to floor down to a solid foundation is used.

Where baulk timbers of the required length are not available, 175 mm by 75 mm timbers of random lengths may be bolted together so that the joints end on are staggered. 230 mm by 75 mm timbers with 25 mm packing pieces bolted between may be used as an alternative shore. If this method is adopted packings must be placed at the head and foot of the shores to provide a good solid bearing.

Shoring for Column Removal

Sheet 6.11 shows a method of shoring when it is necessary to remove and replace a column supporting a series of arches in the cloisters of a church. Strong centres are constructed using 100 mm thick ribs, cut from the plank or laminated, and fitted accurately to the contour of the arch on each side of the

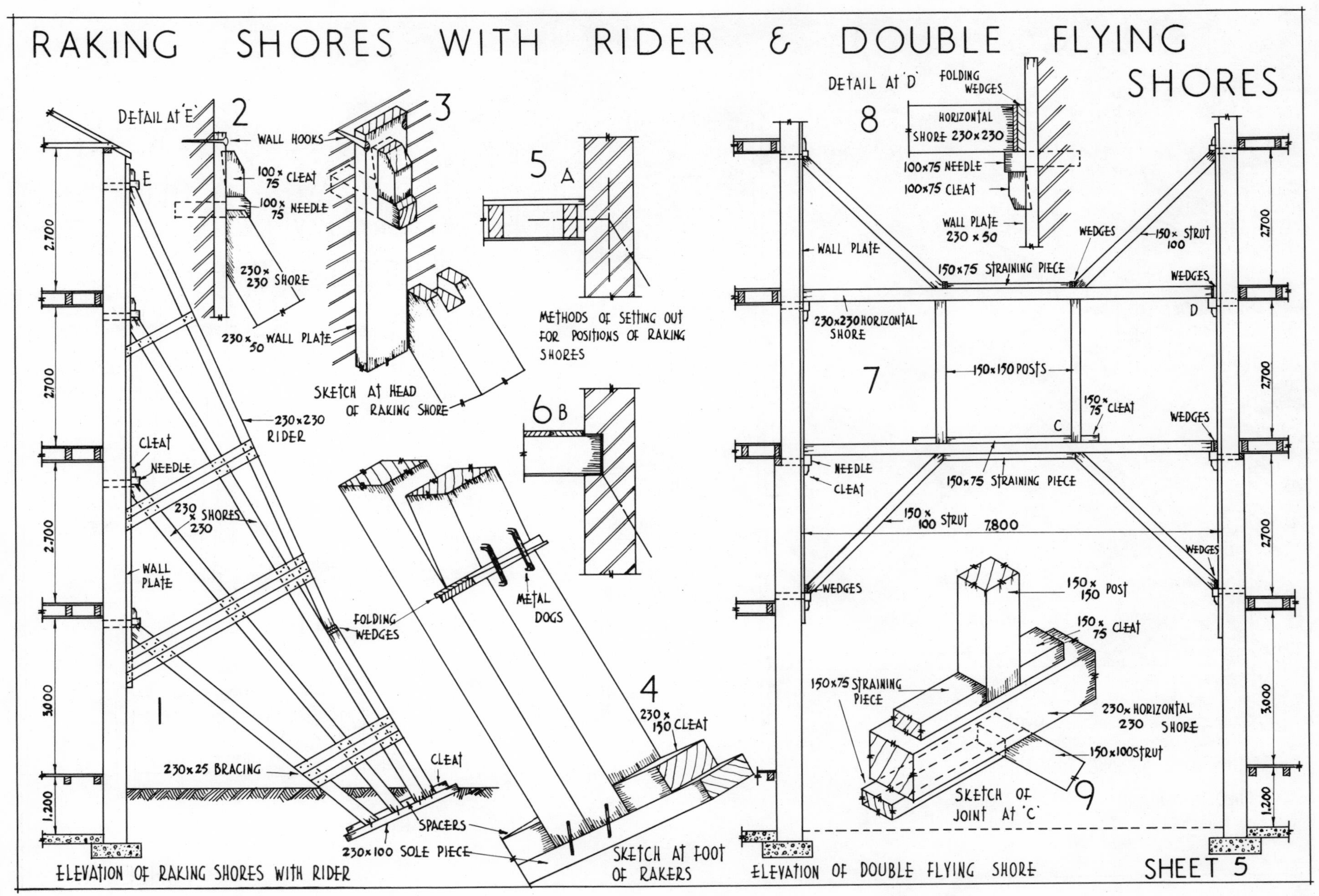
RAKING SHORES WITH RIDER & DOUBLE FLYING SHORES
DETAIL AT 'E'
2
WALL HOOKS
100 × 75 CLEAT
100 × 75 NEEDLE
230 × 230 SHORE
230 × 50 WALL PLATE
E
2.700
2.700
2.700
3.000
1.200
3
SKETCH AT HEAD OF RAKING SHORE
5 A
6 B
METHODS OF SETTING OUT FOR POSITIONS OF RAKING SHORES
230×230 RIDER
CLEAT
NEEDLE
230 × 230 SHORES
WALL PLATE
FOLDING WEDGES
METAL DOGS
1
4
230 × 150 CLEAT
230×25 BRACING
CLEAT
SPACERS
230×100 SOLE PIECE
SKETCH AT FOOT OF RAKERS
ELEVATION OF RAKING SHORES WITH RIDER
DETAIL AT 'D'
8
FOLDING WEDGES
HORIZONTAL SHORE 230×230
100×75 NEEDLE
100×75 CLEAT
WALL PLATE 230 × 50
WALL PLATE
WEDGES
150 × 100 STRUT
150×75 STRAINING PIECE
WEDGES
D
230×230 HORIZONTAL SHORE
7
150×150 POSTS
150 × 75 CLEAT
C
WEDGES
NEEDLE
CLEAT
150×75 STRAINING PIECE
150 × 100 STRUT
7.800
WEDGES
WEDGES
150 × 150 POST
150 × 75 CLEAT
150×75 STRAINING PIECE
230 × 230 HORIZONTAL SHORE
150×100 STRUT
SKETCH OF JOINT AT 'C'
9
ELEVATION OF DOUBLE FLYING SHORE
SHEET 5

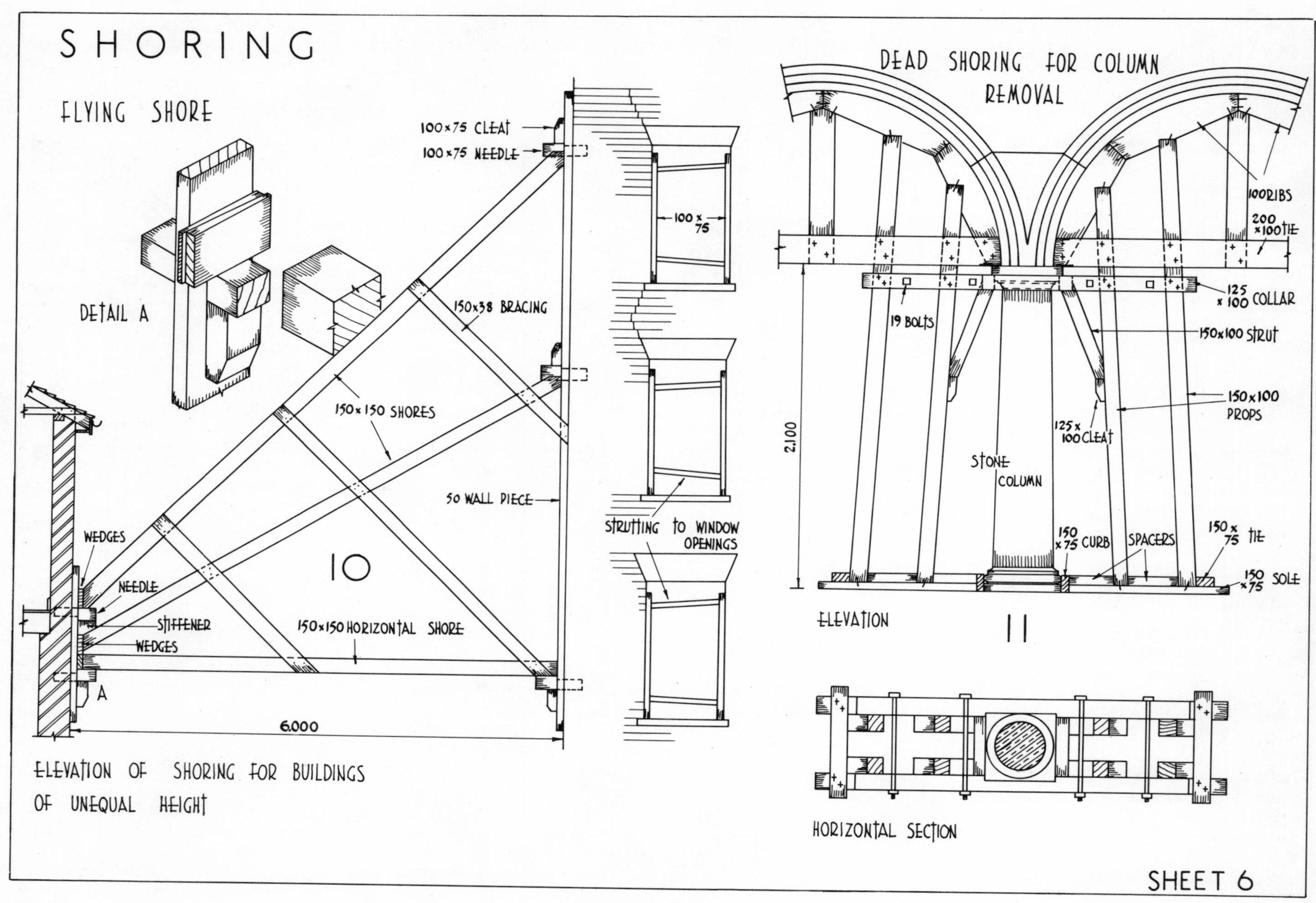
SHORING
FLYING SHORE
DETAIL A
100×75 CLEAT
100×75 NEEDLE
100 × 75
150×38 BRACING
150×150 SHORES
50 WALL PIECE
STRUTTING TO WINDOW OPENINGS
WEDGES
NEEDLE
STIFFENER
WEDGES
150×150 HORIZONTAL SHORE
A
6,000
10
ELEVATION OF SHORING FOR BUILDINGS OF UNEQUAL HEIGHT
DEAD SHORING FOR COLUMN REMOVAL
100RIBS
200 ×100TIE
125 × 100 COLLAR
19 BOLTS
150×100 STRUT
150×100 PROPS
125× 100CLEAT
2,100
STONE COLUMN
150 ×75 CURB
SPACERS
150× 75 TIE
150 ×75 SOLE
ELEVATION
11
HORIZONTAL SECTION
SHEET 6

defective column. The butt joints between the ribs, cut radiating with the centre, should fit perfectly and are held together by iron dogs. Shores 150 mm by 100 mm are then cut as shown fitted into the ribs and resting on sole pieces. Folding wedges are not used, the shores being pinched up tight with a crow-bar with spacers spiked in position. Ties are then placed across the sole pieces and fixed. Horizontal ties are bolted on each side of the rakers and 125 mm by 100 mm collars bolted together and to the rakers, keep the capital in position. Struts are then cut tight under the collar. The rakers are pitched to about 85 degrees, and sufficient room is allowed to enable workmen to carry out the removal and reinstatement of the column.

Although the centres are much more substantial than the ones used originally to construct the arches, they have to support, in addition to the arches, the weight above them. The timber sizes would therefore depend on these factors.

Dead Shoring

The shoring of a wall and floors while the lower part of the wall is removed, is termed dead shoring. Two vertical timbers, one inside the building and one outside, support a needle which passes through the wall. Window openings are strutted and floors propped separately, to avoid the weight of the floors being transferred to the main shores. These measures are taken to ensure the safety of the building while the operation of shoring and work on the main wall is carried out.

Dead shores are placed at about 1.800 m intervals and are of square section, usually 230 mm by 230 mm with needles the same size.

Centring

Temporary timber structures placed under arches, domes, vaults, etc. during their construction are termed centres. They must be constructed sufficiently strong to support temporarily the load to be imposed, without distortion, and require to be carefully and rigidly put together. The construction of centres differs widely according to span, shape, width of soffit and the material of which the arches are to be constructed. Centres consist of ribs shaped to the curved soffit of the intended arch, together with ties and braces arranged to give the strongest possible form of structure. The chief consideration in the design of a centre must be to ensure that the centre cannot alter its shape, but is kept rigid during the construction of the arch.

A number of the various outlines for the more generally used forms of centres, together with centres for special types of arches are shown in the following illustrations:

Sheet 7.1 shows a centre designed for a semicircular arch in which the jambs and soffit are splayed at 60 degrees. This centre is for a 230 mm brick arch over the entrance to a building. The plan shows the ribs and the splay of the arch at the springing. The centre is close lagged for a gauged brick arch, and supported on props. Because the arch is splayed at the crown as at the springing, the centre is part of the frustum of a cone.

Sheet 7.2 shows the plan and elevation of a centre for an arch where the jambs only are splayed, the soffit being level at the crown. This causes the arch to be elliptical on one face and semicircular on the other. It will be seen that the ribs in this and the previous example are built up.

The following examples deal with centres of various types having solid ribs:

Sheet 7.3 shows a framed centre for a semi-elliptical arch of 9.000 m span. The design is suitable for bridge construction where support for the arch under construction is taken at the sides only, and the main tie has been raised considerably in order to give increased headroom, allowing traffic to pass beneath. The ribs are built up of solid members as shown and held together at the joints by iron bolts, 'dogs' and metal straps. With the advent of timber connectors much of the elaborate jointing and ironwork can be dispensed with, giving a saving both in material and labour. The ribs are placed at 1.200 m centres with suitable bracing freely used to prevent any lateral movement. 125 mm by 75 mm close lagging for gauged brick arches is used. For stone arches the lags are arranged in pairs, one on each side of the position to be occupied by the joints in the arch. The centres are adjusted and eased when the arch is set by means of large hardwood folding wedges on the top of a 230 mm by 75 mm head, carried on the supports.

Provision must be made for the gradual easing of centres so that the arch will take its bearings gradually. If the centres are heavy the wedges should be greased, alternatively screw jacks may be used.

Positioning of such heavy centres on the site is done either by mechanical crane or by derrick and lifting tackle. Care should be taken when fixing the slings to pick up on the tie and not at the crown, otherwise the centre may be deformed.

Sheet 7.4 shows the plan and elevation of a circle-on-circle centre for a window or door opening, in which the jambs are parallel and the soffit of the arch is level at the crown.

The shapes of the inner and outer ribs are identical, but the true shape of these and the soffit mould has to be determined. The method of dealing with this has been covered elsewhere in this book and will not be repeated here. When the ribs are sawn out, the edge should be planed in such a way that it is level at the top, but gradually bevels until at the springing it is at the angle shown in the plan. Sheet 7.5 shows the construction of the centre with part of the lagging removed.

Sheet 8.6 shows a sketch of framed centring suitable for arch construction

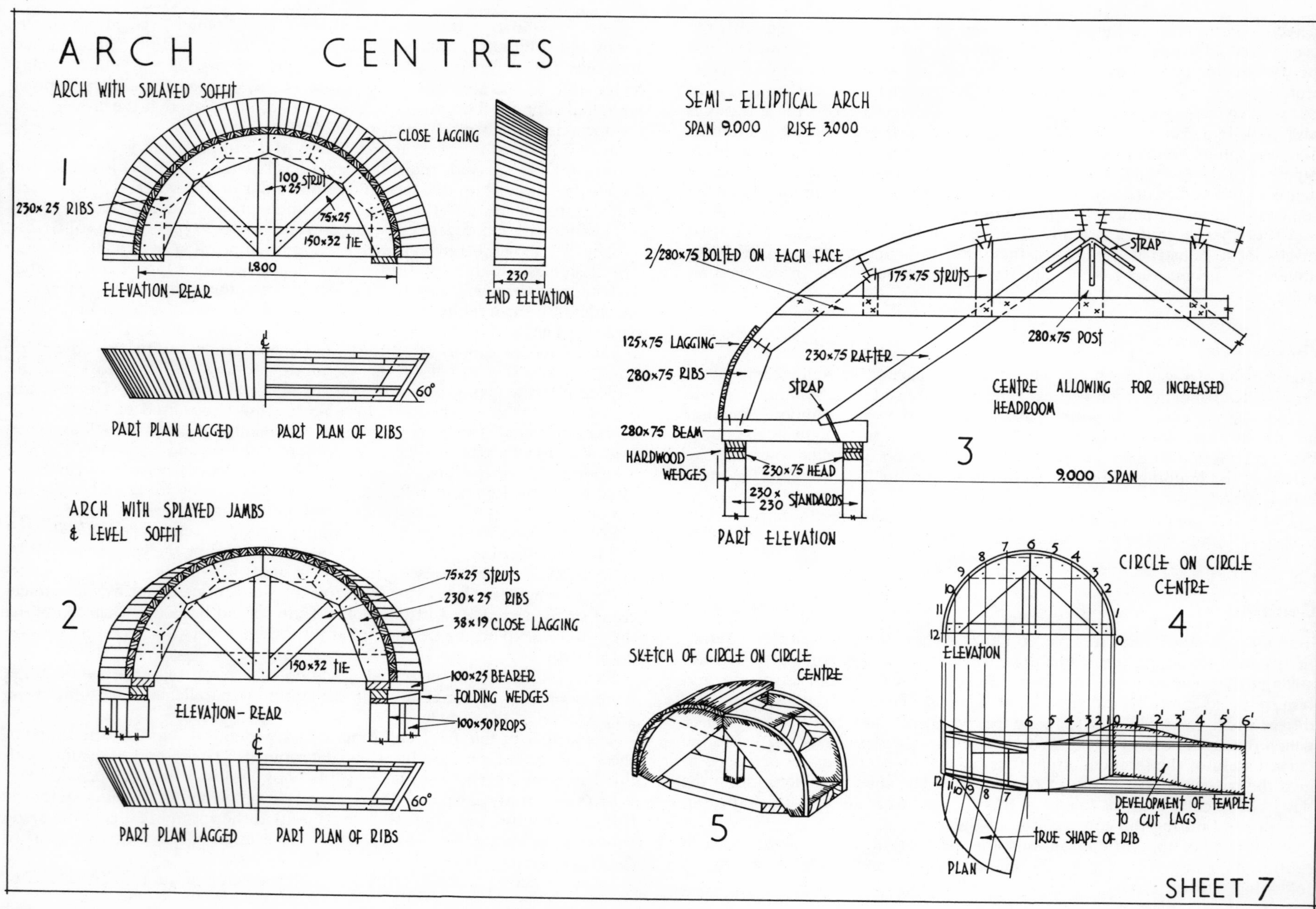

ARCH CENTRES
ARCH WITH SPLAYED SOFFIT
1
CLOSE LAGGING
230×25 RIBS
100 ×25 STRUT
75×25
150×32 TIE
1.800
ELEVATION–REAR
230
END ELEVATION
PART PLAN LAGGED
PART PLAN OF RIBS
60°
SEMI-ELLIPTICAL ARCH
SPAN 9.000 RISE 3.000
2/280×75 BOLTED ON EACH FACE
175×75 STRUTS
STRAP
280×75 POST
125×75 LAGGING
280×75 RIBS
230×75 RAFTER
STRAP
CENTRE ALLOWING FOR INCREASED HEADROOM
280×75 BEAM
HARDWOOD WEDGES
230×75 HEAD
3
9.000 SPAN
230×230 STANDARDS
PART ELEVATION
ARCH WITH SPLAYED JAMBS & LEVEL SOFFIT
2
75×25 STRUTS
230×25 RIBS
38×19 CLOSE LAGGING
150×32 TIE
100×25 BEARER
FOLDING WEDGES
ELEVATION–REAR
100×50 PROPS
PART PLAN LAGGED
PART PLAN OF RIBS
60°
SKETCH OF CIRCLE ON CIRCLE CENTRE
5
CIRCLE ON CIRCLE CENTRE
4
ELEVATION
DEVELOPMENT OF TEMPLET TO CUT LAGS
TRUE SHAPE OF RIB
PLAN
SHEET 7

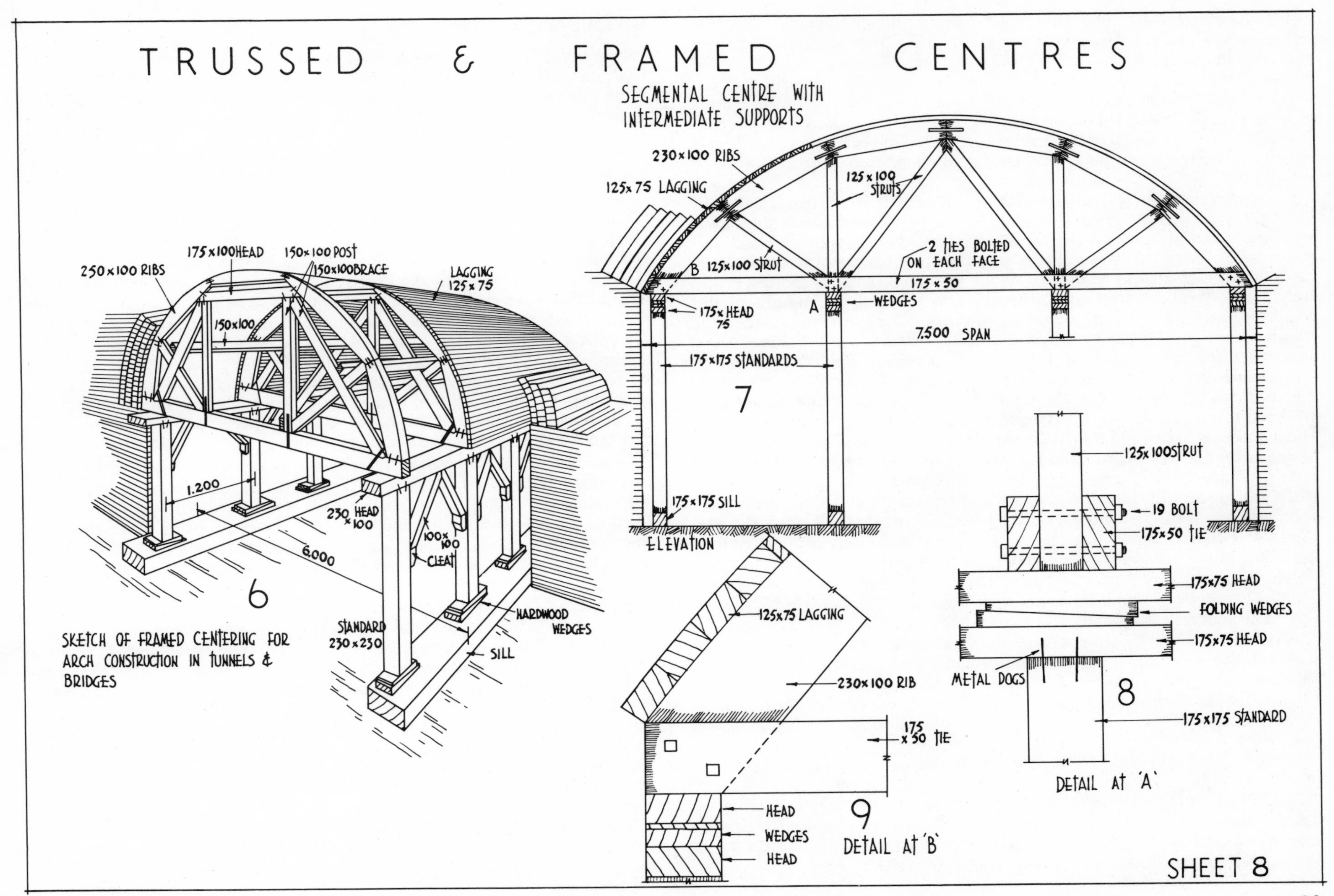
TRUSSED & FRAMED CENTRES
SEGMENTAL CENTRE WITH INTERMEDIATE SUPPORTS
230×100 RIBS
125×75 LAGGING
125×100 STRUTS
B 125×100 STRUT
2 TIES BOLTED ON EACH FACE
175×50
WEDGES
175×HEAD 75
A
7.500 SPAN
175×175 STANDARDS
7
175×175 SILL
ELEVATION
175×100HEAD
150×100 POST
150×100BRACE
250×100 RIBS
LAGGING 125×75
150×100
1.200
230 HEAD ×100
6.000
100×100 CLEAT
HARDWOOD WEDGES
STANDARD 230×230
SILL
6
SKETCH OF FRAMED CENTERING FOR ARCH CONSTRUCTION IN TUNNELS & BRIDGES
125×75 LAGGING
230×100 RIB
175 ×50 TIE
HEAD
WEDGES
HEAD
9
DETAIL AT 'B'
125×100STRUT
19 BOLT
175×50 TIE
175×75 HEAD
FOLDING WEDGES
175×75 HEAD
METAL DOGS
8
175×175 STANDARD
DETAIL AT 'A'
SHEET 8

in tunnels and bridges. The centres are trussed and the members held together at the joints using iron straps, dogs and bolts. They are placed at 1.200 m centres and supported on a 230 mm by 100 mm head with 230 mm by 230 mm standards braced as shown with angle braces. It will be seen that the wedges for easing the centres are placed at the sill and not under the centres as in the previous and following examples. This method is seldom used in practice, a gradual easing of the centres before striking being more easily and readily carried out when the wedges are placed under each centre, easing each one in turn.

A centre suitable for a segmental arch or bridge of 7.500 m span is shown in Sheet 8.7. The centre is supported using 175 mm by 175 mm standards at each end, and at two intermediate points. When intermediate supports are used the construction of the centre can of course be much lighter than a centre without these supports. It should be noted that the load on a centre is greatest at the crown, but during the actual construction of the arch the weight of the voussoirs bears most of their weight downwards at the springing. As the arch rises more weight is taken by the centre, while at the crown the voussoirs are almost bearing their weight on the centre at that point. The arrangement of the members at the crown requires special care. The construction of the centre shown shows solid ribs out of 230 mm by 100 mm, struts 125 mm by 100 mm arranged to give an outline of triangulation as the strongest form of centre. The tie consists of two 175 mm by 50 mm timbers bolted one on each face with 19 mm bolts through the ribs and the struts. Timber connectors may be used at these points. The ties serve to stop the centre spreading under load while the struts and braces stiffen the ribs and ties, and transmit loads from one point to another, so that the stress is evenly distributed throughout. The centres are placed at 1.200 m centres and fixed to a 175 mm by 75 mm head, extending the width of the bridge; these rest upon pairs of folding hardwood wedges, which in their turn are supported by a similar horizontal plate or head supported by 175 mm by 175 mm standards. These rest on a 175 mm by 175 mm sill and are well braced diagonally.

A detail at the foot of one centre showing the wedging and supports is shown in Sheet 8.8 with a detail at the springing shown in Sheet 8.9.

Sheet 9.10 shows a centre for a segmental arch of 12.000 m span where support for the arch is taken at the sides only. Here again the tie has been raised considerably in order to give increased headroom, allowing traffic to pass beneath.

Sheet 9, figs. 11, 12 and 13 show the details of construction and supporting of the centre.

Sheet 9.14 shows the elevation of a segmental centre where alternative designs and construction have been shown. On the left, the centre is self-supporting with side supports only used, whilst to the right, the centre is shown with intermediate supports over a span of 9.000 m.

Spectators' Stands

Where sports meetings, street processions, royal occasions, etc., are to be held, elevated timber spectator stands are often used. The construction of these will vary slightly, as the stand may be required for one occasion only, or for a longer period. As they are of a temporary character, they are termed temporary stands.

For football and cricket grounds, race-courses, etc., permanent roofed-in structures are required which may be constructed in timber, steel or concrete, or a combination of these. This type of stand requires careful design, and where timber is used they are usually prefabricated in large sections for erection on the site by mechanical means. These are termed permanent stands.

With temporary stands, the construction consists of an undercarriage of triangulated construction with plain butted joints, spiked or dogged with bolted diagonal bracing. This is most important where the stand is erected in the open and exposed to the force of the wind. Sleepers may be used to form a foundation and the stand securely anchored to the ground by driving stakes on each side of the runner piece, and connecting them by a tie, over the top of the runner. The inclination of the stringer is usually determined by the area of the site or the height of the stand. An allowance of 400 mm per person is allowed for seating with a gangway of 300 mm.

With permanent structures, frame construction using timber connectors, iron straps and bolts is adopted. A high factor of safety must be allowed in calculating the timber sizes. When stands are loaded with people who are likely to become excited at some incident, stresses are likely to be great and not easily determined. Hence, stands are always made strong enough and adequately braced to prevent any collapse which might result in injury or loss of life.

Frames are placed at 1.500 m centres with cross-bracing between them and anchored with steel angles and rag bolts to concrete foundations. Timber connectors are used at the joints between tier frames, stringers and bracing members. The ends of the stand are boarded up to a height of 900 mm on 100 mm by 50 mm framing and finished with a capping. The back also is boarded. When the stand covers a large area, stairs at each end as well as at the centre should be used. These should be much stronger than the seating part of the stand, because the load is frequently concentrated at these points, whilst it is equally distributed elsewhere. Handrails must also be provided at the stairs and in large stands, at other positions, to divide the stand up to help control in case of panic.

Sheet 10.1 shows the section of a temporary stand, the construction being similar to that already described. Each tier is 600 mm wide with a rise of 375 mm, a section of a tier is shown in Sheet 10.2. This shows timber connectors

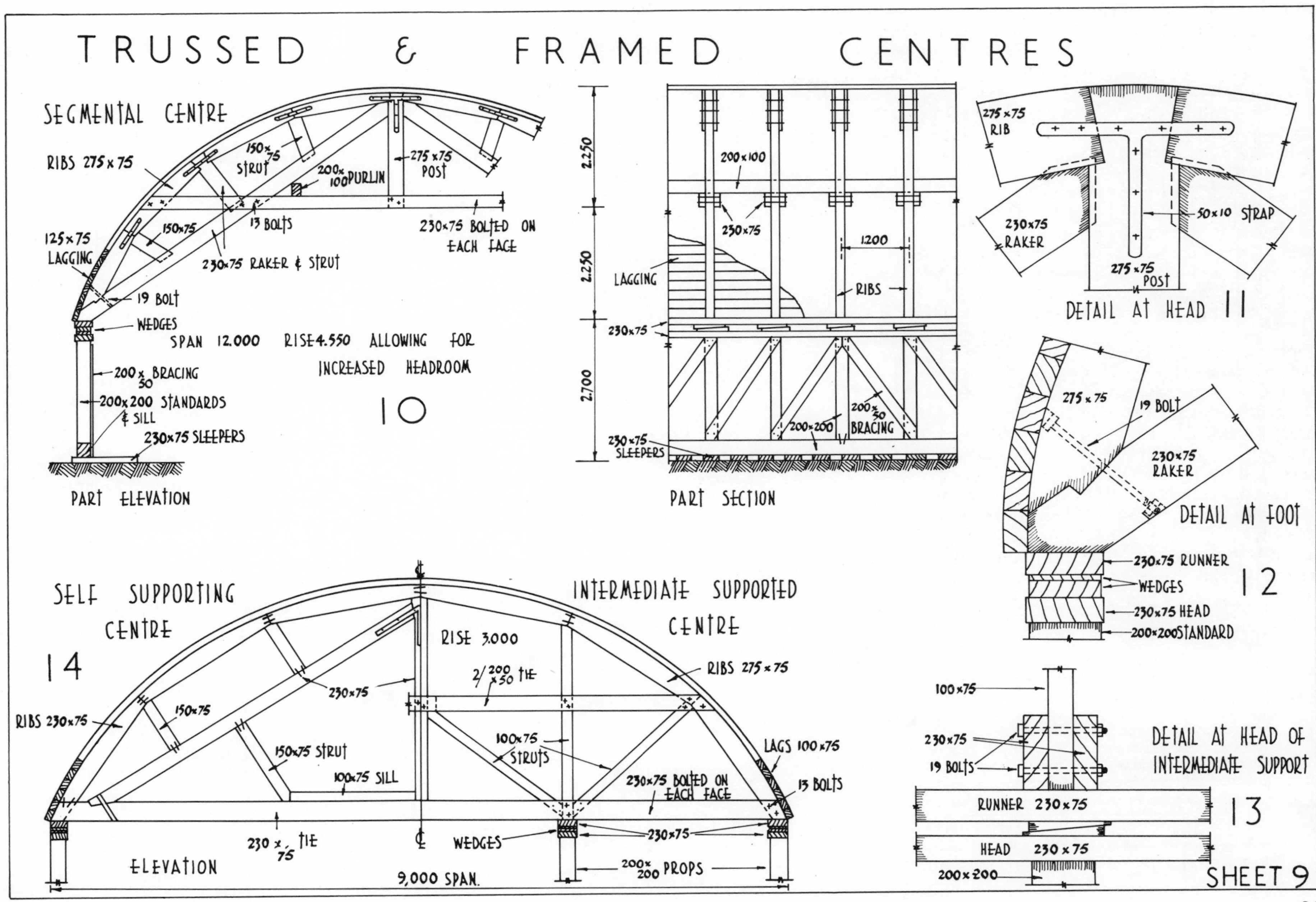
TRUSSED & FRAMED CENTRES
SEGMENTAL CENTRE
RIBS 275 x 75
150x75 STRUT
200x100 PURLIN
275 x 75 POST
13 BOLTS
230x75 BOLTED ON EACH FACE
125x75 LAGGING
150x75
230x75 RAKER & STRUT
19 BOLT
WEDGES
SPAN 12.000 RISE 4.550 ALLOWING FOR INCREASED HEADROOM
200x50 BRACING
200x200 STANDARDS & SILL
230x75 SLEEPERS
10
PART ELEVATION
2.250
2.250
2.700
200x100
230x75
1200
LAGGING
RIBS
230x75
200x200
200x50 BRACING
230x75 SLEEPERS
PART SECTION
275x75 RIB
50x10 STRAP
230x75 RAKER
275x75 POST
DETAIL AT HEAD 11
275x75
19 BOLT
230x75 RAKER
DETAIL AT FOOT
230x75 RUNNER
WEDGES
230x75 HEAD
200x200 STANDARD
12
SELF SUPPORTING CENTRE
14
INTERMEDIATE SUPPORTED CENTRE
RISE 3.000
2/200x50 TIE
RIBS 275x75
230x75
RIBS 230x75
150x75
150x75 STRUT
100x75 SILL
100x75 STRUTS
LAGS 100x75
230x75 BOLTED ON EACH FACE
13 BOLTS
230x75 TIE
WEDGES
230x75
ELEVATION
200x200 PROPS
9,000 SPAN.
100x75
230x75
19 BOLTS
DETAIL AT HEAD OF INTERMEDIATE SUPPORT
RUNNER 230x75
13
HEAD 230x75
200x200
SHEET 9

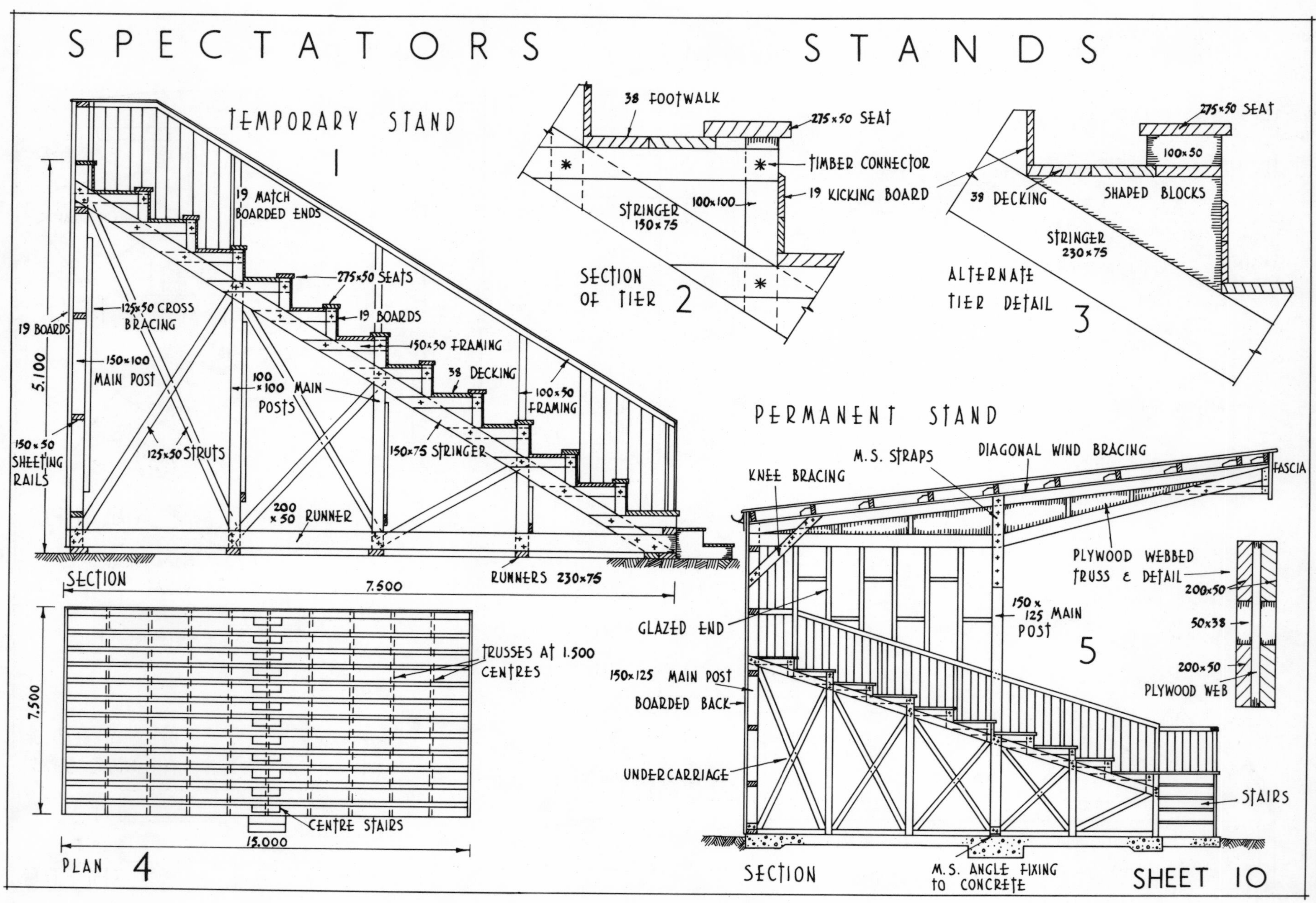

SPECTATORS STANDS
TEMPORARY STAND
1
19 MATCH BOARDED ENDS
275×50 SEATS
19 BOARDS
150×50 FRAMING
38 DECKING
100×50 FRAMING
125×50 CROSS BRACING
19 BOARDS
5.100
150×100 MAIN POST
100 ×100 MAIN POSTS
150×50 SHEETING RAILS
125×50 STRUTS
150×75 STRINGER
200 ×50 RUNNER
RUNNERS 230×75
SECTION
7.500
38 FOOTWALK
275×50 SEAT
TIMBER CONNECTOR
19 KICKING BOARD
STRINGER 150×75
100×100
SECTION OF TIER 2
275×50 SEAT
100×50
38 DECKING
SHAPED BLOCKS
STRINGER 230×75
ALTERNATE TIER DETAIL 3
PERMANENT STAND
M.S. STRAPS
DIAGONAL WIND BRACING
KNEE BRACING
FASCIA
PLYWOOD WEBBED TRUSS & DETAIL
200×50
50×38
200×50
PLYWOOD WEB
150× 125 MAIN POST
GLAZED END
5
150×125 MAIN POST
BOARDED BACK
UNDERCARRIAGE
STAIRS
SECTION
M.S. ANGLE FIXING TO CONCRETE
TRUSSES AT 1.500 CENTRES
7.500
CENTRE STAIRS
15.000
PLAN 4
SHEET 10

used at the jointing of the framework and the stringer. A planked footwalk 38 mm thick is nailed to the 100 mm by 50 mm frames with a 50 mm seat on top. A kicking board nailed to the frames completes the tiers.

An alternative tier detail is shown in Sheet 10.3, shaped triangular blocks are fixed to the stringers with the decking nailed to them. The seat is raised above the decking on 100 mm by 50 mm blocks, allowing room for the feet of the people sitting behind to pass beneath. Sheet 10.4 shows the plan of the stand.

A permanent stand requires a roof covering of some kind. Whatever form this may take, careful design by a competent person is necessary and most important. Since fabrication and construction are the concern of the joiner and not design, a possible roof arrangement is shown in Sheet 10.5.

Plywood web cantilever trusses of nailed construction support the roof members with the covering of tongue and grooved boards and felt finish, or aluminium sheeting, nailed to purlins. For protection against the weather, the ends of the stand are usually partly boarded and the remainder glazed, for good visibility. Trusses are spaced at 4.500 m centres with diagonal wind bracing between each truss. Knee bracing is fixed to each side of the trusses as shown. A detail of the truss is shown in Sheet 10.6.

Chapter Two: Carpentry

Formwork

The preparation of formwork or shuttering for steel frame and reinforced concrete construction has become a most important aspect of the carpenter's work. In the setting out, making and erecting of the formwork, care is needed to see that it can be easily stripped after concreting, otherwise added expense to the contractor is inevitable.

Concrete work is divided into two main groups known as precast and cast *in situ*. Examples have been selected of work classed in both groups to show the preparation and erection of the various mould boxes and forms.

The appearance of formwork is of little importance, and in many cases the timber used need not be cut to exact lengths, the ends being left long to avoid waste. The timbers generally used are softwoods, such as spruce and deal and are best wrought on all four sides. They will then be of uniform size and more easily adaptable for different jobs. The thickness of boards used in any one size is most important, a uniform thickness allows the boards to match up. Much time is wasted in having to plane down joints if boards are not thicknessed.

Timber partially seasoned is often regarded as best for formwork, as if it is too dry it will swell from moisture absorption. The reverse is the case in hot weather if the timber is green. Seasoned timber will require well wetting before the concrete is placed, so that moisture will not be absorbed from the wet concrete.

When a high finish to the concrete is required, the inside faces of the forms are lined with exterior grade plywood or other suitable kinds of hardboard. If the concrete is not to be plastered, the surfaces of all forms coming into contact with the concrete should be oiled with a mould oil to prevent the concrete from sticking to the face. Soft soap solution, lime wash and diesel oil are alternative treatments.

Formwork, in addition to being sufficiently strong and rigid enough to support the dead load of the concrete, must be designed to allow for the temporary live load of workmen wheeling barrows and the tamping of concrete. In all formwork wire nails should be used and in some cases their heads left protruding for easy withdrawal.

Wire ties are used for holding shutters through walls by being strained against distance pieces, Nos. 8, 9, 10, 11 and 12 s.w.g. soft black iron wire being common sizes.

Bolts 13 mm to 19 mm are generally used in heavy work. They are always withdrawn after striking the formwork, and should be greased for this purpose, or fitted with cardboard sleeves.

Precast Units

The requirements of moulds for manufacturing precast units is that they should each be used many times to produce units of the required size and shape. The work is either carried out on the site or in the shop.

Sheet 11.1 shows a mould box suitable for the repetitive casting of beams, posts or lintels. The boxes are constructed to take to pieces easily for stripping and jointed so that they can be quickly reassembled. The ends of the mould box are housed into the sides and the divisions housed into the ends. 19 mm bolts are used to tie the box and purpose-made wooden straps fitted over the top of the box as shown, to prevent spreading. An alternative corner jointing of the box is shown in Sheet 11.2.

An alternative mould box for precast beams is shown in Sheet 11.3.

Sheet 11.4 shows a mould box for the repetition casting of concrete panels.

A small-scale drawing, Sheet 11.5, shows the section of a concrete sill. The mould box for casting the sills is shown in the accompanying sketch.

Precast Cornice

Sheet 12.6 shows the section of a moulded cornice. In preparing a mould box for a unit of this form to stand the same way as the section, a shaped screed is necessary to produce the weathering on the top side. It is better to design a box to enable the casting to be removed by inverting the mould, permitting the sections to be drawn away at right angles to the faces of the moulded concrete without damage. This is shown in the detailed section, Sheet 12.7. The box is constructed as for plain beams, except for the triangular fillet forming the weathering. Thinner sheeting boards are used at this point in order that the fillet can be recessed and not run out to a feather edge. The units of the moulded portion of the cornice are built up out of separate pieces as shown and nailed together.

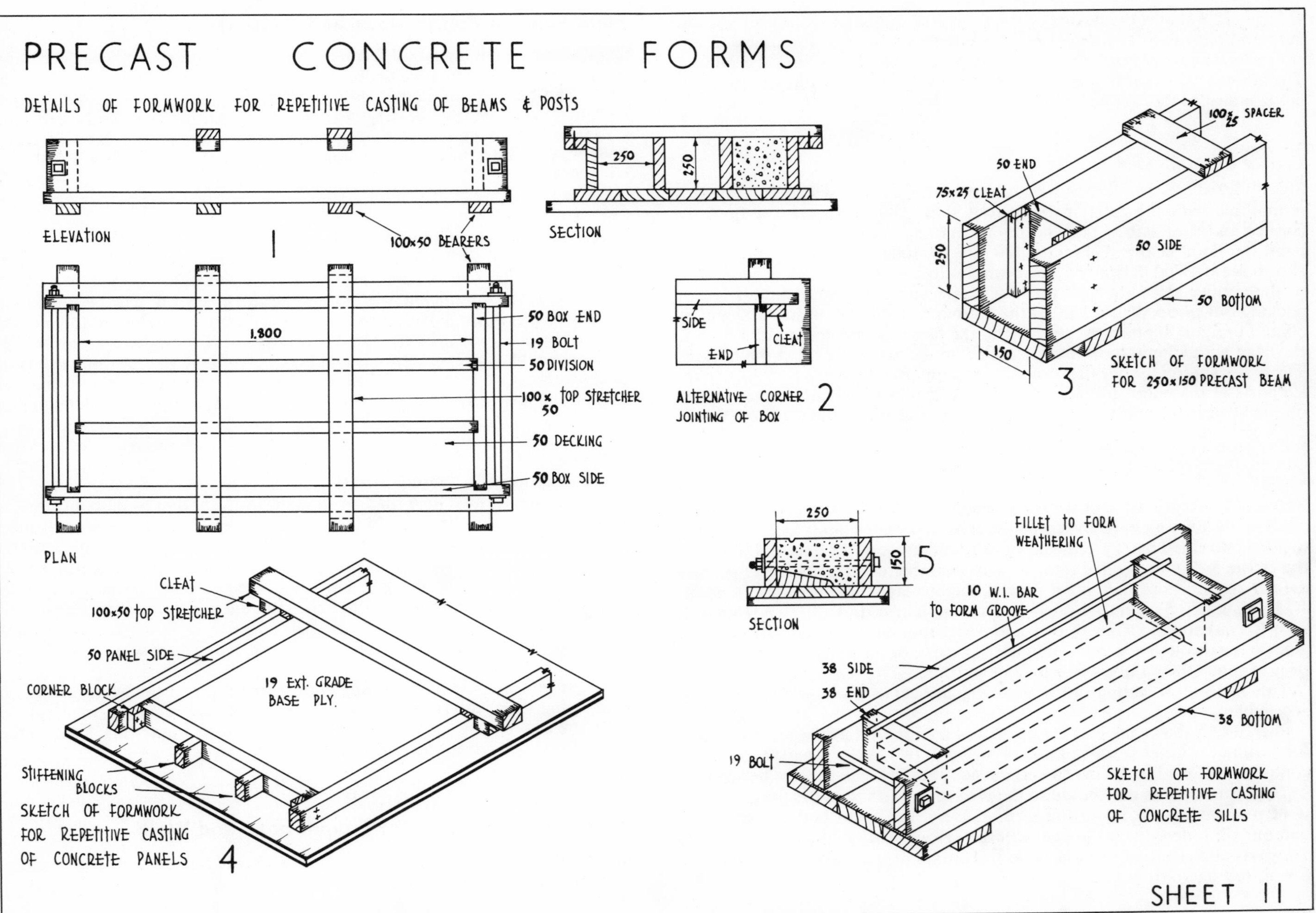
PRECAST CONCRETE FORMS
DETAILS OF FORMWORK FOR REPETITIVE CASTING OF BEAMS & POSTS
ELEVATION
1
100×50 BEARERS
SECTION
250
250
PLAN
1.800
50 BOX END
19 BOLT
50 DIVISION
100×50 TOP STRETCHER
50 DECKING
50 BOX SIDE
SIDE
CLEAT
END
ALTERNATIVE CORNER JOINTING OF BOX
2
100×25 SPACER
50 END
75×25 CLEAT
250
50 SIDE
50 BOTTOM
150
3
SKETCH OF FORMWORK FOR 250×150 PRECAST BEAM
CLEAT
100×50 TOP STRETCHER
50 PANEL SIDE
CORNER BLOCK
19 EXT. GRADE BASE PLY.
STIFFENING BLOCKS
SKETCH OF FORMWORK FOR REPETITIVE CASTING OF CONCRETE PANELS
4
250
150
5
SECTION
FILLET TO FORM WEATHERING
10 W.I. BAR TO FORM GROOVE
38 SIDE
38 END
38 BOTTOM
19 BOLT
SKETCH OF FORMWORK FOR REPETITIVE CASTING OF CONCRETE SILLS
SHEET 11

Sheet 12.8 shows the details of the formwork required to cast an upstand curb and base slab. The outer sheeting boards are fixed to 75 mm by 50 mm stiffeners which are strutted back to support stakes driven into the ground. The haunch is formed by sheeting boards held in position by the cleats shown, being nailed to the inclined strut.

Cornice Cast with Slab

Sheet 12.9 shows the formwork required for a cornice which is to be cast monolithic with the 150 mm thick roof slab. The formwork for both the cornice and roof slab is built up on a head and shores or props supported from the floor below. These are 100 mm by 100 mm with an outrigger as shown set at 1.200 m centres. The beam bottom is 50 mm thick. The shutters to the cornice sides are built up as separate units, the outer shutter having packing out pieces placed behind the battens to form the offsets. 100 mm by 50 mm uprights support the shutters on the face side and are strutted back to the head with 75 mm by 50 mm struts. 32 mm cleats are used at the bottom. The inside shutters above the roof slab are supported by struts tied back to the supports of the outer shutters as shown. A 75 mm by 75 mm runner carried on the headtree supports the joists and the shuttering of the roof slab.

Curved Lintel

Sheet 12.10 shows the geometrical setting out of a little more than half of a 500 mm by 300 mm lintel over 3.000 m span. The half span is first drawn and divided into equal parts 0–5 as shown. The curvature of the lintel is 150 mm at the centre 5–5′ and divided into the same number of equal parts. A perpendicular to the chord 0–5′ is set up and the increased span divided in points 1′, 2′, 3′, 4′ and 5′. Points 1–1′, 2–2′, etc., are joined. Lines drawn from 0, 1, 2, 3, 4, 5 radiating to 5′ are drawn, and where they cut their opposite numbers give points through which to draw the fair curve of the lintel. A lath carefully bent to touch the points is best for this purpose.

This method of setting out is used for any curve where the centre is inaccessible.

Sheet 12.11 shows the part elevation and section of the formwork required for casting the lintel *in situ*. 10 mm plywood in two layers is nailed to shaped contour ribs forming the inside and outside shutters with spacers between.

The beam bottom may be cut to the plan shape with the sides fitting against it or prepared as shown in the section. Here, the beam bottom and lower contour rib follow the same line with the vertical struts fitting against them. Supports and strutting are similar to those used in the previous example and for an ordinary straight cast *in situ* lintel.

Formwork for Foundations and Footings

These are the simplest of all forms to build where rough timber is generally used. Sheet 13.12 shows the details of the formwork for foundations 530 mm deep. The sides are of 25 mm sheeting built into panels about 2.400 m long with battens at 600 mm centres; 100 mm by 50 mm walings are braced, one side from the trench side, the other from stakes driven at 1.500 m intervals. Stripping can be done after a few hours and the forms moved forward.

Square Footings

Sheet 13.13 shows the square or box form, consisting of two ends and sides. These are made from 150 mm by 25 mm sheeting boards cleated with 100 mm by 32 mm cleats. The two ends are cut to the exact width of the footing with the cleats set in the thickness of the side cleats as shown. The two sides are allowed to run through with the cleats nailed on the inside against the outside face of the end forms. Intermediate cleats are used to stiffen the shuttering. When assembled, the forms are nailed together and the wire ties, consisting of double strands of wire, inserted through the sides and ends of the forms and around the intermediate cleats before being drawn taught by windlassing.

Sloping Side Footings

A cross-section of a footing with sloping sides is shown in Sheet 13.14. The splayed box is formed on a shallow-sided base of 230 mm by 32 mm sheeting supported by stakes driven into the ground, Sheet 13.15. The ends and sides are built up as in the previous example, the two ends being cut to the exact dimensions of the footing. It is not necessary to cut the sides to the slope; they are allowed to run through. The stakes are allowed to project above the top of the base sheeting to hold the bottom edges of the sloping sides.

The inside or end shutters require their edges to be bevelled to fit against the side shutters. The outside cleats will also require to be bevelled at the same angle.

To prevent the form from lifting with the pressure of the wet concrete on the sloping sides, the forms may be wired to the reinforcement or weighted with any heavy material or secured through a rough framework fixed to the forms to stakes driven into the ground on each side.

Basement Timbering for Foundations and Walls

Sheet 13.16 shows the first stage of a trench excavation with the required timbering to allow the foundation concrete for a basement to be laid. A

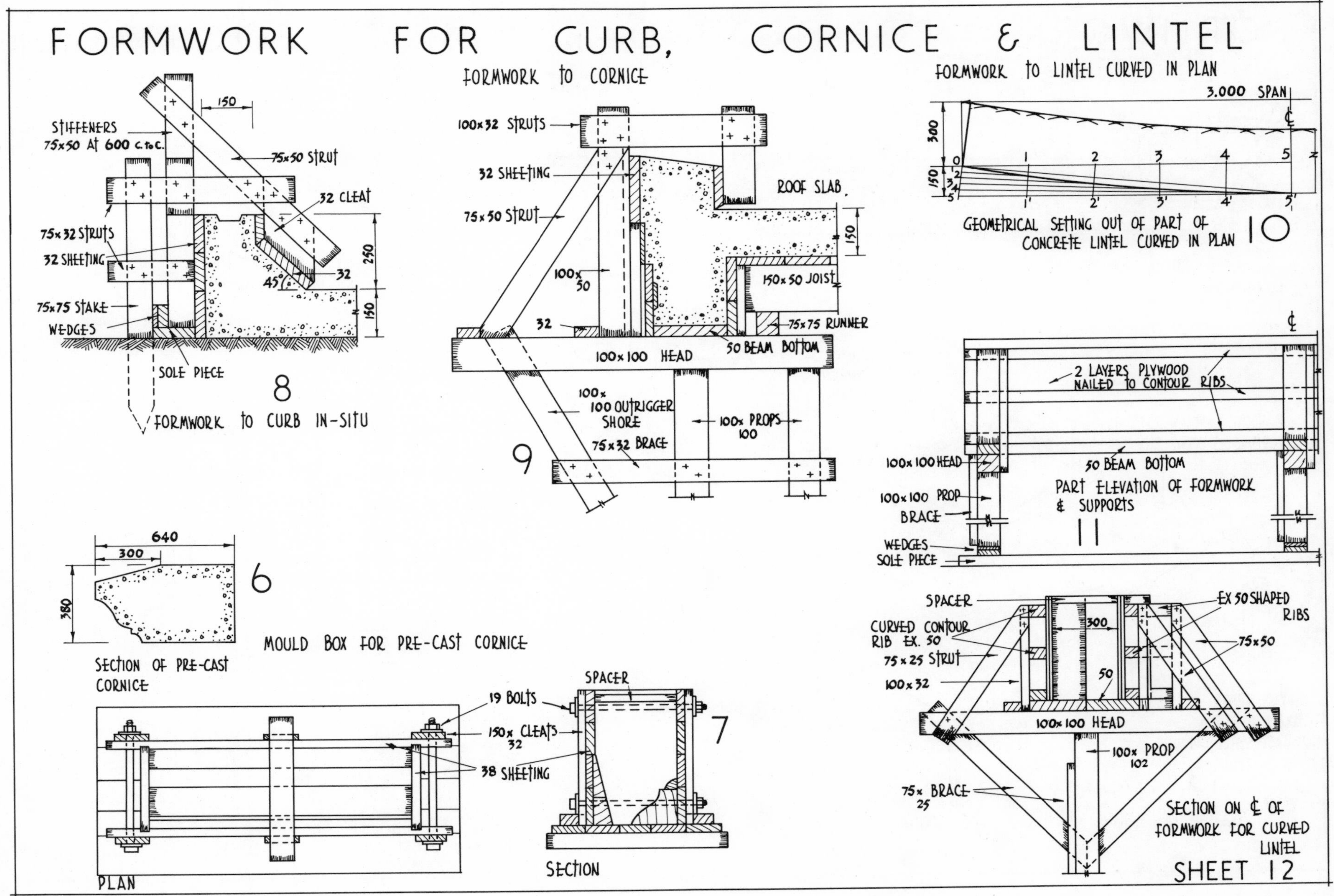

FORMWORK FOR CURB, CORNICE & LINTEL
STIFFENERS 75x50 AT 600 C. to C.
150
75x50 STRUT
32 CLEAT
75x32 STRUTS
32 SHEETING
45°
32
250
150
75x75 STAKE
WEDGES
SOLE PIECE
8
FORMWORK TO CURB IN-SITU
FORMWORK TO CORNICE
100x32 STRUTS
32 SHEETING
75x50 STRUT
ROOF SLAB
150
100x50
150x50 JOIST
32
75x75 RUNNER
50 BEAM BOTTOM
100x100 HEAD
100x100 OUTRIGGER SHORE
100x100 PROPS
75x32 BRACE
9
FORMWORK TO LINTEL CURVED IN PLAN
3.000 SPAN
300
150
0 1 2 3 4 5
1' 2' 3' 4' 5'
GEOMETRICAL SETTING OUT OF PART OF CONCRETE LINTEL CURVED IN PLAN
10
2 LAYERS PLYWOOD NAILED TO CONTOUR RIBS
100x100 HEAD
50 BEAM BOTTOM
100x100 PROP
BRACE
WEDGES
SOLE PIECE
PART ELEVATION OF FORMWORK & SUPPORTS
11
640
300
380
6
SECTION OF PRE-CAST CORNICE
MOULD BOX FOR PRE-CAST CORNICE
19 BOLTS
SPACER
150x32 CLEATS
38 SHEETING
7
PLAN
SECTION
SPACER
EX 50 SHAPED RIBS
CURVED CONTOUR RIB EX. 50
300
75x50
75x25 STRUT
100x32
50
100x100 HEAD
100x102 PROP
75x25 BRACE
SECTION ON ℄ OF FORMWORK FOR CURVED LINTEL
SHEET 12

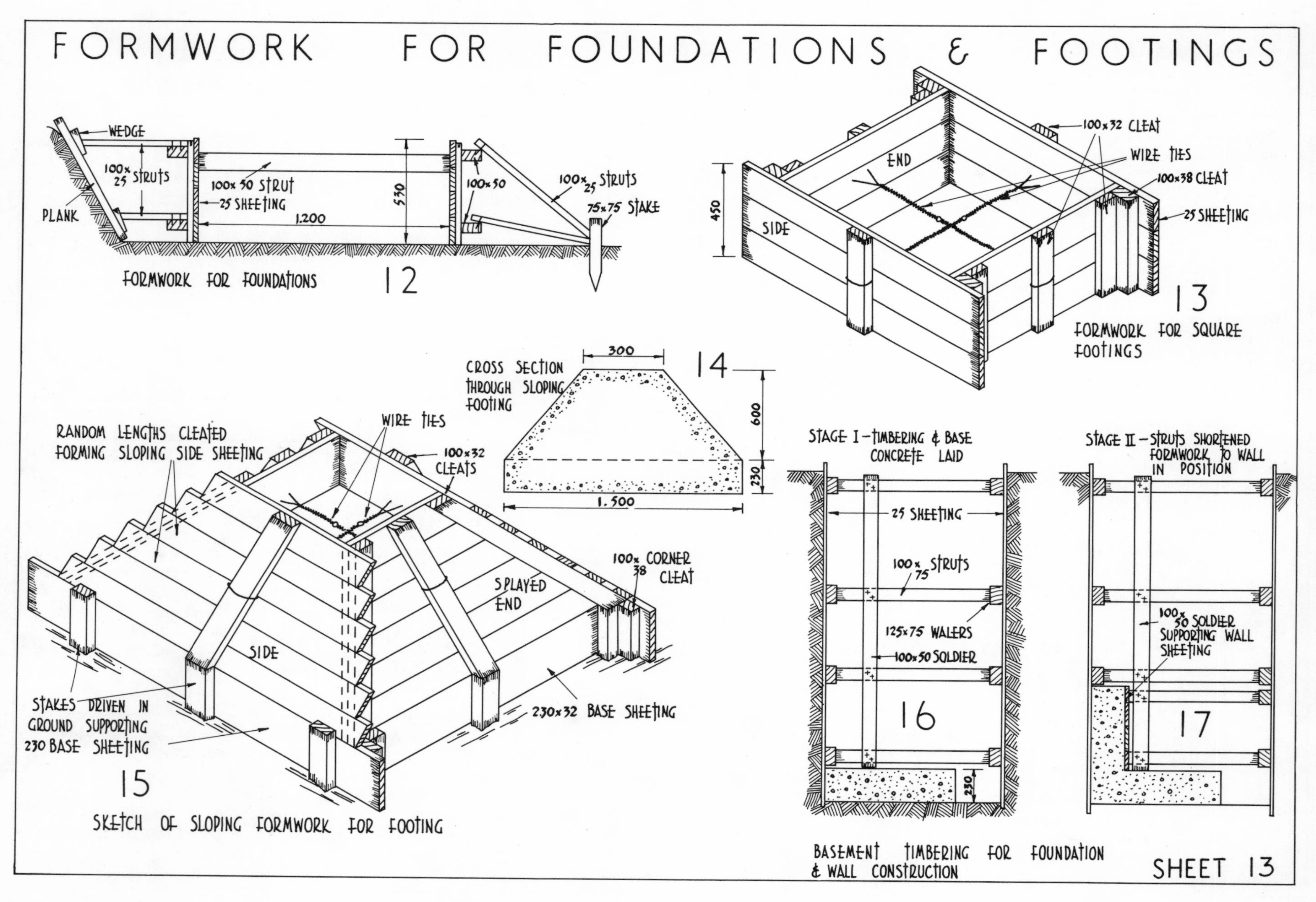
FORMWORK FOR FOUNDATIONS & FOOTINGS
WEDGE
100x25 STRUTS
PLANK
100x50 STRUT
25 SHEETING
1.200
530
100x50
100x25 STRUTS
75x75 STAKE
FORMWORK FOR FOUNDATIONS
12
100x32 CLEAT
WIRE TIES
END
100x38 CLEAT
25 SHEETING
450
SIDE
13
FORMWORK FOR SQUARE FOOTINGS
CROSS SECTION THROUGH SLOPING FOOTING
300
14
600
230
1.500
RANDOM LENGTHS CLEATED FORMING SLOPING SIDE SHEETING
WIRE TIES
100x32 CLEATS
100x38 CORNER CLEAT
SPLAYED END
SIDE
STAKES DRIVEN IN GROUND SUPPORTING 230 BASE SHEETING
230x32 BASE SHEETING
15
SKETCH OF SLOPING FORMWORK FOR FOOTING
STAGE I – TIMBERING & BASE CONCRETE LAID
25 SHEETING
100x75 STRUTS
125x75 WALERS
100x50 SOLDIER
16
230
STAGE II – STRUTS SHORTENED FORMWORK TO WALL IN POSITION
100x50 SOLDIER SUPPORTING WALL SHEETING
17
BASEMENT TIMBERING FOR FOUNDATION & WALL CONSTRUCTION
SHEET 13

timbered trench is excavated round the site wide enough to allow working room and to position reinforcement. The second stage, Sheet 13.17, requires the lower walers to be removed and the strut shortened. The sheeting on the inside face for the retaining wall is placed in position and packed off the vertical soldier. This is bolted to each strut as shown, successive lifts are prepared and concreted in the same way with each series of walers and struts removed as the work proceeds. After the earth in the centre of the site has been removed, the basement floor can be completed. As the earth from the centre area is excavated and the supports are removed, raking struts should be inserted at intervals as temporary wall supports from stakes driven in the ground. These are removed as the floor concrete is laid.

Stripping of Forms

Great care is required in the stripping or striking of forms and the responsibility for deciding when this should take place should be taken by the architect or engineer. Each job has to be considered on its merits, the setting qualities of the cement, the temperature since the placing of the concrete, and the nature and purpose of the member used. Beams and slabs should be left longer than columns and walls, and long spans longer than short spans.

The following times are usually accepted as safe, but the conditions already mentioned should be borne in mind.

Slabs

Soffit boards can be removed in seven days where occasional props are used for spans up to 3.000 m. Longer spans will require shutters to be left in place for two weeks. These times will be reduced where rapid-hardening cement is used.

Beams

Beam sides can be removed in three days and the beam bottom boards left as long as the slab shuttering, the props remaining for two to three weeks.

Columns

The side sheeting may be removed at the same time as the beam bottoms.

Walls

Shutters can usually be stripped in three days or even earlier where they are not subject to horizontal or vertical loading.

Columns

Sheet 14.18 shows the formwork required for a square or rectangular column. This consists of two ends and two sides of vertical boards held in position by nailing to yokes. These are held at the ends by 19 mm bolts. Wedges about 150 mm long cut out of 100 mm by 50 mm are driven in at each end between the bolts and yokes to bring the ends tight up to the side sheeting. A clean-out trap is formed at the foot of the column box for the cleaning out of rubbish, and is positioned behind the bottom yoke secured by a cleat screwed to the box as shown.

Octagonal Column

Sheet 14.19 shows the required formwork which is a modification of the form in the previous example. Corner pieces with their edges bevelled at 45 degrees are used to form the additional sides. An alternative construction to that shown is to reduce in width the sides of the normal column box and use spacers to make up the width.

Spacing of Yokes

The formwork for columns must be designed to resist the hydrostatic pressures of the wet concrete. This pressure is influenced by the method of placing the concrete, the consistency of the mix, the rate of filling, temperature and the height of the column. Columns are seldom less than 2.700 m high, and should never be filled to the top without a break. The weight of concrete may be taken as 2.240 kg/m^3, and the spacing of the yokes may be governed by the capacity of the shuttering to take the pressure of the wet concrete without bending, and the strength of the yoke to take the pressure from the shuttering. For a column 3.000 m high the spacing of the yokes is shown in Sheet 14.20.

The formwork for a circular column is shown in Sheet 14.21. The casing is made up in sections using horizontal or contour ribs, shaped out of 38 mm thick timber as shown. These are stiffened by 100 mm by 38 mm vertical ribs through which the contour ribs are bolted. 25 mm by 25 mm laggings are fitted between the vertical ribs and fixed for the contour ribs.

The section of the timber formwork for an L-shaped column is shown in Sheet 14.22. This form of column is usually an external corner column in a building, with the construction similar to that for square or rectangular

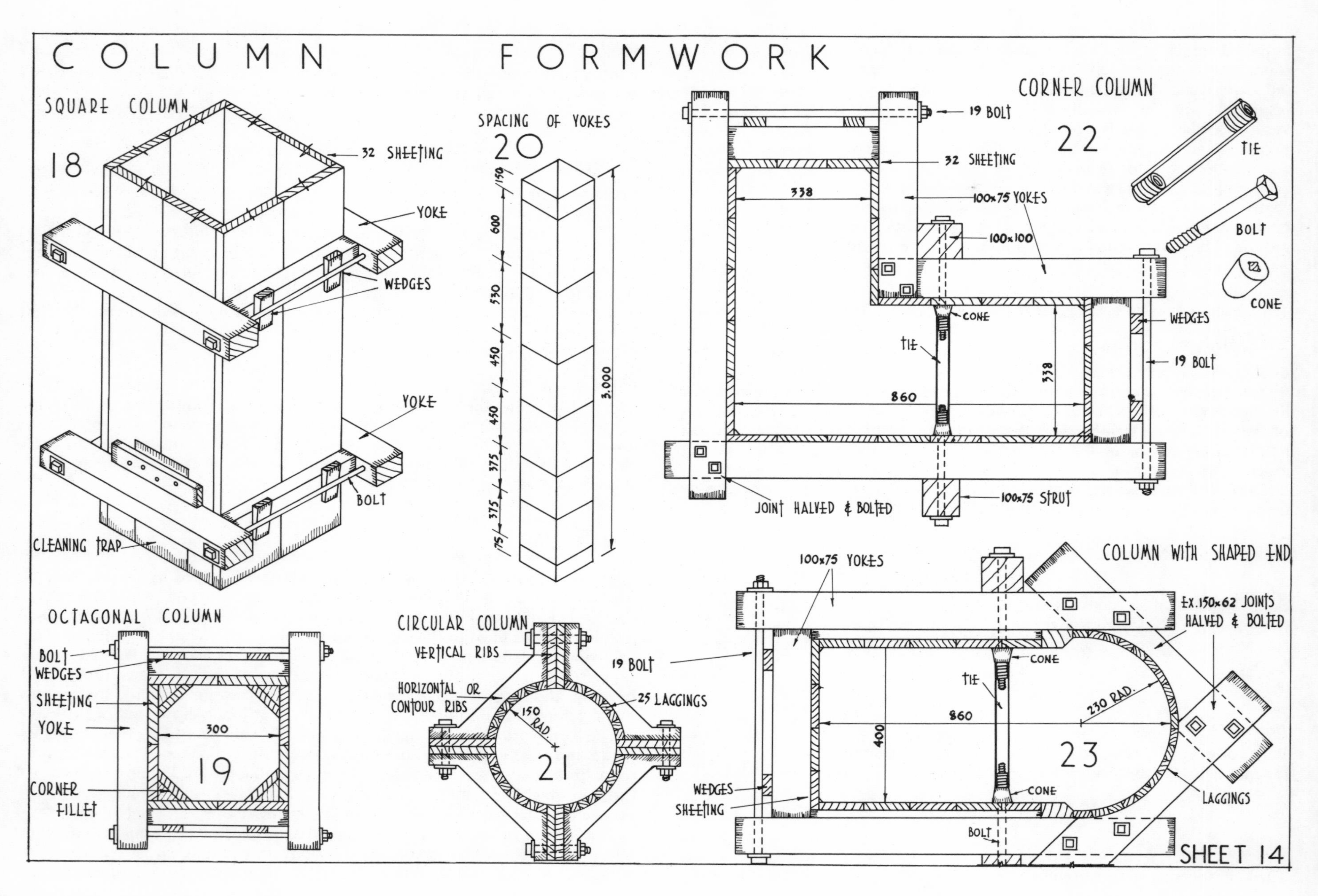
COLUMN FORMWORK
SQUARE COLUMN
18
32 SHEETING
YOKE
WEDGES
YOKE
BOLT
CLEANING TRAP
SPACING OF YOKES
20
150
600
530
450
450
375
375
75
3,000
CORNER COLUMN
22
19 BOLT
32 SHEETING
338
100x75 YOKES
100x100
CONE
TIE
338
860
WEDGES
19 BOLT
JOINT HALVED & BOLTED
100x75 STRUT
TIE
BOLT
CONE
OCTAGONAL COLUMN
BOLT
WEDGES
SHEETING
YOKE
300
19
CORNER FILLET
CIRCULAR COLUMN
VERTICAL RIBS
HORIZONTAL OR CONTOUR RIBS
150 RAD.
21
19 BOLT
25 LAGGINGS
100x75 YOKES
COLUMN WITH SHAPED END
EX.150x62 JOINTS HALVED & BOLTED
CONE
TIE
860
230 RAD.
400
23
CONE
WEDGES
SHEETING
BOLT
LAGGINGS
SHEET 14

columns. At the internal and external angles the yokes are halved together and bolted. A brace may be used across the internal corner to square up the column. Bolts or ties may be used passing through the vertical 100 mm by 100 mm to stiffen the form. These do not pass through the yokes. *Acrow* form ties are shown in Sheet 14, figs. 22 and 23.

Sheet 14.23 shows the formwork required for a column with a shaped end. The arrangement of the shuttering is similar to that used in the two previous examples.

The formwork for encasing steel stanchions in concrete differs slightly from that illustrated for casting reinforced columns. The two ends and sides of vertical boards are formed by using battens to hold the boards together. The shutters are then assembled round the steel and strutted from it by inserting spacers or small concrete blocks between the steelwork and the shutters. These ensure that the correct margin of concrete is formed on all the sides. The shutters are nailed together along the edges or, if the battens are left long on two sides, nailed through the battens. As the pressure from the concrete is much lower in this type of construction yokes are not necessary, the fixing being by nailing with the heads of the nails left proud for easy striking.

The casing of all steelwork is intended to increase strength and stiffness, improve the appearance and reduce the fire hazard.

Column with Splayed Cap

Sheet 15.24 shows the formwork for a square column with a splayed cap which is sometimes used at the top of the column to give an increased bearing area. The formwork for the column is the same as that already described. The cap is made up of shutters prepared in a similar way to those for sloping footings already described and illustrated on Sheet 13. Support for the shutters is taken from the top column yoke with struts to the floor slab joists as shown. An alternative method of support is by a yoke, having the inside faces bevelled to suit the splay of the cap, bolted and wedged in the usual way near to the top of the splay. The shape of the splayed shutters will have to be developed and the angle to which the edges are bevelled obtained.

Beam and Girder Floors

The formwork required for a reinforced concrete floor, having 500 mm by 250 mm girders or main beams and 400 mm by 200 mm secondary beams is shown in Sheet 15.25. In beam and girder construction, the slab panels, beam and girder sides and bottoms are prepared to the correct dimensions in advance ready for erection. The formwork for the columns may be cast first and stripped, or erected with the beams and floor shuttering. Whichever method is used, the columns should only be filled to just below the lowest beam bottom at the first pour. A most important item in column and beam work is the corner fillet. When stripping of formwork is carried out the sharp corners in concrete are easily damaged, so a small triangular fillet is nailed at the corners.

Sheet 15.26A shows a sketch of the formwork at the intersection of the various members. This clearly shows how the column sides are prepared at the top to correspond with the formwork for the various beams. Support for these is obtained from 75 mm by 50 mm ledgers nailed as shown.

In Sheet 15.26 the section of the secondary beam is shown with the formwork for the main beam and slab shown in Sheet 15.27. The 50 mm beam bottom rests on 125 mm by 100 mm headtrees which are supported by props. Timber or adjustable steel or steel scaffolding props may be used from a sole-piece and spaced at 1.500 m centres. The sides are formed from 32 mm boards battened together with 25 mm battens at about 900 mm intervals. Temporary spacers are used to keep the sides apart at the top.

The slab sheeting is formed by 25 mm thick boards laid and carried on joists. These should not be less than 100 mm by 50 mm in section. A continuous ledger is nailed to the cleats on the sides of the beam boxes to support the ends of 100 mm by 50 mm joists spaced at 750 mm centres. The boards are laid on the joists without nailing but should this be necessary to keep the boards in position, or with badly warped boards, as few nails as possible should be used. The ends of the joists should be bevelled inwards at the bottom about 6 mm so that they can easily be stripped without binding. The pressure on the beam sides is not great except where deep beams are used, but if necessary a cleat nailed to the headtree will secure the bottom of the beam sides.

Supports for the secondary beams have been omitted in the details, but it will be noticed in the sketch that the propping and headtree only supports the concrete of the beam itself while the propping of the secondary beam has to support the joisting, decking and the slab.

Wall Forms

There are various ways of constructing wall shuttering, but the same general methods employed will apply to any kind of wall whatever the height and situation. It is usual to use horizontal boards or sheeting nailed to vertical studs if the formwork is to be built in place, with the assembly and erection being done in one operation. In thin walls the most important shutter is erected first, and the other connected by bolts or wire ties or patent ties to this with spacers between. Braces nailed from the studs to stakes driven in the ground at intervals keep the shutters in line and hold them vertical.

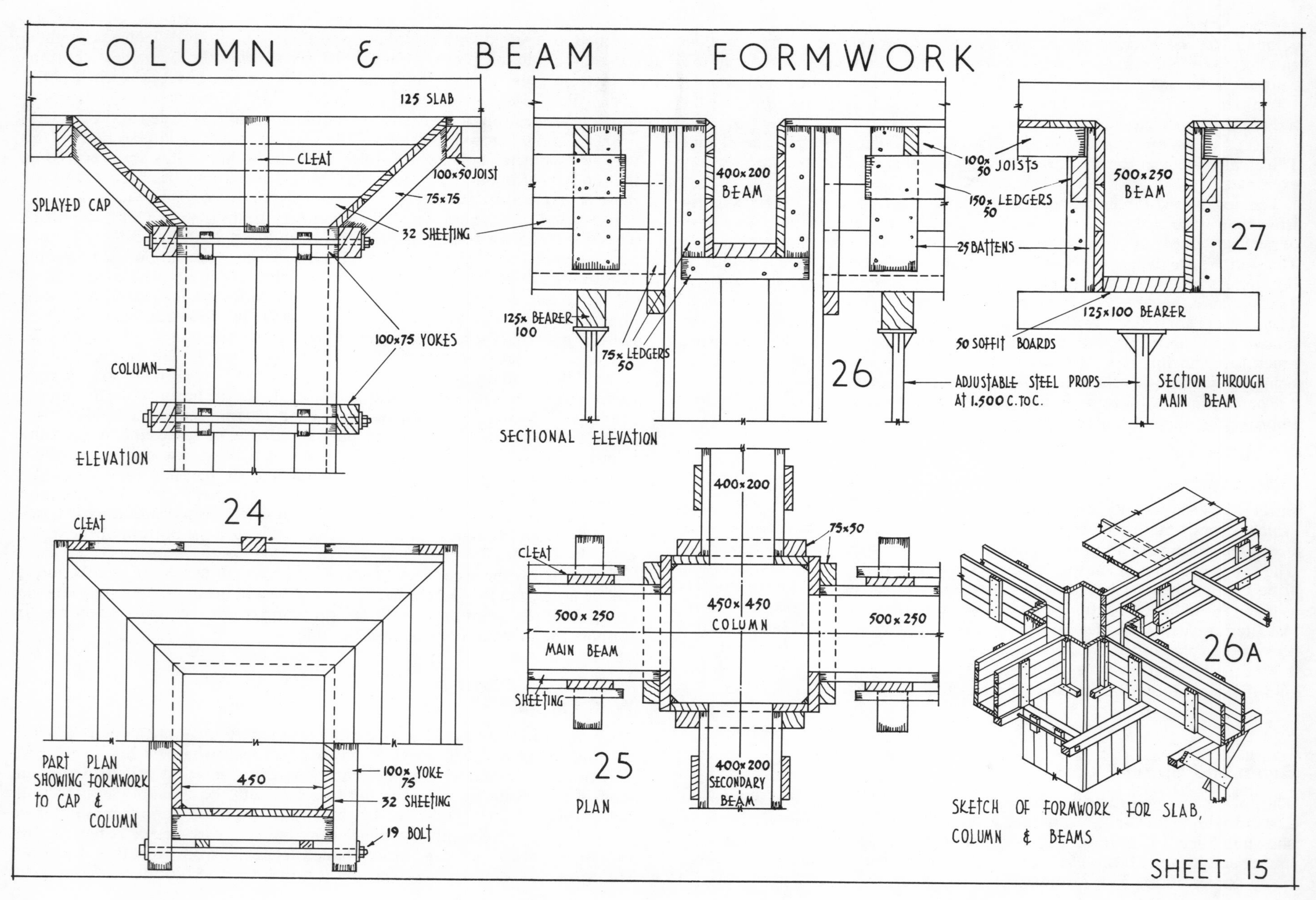
COLUMN & BEAM FORMWORK
125 SLAB
CLEAT
100×50 JOIST
SPLAYED CAP
75×75
32 SHEETING
100×75 YOKES
COLUMN
ELEVATION
24
400×200 BEAM
100× 50 JOISTS
150× 50 LEDGERS
25 BATTENS
125× 100 BEARER
75× 50 LEDGERS
26
ADJUSTABLE STEEL PROPS AT 1.500 C. TO C.
SECTIONAL ELEVATION
500×250 BEAM
27
125×100 BEARER
50 SOFFIT BOARDS
SECTION THROUGH MAIN BEAM
CLEAT
PART PLAN SHOWING FORMWORK TO CAP & COLUMN
450
100× 75 YOKE
32 SHEETING
19 BOLT
400×200
75×50
CLEAT
500 × 250
MAIN BEAM
450×450 COLUMN
500×250
SHEETING
25
PLAN
400×200 SECONDARY BEAM
26A
SKETCH OF FORMWORK FOR SLAB, COLUMN & BEAMS
SHEET 15

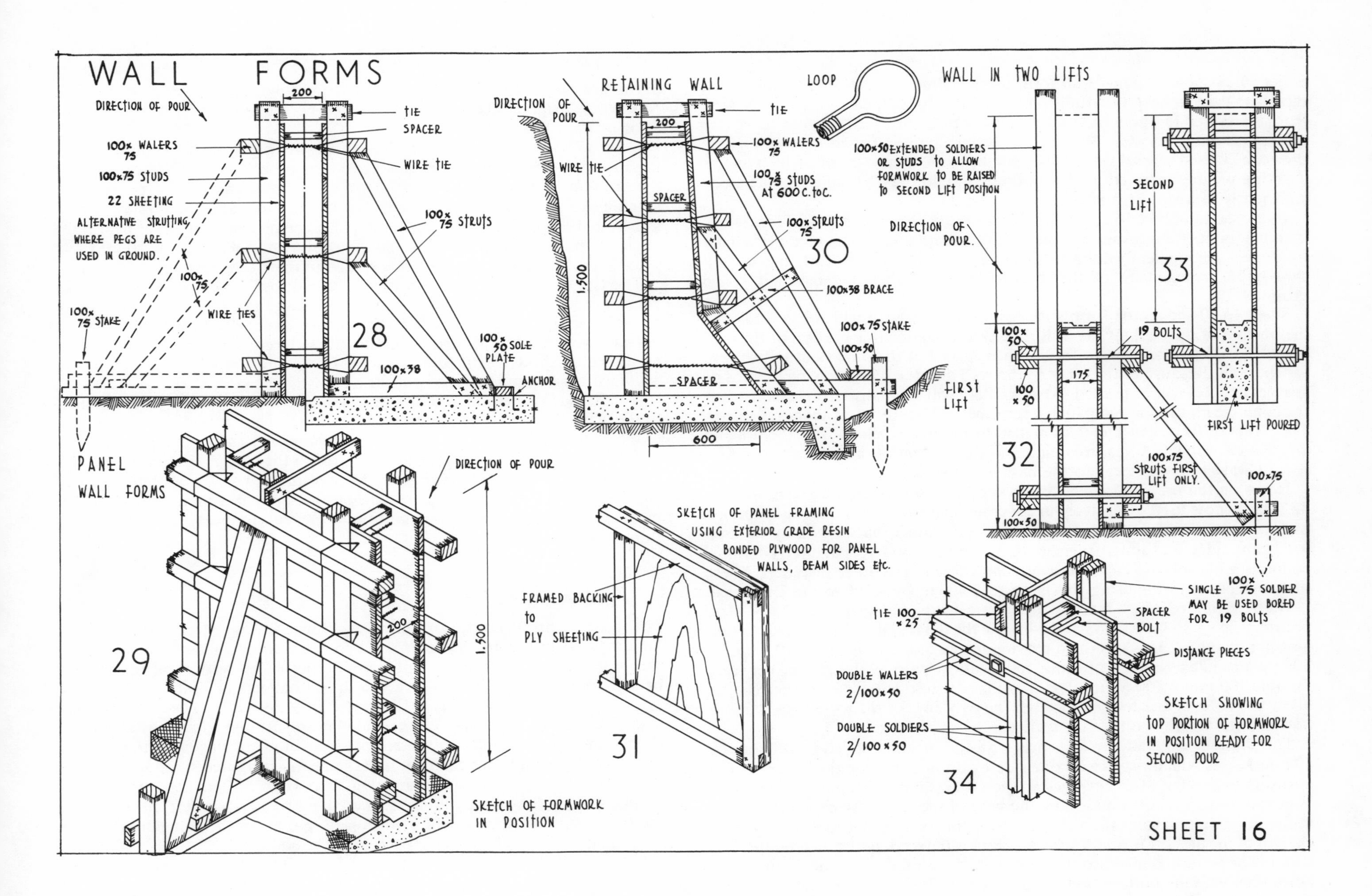

WALL FORMS
DIRECTION OF POUR
200
TIE
SPACER
100× 75 WALERS
100×75 STUDS
22 SHEETING
WIRE TIE
ALTERNATIVE STRUTTING WHERE PEGS ARE USED IN GROUND.
100× 75 STRUTS
100× 75
WIRE TIES
100× 75 STAKE
28
100× 50 SOLE PLATE
100×38
ANCHOR
RETAINING WALL
DIRECTION OF POUR
WIRE TIE
200
SPACER
100× 75 WALERS
100× 75 STUDS AT 600 C. to C.
100× 75 STRUTS
30
100×38 BRACE
1.500
100× 75 STAKE
100×50
SPACER
600
FIRST LIFT
LOOP
WALL IN TWO LIFTS
100×50 EXTENDED SOLDIERS OR STUDS TO ALLOW FORMWORK TO BE RAISED TO SECOND LIFT POSITION
SECOND LIFT
DIRECTION OF POUR.
33
100× 50
19 BOLTS
100 × 50
175
FIRST LIFT POURED
32
100×75 STRUTS FIRST LIFT ONLY.
100×75
100×50
PANEL WALL FORMS
DIRECTION OF POUR
200
1.500
29
SKETCH OF FORMWORK IN POSITION
SKETCH OF PANEL FRAMING USING EXTERIOR GRADE RESIN BONDED PLYWOOD FOR PANEL WALLS, BEAM SIDES ETC.
FRAMED BACKING TO PLY SHEETING
31
100× 75 SOLDIER
SINGLE MAY BE USED BORED FOR 19 BOLTS
TIE 100 ×25
SPACER
BOLT
DISTANCE PIECES
DOUBLE WALERS 2/100×50
DOUBLE SOLDIERS 2/100×50
SKETCH SHOWING TOP PORTION OF FORMWORK IN POSITION READY FOR SECOND POUR
34
SHEET 16

Wall Forms with Walings

Sheet 16.28 shows a panel wall form using walings. These are the horizontal members used to keep the studs in line, and to stiffen the form. It will be seen from the sketch in Sheet 16.29 that where walings are used it is not necessary to wire or bolt every stud. Wall forms are best built in panels for work where they can be used several times. They can be made up in advance which means a saving of time in erection. The walings should only be nailed to the studs sufficiently to keep them in place.

In the section Sheet 16.28, to the right of the centre line the panel formwork is shown in position, using wire ties with the strutting from a sole piece fixed by a steel anchor, embedded in the base concrete. To the left, the section shows the arrangement of the strutting where stakes are driven into the ground.

Sheet 16.30 shows the formwork for a retaining wall which has a fillet at the bottom of the back of the wall. The shuttering for this portion of the wall is built as a separate panel with the studs spaced at the same centres as those for the wall. Because of the taper in the wall and the fillet, varying lengths of spacers and ties will be needed. Also, as there will be an upward pressure from the concrete on the splayed panel, it must be well strutted and anchored down. The struts of the upper panel are nailed to the struts below, with the bracing arranged as shown. A spacer is placed at the bottom of the wall with the first row of walers kept as near to the base as possible.

Notes on the use of Douglas fir plywood for formwork are given on page 42. This is now used extensively in concrete formwork with many advantages over boards. Panels prepared with softwood framed backings may be used for beam sides as shown in sheet 16.31. Square and rectangular columns, columns with splayed caps, stair and wall forms are other instances where plywood panel forms using sheets up to 2.400 m by 1.200 m are employed.

High walls are often built in lifts by raising the forms vertically after pouring. Sheet 16 figs. 32, 33 and 34 show the details of the formwork to cast a wall in two lifts. This is termed moving formwork, consisting of 100 mm by 50 mm extended soldiers or studs made up in pairs with spacers between. Double 100 mm by 50 mm walers may also be used with spacers between, to allow the bolts or ties to pass through the two sides of the formwork without boring holes for the bolts in the walers.

The shutters are made up in panels with the sheeting nailed to 100 mm by 50 mm ledges which project on one side by an amount equal to the lift. The strutting to the first lift is arranged as shown to stakes driven into the ground. Where foundation concrete is laid, the base of the wall may be raised 75 mm at the same time to form the starting points for the side shutters. The bolts at the top of the first lift form the centre holes for the second lift; this requires the shutters to be turned upside down and refixed with the bolts, the middle bolt-holes then take the bottom bolts of the second lift.

Stair Forms

Stairs are not normally poured until all the floors have been cast and the formwork stripped. They are designed to be self-supporting from floor to floor or floor to landing.

Sheet 17.35 shows the section through a lower flight of stairs and the landing with one side supported by a wall and the other open. The formwork for the landing is prepared first supported by joists, runners and props. The sloping slab is supported on 100 mm by 75 mm joists and 150 mm by 75 mm stringers. 100 mm by 75 mm struts are placed below the stringers as near right angles as possible and cleated at the top as shown. A 150 mm by 50 mm sole piece is used at the floor with cleats spiked to this at the foot of each strut.

The side forms or strings consist of 275 mm by 38 mm or 50 mm planks which are set out and cut to the required tread and rise. The risers are usually 50 mm thick with the lower edge bevelled to allow the full width of the tread to be trowelled. The wall string is cut to allow the ends of the risers to be fixed as shown in Sheet 17.36. Allowance for trowelling the tread right up to the wall face must be made at the wall string.

An alternative method of fixing the risers at the wall end is by cleats nailed to the wall string, as shown in Sheet 17.38. A further method is to fix a plank to the wall above the line of the risers with hangers to support the risers fixed to this. The risers at the open end are nailed to the cut string. For wide stairs an intermediate support is necessary to prevent the risers from bulging. The pieces cut from the outer string are nailed to a 100 mm by 75 mm support which must be fastened at the top and bottom to prevent it moving forward. This is shown in Sheet 17.37.

Suspended Forms

The general arrangement of formwork for encasing steel framework in buildings is similar to that employed in reinforced concrete construction, except that the runners, instead of being supported by vertical members from below, are suspended from above with metal hangers or bolts.

The *Acrow* hanger is the type illustrated in Sheet 17.39. The rod is shaped to enable a concrete casing to be provided *in situ* around a haunched beam by supporting the formwork in the required position. In this case, the bottom flange of the beam is not cased so that the bearer is tightened up against this, which in turn supports the remaining formwork. When the formwork is stripped by withdrawing the bolts, the coiled sockets are left in the beam; these can be subsequently used as anchorage points for the fixing of fittings or equipment after the holes have been made good.

Sheet 17.40 shows a similar hanger used on a larger beam cased on all sides. Concrete spacers are used below the beam to give the necessary cover of

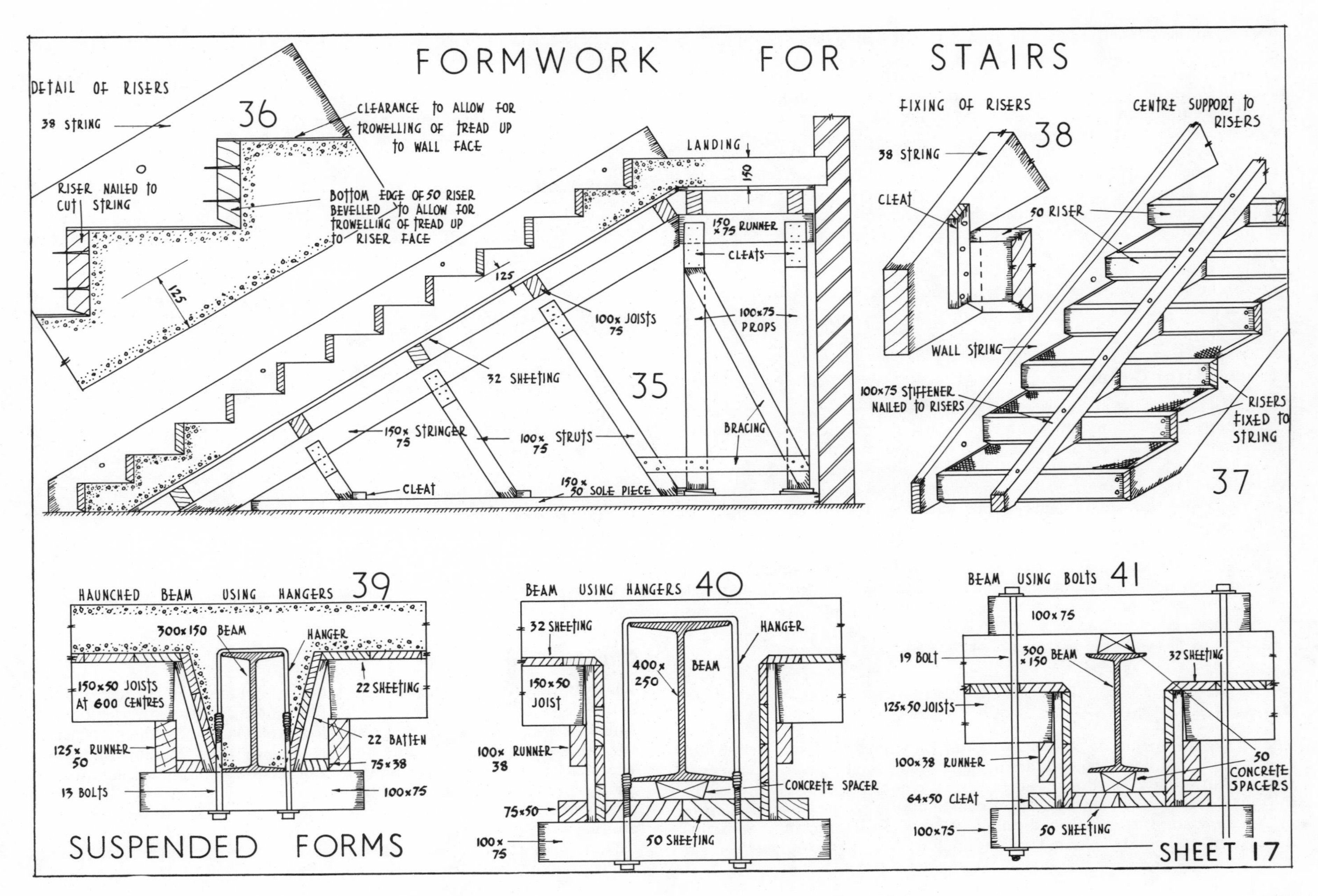
FORMWORK FOR STAIRS
DETAIL OF RISERS
36
38 STRING
CLEARANCE TO ALLOW FOR TROWELLING OF TREAD UP TO WALL FACE
RISER NAILED TO CUT STRING
BOTTOM EDGE OF 50 RISER BEVELLED TO ALLOW FOR TROWELLING OF TREAD UP TO RISER FACE
125
LANDING
150
150 × 75 RUNNER
CLEATS
125
100× JOISTS 75
100×75 PROPS
32 SHEETING
35
BRACING
150× STRINGER 75
100× STRUTS 75
CLEAT
150× 50 SOLE PIECE
FIXING OF RISERS
38
38 STRING
CLEAT
50 RISER
WALL STRING
CENTRE SUPPORT TO RISERS
100×75 STIFFENER NAILED TO RISERS
RISERS FIXED TO STRING
37
HAUNCHED BEAM USING HANGERS 39
300×150 BEAM
HANGER
150×50 JOISTS AT 600 CENTRES
22 SHEETING
22 BATTEN
125× RUNNER 50
75×38
13 BOLTS
100×75
SUSPENDED FORMS
BEAM USING HANGERS 40
32 SHEETING
HANGER
150×50 JOIST
400× 250
BEAM
100× RUNNER 38
CONCRETE SPACER
75×50
100× 75
50 SHEETING
BEAM USING BOLTS 41
100×75
19 BOLT
300 ×150 BEAM
32 SHEETING
125×50 JOISTS
100×38 RUNNER
50 CONCRETE SPACERS
64×50 CLEAT
100×75
50 SHEETING
SHEET 17

concrete to the steel. The beam bottom is pulled tight against the spacer by the bolts in the hangers.

Sheet 17.41 shows the details of suspended formwork using 19 mm diameter bolts. In this example, concrete spacers are used on the top of the steel beam to carry the top runners and below, between the lower flange and the 50 mm beam bottoms. 100 mm by 75 mm bottom runners are connected by the bolts from below.

The beam sides are made up as shutters in the usual way, the bottom edges held tight against the beam bottom by cleats or ribbons spiked to the runners. To the battens of the side shutters, ledgers or runners are nailed to support the joists with the slab decking laid across them. The outside boards are bevelled on the edges as shown.

When stripping, the bolts are slackened to allow the bearers to drop, when most of the joists can be removed, a few being left flat to catch the decking as it is stripped.

Formwork for Canopy

Sheet 18.42 shows the key section and plan of a canopy having shaped ends. The canopy is cast monolithic with the floor slab between the two columns shown, with the necessary supports taken from the floor below.

Sheet 18.43 shows the details of the necessary formwork. The headtrees are formed using two 150 mm by 50 mm members placed on each side of the 100 mm by 75 mm props and struts. Cleats nailed on these supports assist in positioning the headtrees in assembly. A 150 mm by 50 mm horizontal tie is used spanning across the faces of the columns.

The foot of the strut is supported on a 125 mm by 75 mm runner, bolted to the lower slab as shown. Double joisting is used to support the decking which is framed, using 19 mm exterior grade resin-bonded ply with a 50 mm backing.

The shaped ends are formed with two layers of 6 mm ply secured to battens and struts as shown in the part plan, Sheet 18.44.

Formwork for Balcony

The section through a balcony showing the required formwork is shown in Sheet 18.45. The floor and wall are to be poured in one operation. Support for the shuttering to the floor is taken from the floor below with strutting to the outside face of the balustrade as shown. Supports for the inside shuttering of the balustrade are taken from the main wall where a 100 mm by 75 mm runner is fixed. 100 mm by 50 mm struts are tied to this and the outside vertical studs.

The shaped end to the balustrade is formed by two layers of 6 mm exterior grade plywood which is fixed to a series of 50 mm contour ribs, Sheet 18.46. These are cut to the required radius and well nailed at the crossings. Vertical 100 mm by 50 mm studs are used to join the curved portion to the straight by bolts, and to provide a fixing at the wall.

Formwork for Dome

Sheet 19.47 shows the plan and section of the formwork and supports to a dome octagonal in plan. The elevation of the rib A–B shows two methods of constructing the truss or main rib, which is similar to the construction used in centring.

Main ribs are placed at the intersections in the plan with the spaces between each rib filled with smaller ribs which are housed in purlins at the top and fixed to the curb at the bottom. The ribs are supported by a head piece and props at the centre.

The covering of a dome of this shape is straightforward. 75 mm by 25 mm horizontal boards are nailed to the main ribs, as shown in the lower half of the plan in Sheet 19.48.

A hemispherical dome which is curved in two directions is much more difficult and expensive to cover. The sheeting in this case may consist of 50 mm by 13 mm strips arranged circumferentially, this size allowing the strips to be laid to the radius of the dome where the diameter is not too large.

Shuttering for the outer face of the domed surface is necessary to support the wet concrete and consists of radial ribs similar to those for the soffit, the boards being attached to the underside of the ribs as the concreting proceeds. These forms are called back forms and are bolted together as shown in the upper part of the plan. The part elevation is shown in Sheet 19.49.

Circular Tank

The formwork for a circular tank 2.700 m diameter and 1.500 m deep is shown in the plan and elevation. Sheet 19.50. The shutters consist of vertical boards nailed to contour or horizontal ribs. These are concave for the outer panels and convex for the inner ones and are formed from 230 mm by 38 mm. The panels are 750 mm deep and are connected by bolting through the 230 mm by 38 mm vertical ribs and the horizontal ribs as shown. All the bolt-holes are set out to a standard rod, so that the panels can be used in any position.

Horizontal strutting is cut between the vertical ribs on the inside and fixed to the horizontal ribs.

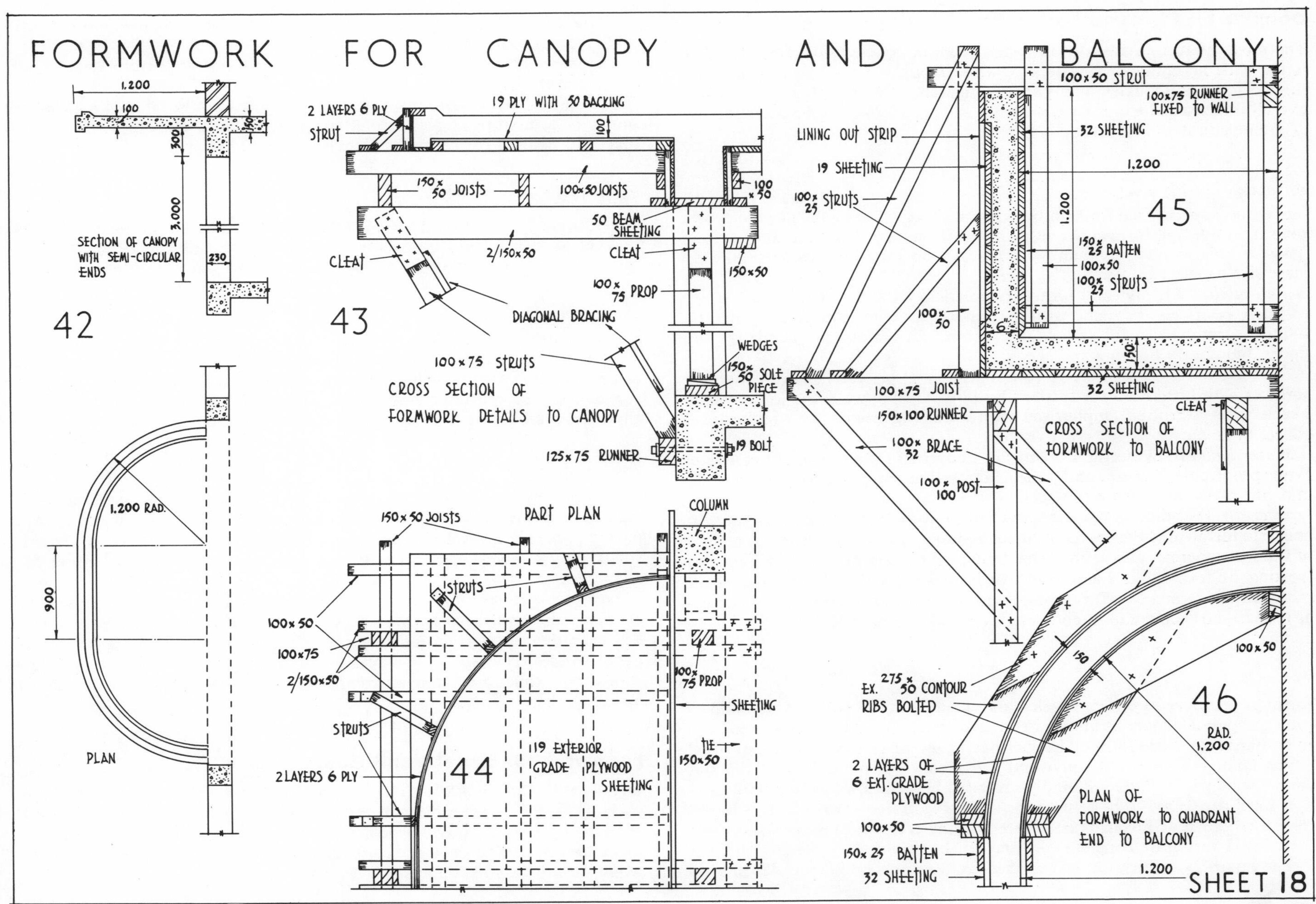

FORMWORK FOR CANOPY AND BALCONY
1.200
100
300
3.000
50
230
SECTION OF CANOPY WITH SEMI-CIRCULAR ENDS
42
2 LAYERS 6 PLY
STRUT
19 PLY WITH 50 BACKING
100
150 x 50 JOISTS
100x50 JOISTS
100 x 50
50 BEAM SHEETING
CLEAT
2/150x50
150x50
100 x 75 PROP
43
DIAGONAL BRACING
WEDGES
150 x 50 SOLE PIECE
100 x 75 STRUTS
CROSS SECTION OF FORMWORK DETAILS TO CANOPY
125 x 75 RUNNER
19 BOLT
1.200 RAD.
900
PLAN
150 x 50 JOISTS
PART PLAN
COLUMN
STRUTS
100x 50
100x75
2/150x50
100 x 75 PROP
SHEETING
STRUTS
19 EXTERIOR GRADE PLYWOOD SHEETING
TIE 150x50
2 LAYERS 6 PLY
44
100 x 50 STRUT
100x75 RUNNER FIXED TO WALL
LINING OUT STRIP
32 SHEETING
19 SHEETING
1.200
100 x 25 STRUTS
1.200
45
150 x 25 BATTEN
100x50
100 x 25 STRUTS
100 x 50
150
100 x 75 JOIST
32 SHEETING
150x 100 RUNNER
CLEAT
CROSS SECTION OF FORMWORK TO BALCONY
100 x 32 BRACE
100 x 100 POST
100 x 50
EX. 275 x 50 CONTOUR RIBS BOLTED
150
46
RAD. 1.200
2 LAYERS OF 6 EXT. GRADE PLYWOOD
PLAN OF FORMWORK TO QUADRANT END TO BALCONY
100x50
150x 25 BATTEN
32 SHEETING
1.200
SHEET 18

Douglas Fir Plywood Formwork

The most interesting advance in recent years in concrete form construction is the wide acceptance of waterproof-glue Douglas fir plywood panels for form claddings. Douglas fir plywood's inherent properties together with its overall job cost reduction have played a major part in the ever-widening use of architectural and structural concrete.

Advantages of Plywood

Assembly speed. One 2.400 m by 1.200 m sheet will take the place of seven boards in average formwork with saving on sawing, handling and nailing. There is little waste if the building is planned on a modular basis to take the large size panels.

Re-usability. Fir plywood forms are practically 100 per cent re-usable. Up to thirty pours can be made with the same form with careful handling and oiling. Form panels are re-used in other construction, as sheathing, sub floors, temporary flooring and sheds.

Smooth surfaces. Fir plywood panels give large, perfectly smooth concrete surfaces with little need for further finishing. On big walls the joints where panels meet are often emphasized with moulding inserts for architectural effect.

Watertightness. Watertight forms are essential in pouring good concrete. Waterproof glue qualities and panels attached at the edges to solid backing make fir plywood forms watertight.

Strength. Two-directional strength of fir plywood is of the highest value in concrete formwork. Panels resist direct bending caused by outward pressure of the wet concrete, and at the same time tend to stop diagonal racking and twisting of formwork. Large flat panel surfaces and the fact that small pieces can be nailed close to the edge without splitting are assets architecturally when intricate detail in concrete formwork is required.

Form Oils

Form panels may be ordered sealed and oiled from the mills. Some builders prefer a special lacquer as first coat. Often manufactured form oil is preferred to crankcase oil since form oil keeps wood clean for re-use. A thin coat of oil —just enough to make the plywood feel greasy—is sufficient and will not stain the concrete. Panels should be re-oiled between uses as required.

If form panels are to be re-used for sheathing, a simple expedient is to wet the forms down with water just before pouring concrete. This is a substitute for oil and very little concrete will stick to the plywood.

The number of re-uses to be expected from fir plywood forms depends on the care used in handling panels. Careful use and handling give more re-use and greater savings in cost.

Panels should be handled with reasonable care to prevent damage to corners and edges.

For completely smooth surfaces, joints in forms can be pointed with water putty, or a mixture of plaster of Paris and Portland cement. This mixture may also be used for repairs to damaged areas of panels.

Form Ties

Self-spacing form ties are more efficient than wire and spacer blocks. They are quicker to install and do not stretch. A good form tie can be snapped off inside the concrete, leaving a small hole to be grouted in afterwards.

Form Construction

Construction of Douglas fir plywood concrete forms is similar to formwork with boards in that supporting framework, walers, posts and joists must be strong enough to withstand the pressure of the poured concrete.

Smoothly curved concrete surfaces can readily be obtained because fir plywood can be bent to a very small radius, particularly when wet.

Re-usable Forms

Re-usable plywood forms are braced with wood or metal framework sufficient to withstand any concrete load likely to be encountered. Form panels join together side by side along walls with ties across the walls, installed only at panel edges. Spreaders are not required and walers are used for alignment.

To increase re-usability, forms are often attached side-by-side with bolts or clamps. Care with corners and edges in removing plywood concrete forms will keep them free from damage, and will make forms available for subsequent re-uses. Damage to concrete forms usually occurs in handling. They should be cleaned on removal from the concrete and stacked with cleaned faces together.

Formwork for Barrel Vault Roofs

Details of the formwork for a barrel vault roof of 9.000 m span are shown in sheet 20.51. The arrangement of the shuttering and supports for the beam is similar to previous work. Purpose-made trusses built up at 1.500 m centres, in sections of three trusses, are constructed as shown and support 19 mm plywood sheeting carried on joists.

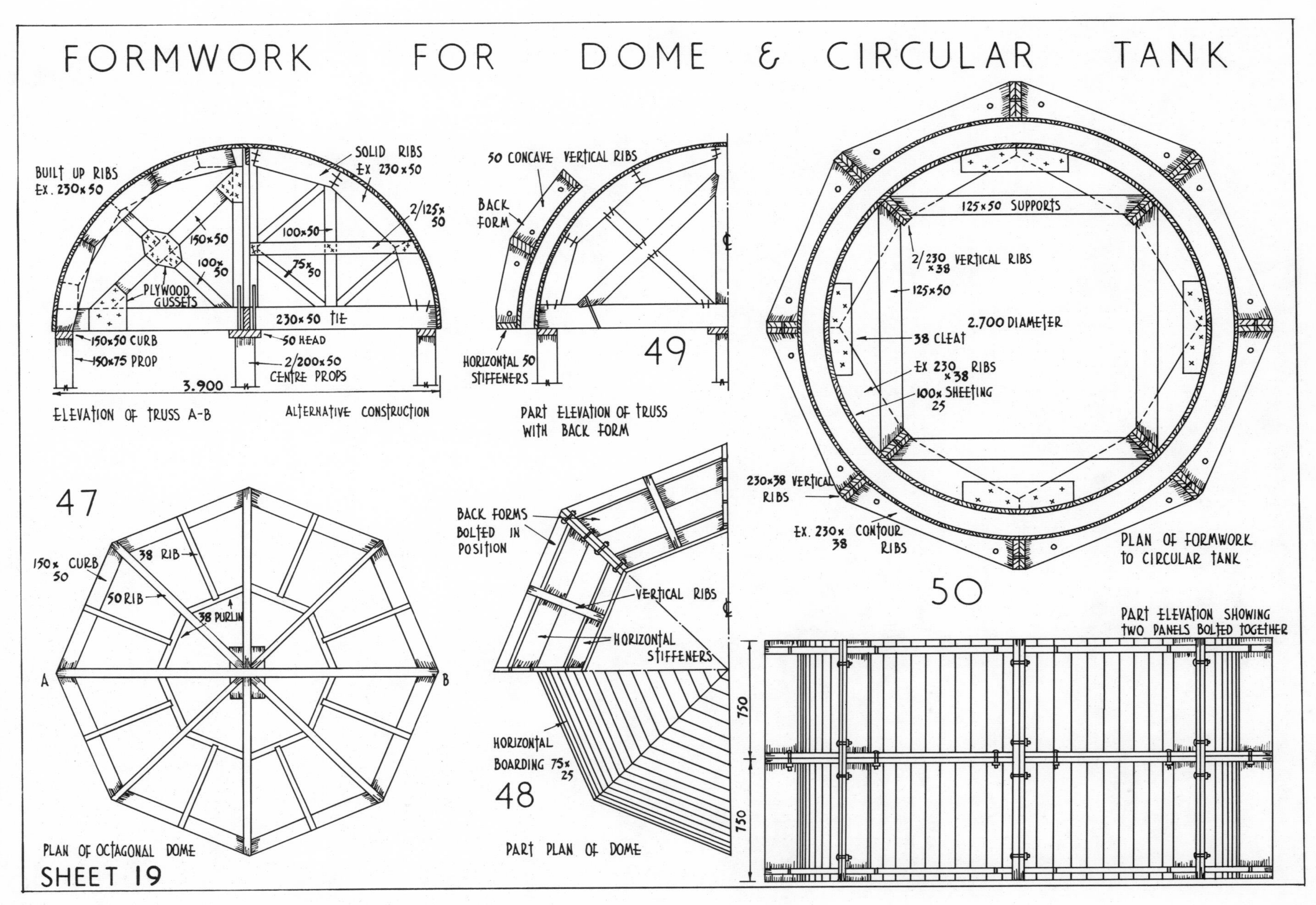
FORMWORK FOR DOME & CIRCULAR TANK
BUILT UP RIBS
EX. 230x50
SOLID RIBS
EX 230x50
2/125x
50
100x50
150x50
100x
50
75x
50
PLYWOOD
GUSSETS
230x50 TIE
150x50 CURB
50 HEAD
150x75 PROP
2/200x50
CENTRE PROPS
3.900
ELEVATION OF TRUSS A-B
ALTERNATIVE CONSTRUCTION
50 CONCAVE VERTICAL RIBS
BACK
FORM
HORIZONTAL 50
STIFFENERS
49
PART ELEVATION OF TRUSS
WITH BACK FORM
125x50 SUPPORTS
2/230 VERTICAL RIBS
x38
125x50
2.700 DIAMETER
38 CLEAT
EX 230 RIBS
x38
100x SHEETING
25
230x38 VERTICAL
RIBS
EX. 230x CONTOUR
38 RIBS
PLAN OF FORMWORK
TO CIRCULAR TANK
50
47
150x CURB
50
38 RIB
50 RIB
38 PURLIN
A
B
PLAN OF OCTAGONAL DOME
BACK FORMS
BOLTED IN
POSITION
VERTICAL RIBS
HORIZONTAL
STIFFENERS
HORIZONTAL
BOARDING 75x
25
48
PART PLAN OF DOME
PART ELEVATION SHOWING
TWO PANELS BOLTED TOGETHER
750
750
SHEET 19

FORMWORK FOR BARREL VAULT ROOFS

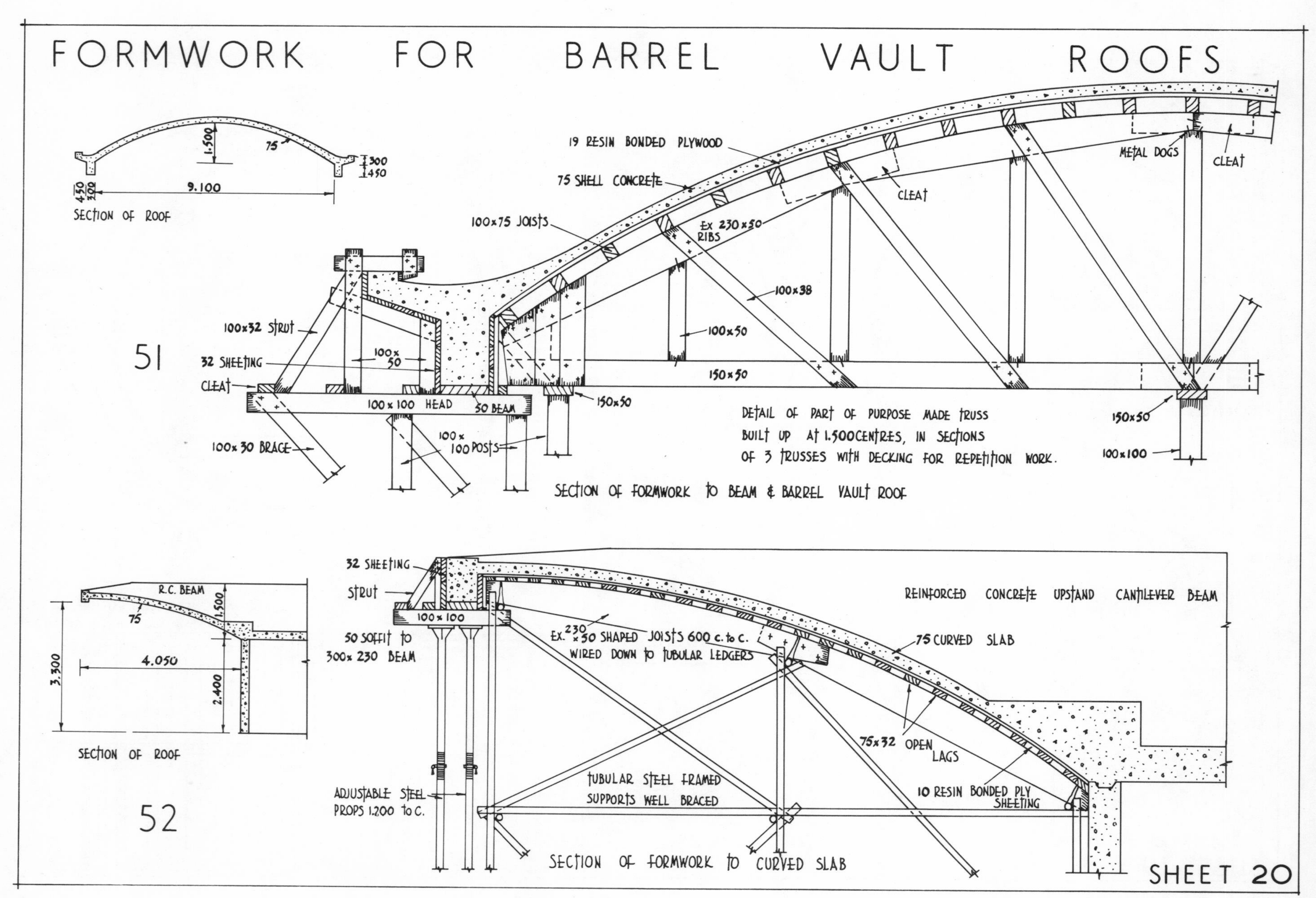

Curved Slab

A suitable arrangement of the shuttering and supports to a curved slab and beam are shown in Sheet 20.52. Adjustable steel props support the beam formwork and tubular steel framed supports used for the curved slab.

Tubular Steel

Tubular steel scaffolding is now used extensively to support formwork for concrete work. An example of this is shown in the work on the curved roof slab. Concrete floors, bridge centring, power stations, etc. are other instances where tubular steel may be used, apart from its normal use of providing platforms.

The main disadvantages of steel scaffolding are:

(1) It is difficult to attach timber to the steel tubes.
(2) The initial cost is high.
(3) When used as horizontal members they cannot be relied upon to take heavy loads without support.
(4) The dangers resulting from couplers not being tightened.

The advantages are:

(1) Ease and speed of erection.
(2) Large stocks are available.
(3) Adaptable for use in formwork, scaffolding and centring.
(4) Tubes are standard from a strength point of view.
(5) Easy to dismantle.
(6) With proper care will last indefinitely.

Patent Devices

A number of patent shutter fixing devices are now available designed to facilitate speed of erection and stripping of formwork for all types of concrete work. They are *Acrow Ties*, *Loops* and *Hangers*. The principle behind all three is the same. A rod is coiled to form a thread and a special bolt is provided to fit. Welded to the coiled thread there is, in the case of the *Loop* shown on Sheet 16, a looped rod; and in the case of the *Tie* two or more straight rods which in turn are welded again at the other end to a second coiled thread shown on Sheet 14. The *Hanger*, instead of having straight rods, has bent rods as shown in Sheet 17.40. The bolt provided fits all three, as the thread is the same in all cases. The bolt passes first through a flat washer, then through the shuttering and through a hardwood cone against the inner face of the shuttering, and is tightened up in the thread. When the bottom lift of the shuttering has been placed in position, with loops or ties, as the case may be, fixed, the concrete is poured.

Column Clamps

Acrow steel adjustable column clamps consist essentially of four arms of flat mild steel 8 mm thick, made in three sizes to suit a wide range of column sizes, Plate 3.1. The four arms are identical in all respects. One end is provided with a series of staggered and lapped slots and the other end is bent over to allow the neighbouring arm to slide through the bend at right-angles. To each bent-over end a steel wedge is firmly secured by a strong length of chain welded to the arm.

When fixing, two nails driven in the timber at the required level provide a temporary rest for the clamp, enabling one man to make the adjustments and fix the wedges in the appropriate slots.

Beam Clamps

Plate 3.3 illustrates the lightweight fixed arm type of *Acrow* beam clamp. This may be used in intermediate positions with the screw type clamp, or when the great strength of the latter is unwarranted, such as in formwork to the casing of steel beams.

Plate 3.4 shows the *Acrow* synchronized-screw type for clamping beam formwork. The adjustment for width is made by turning the handle which operates the screw, clamping the formwork tight. Spigots are provided at the centre and at each end, so that, whether supported by one *Acrow* prop or by two, positioning of the props is instantaneous and secure. A prop is shown in Plate 3.6. Further illustrations of clamps and props in use are shown in Plate 3, figs. 2 and 5.

Chapter Three: Carpentry

Open Roofs

Perhaps the best-known example of this form of ornamental roof to be found in this country is that of Westminster Hall. This 600 years old hammer-beam trussed roof, which has a span of 20.400 m, is a glorious example of the skill of the craftsmen of that period.

Open roofs are steeply pitched and have no tie beams which gives an increased unobstructed height to the interior of the building and adds greatly to the appearance of the roof. Being subject to considerable wind pressure the trusses are usually stiffened from each other or from the trusses and purlins by wind braces.

The principal forms of open roofs are the collar-beam, arched rib and hammer-beam. Although these are only to be found in the older types of public buildings, maintenance work is still required to be done, and for this reason examples of traditional roofs are included in this book.

In these examples it will be noted that the timbers used are of heavy section. These have been largely dispensed with in present-day structures where glued laminated members are used as an alternative to solid timber.

Arched Hammer-Beam

Sheet 21.6 shows the geometrical setting out of this traditional form of truss. In this the main constructional member is the timber arch which is built up from four or more members bolted together and moulded. The horizontal member is called the hammer-beam and projects at the level at the foot of the principal rafter, to which the latter is framed. The upright member, which triangulates the arch at this point, is known as the hammer-post. Additional rigidity is obtained from the two smaller curved ribs.

The pointed main arch is first set out by joining the chord A–B and bisecting it to meet a normal from the springing at point C.

The two smaller curved ribs run into the main rib at points D and E. Chords joining points D–F and F–E are bisected to give the centres G and H to strike the smaller ribs.

Arched Collar-Beam

Sheet 21.5 shows the setting out of part of this truss. As in the previous example, a pointed arch gives the best appearance where the rise must be greater than half the span. The pitch of the principal rafter is first set out with position of the collar drawn in. The width of the rib at points B, C and D are marked and are indicated by the dotted lines as shown. The angle made by the principal rafter and wall post is bisected to meet a normal from the springing at point E. This gives the first centre to strike the arc B–C. The chord C–D is drawn and bisected to give the second centre F in C–E produced.

Sheet 21.1 shows a part elevation of a single hammer-beam truss which has the same outline as the arched hammer-beam, without the main arch.

The 275 mm by 230 mm principal rafters are bridle jointed to the 300 mm by 230 mm hammer-beam and secured by a heel strap. A three-way strap is used at the top to secure the rafters and the key post. The collar-beam, which is tapered, is jointed to the principal rafter and key post and secured by metal straps. The lower edges of the principal rafter and collar and the outer edge of the hammer-post are housed 13 mm deep to receive the 150 mm thick arch ribs. These are scarf jointed in four sections and bolted to the rafter, collar and hammer-post with 19 mm bolts as shown.

The decorative tracery panels are housed 13 mm into the lower edges of the rafter and hammer-beam and the inside edges of the hammer-post, wall post and curved rib.

The purlins and ribs are cogged to the principals and secured by coach screws. The common rafters also are cogged to the purlins, and in this construction are halved together at the ridge and finally covered with roof boarding, two layers of sarking felt, counter battening and slates.

Sheet 21, figs. 2, 3 and 4 show the sections of the principal rafter, collar-beam and wall post.

Arched Collar-Beam Truss

Sheet 22.7 shows the key elevation of this form of traditional truss which is suitable for a 6.000 m span. The pitch of the roof is 45 degrees with a pointed or Gothic arch. As in the previous details of open roof trusses, where there are no main tie beams, thicker walls are necessary. The tendency of an arch when loaded at the crown is to sink at this part and to spread at the haunches and thrust out the walls. To counteract this tendency, the walls are generally

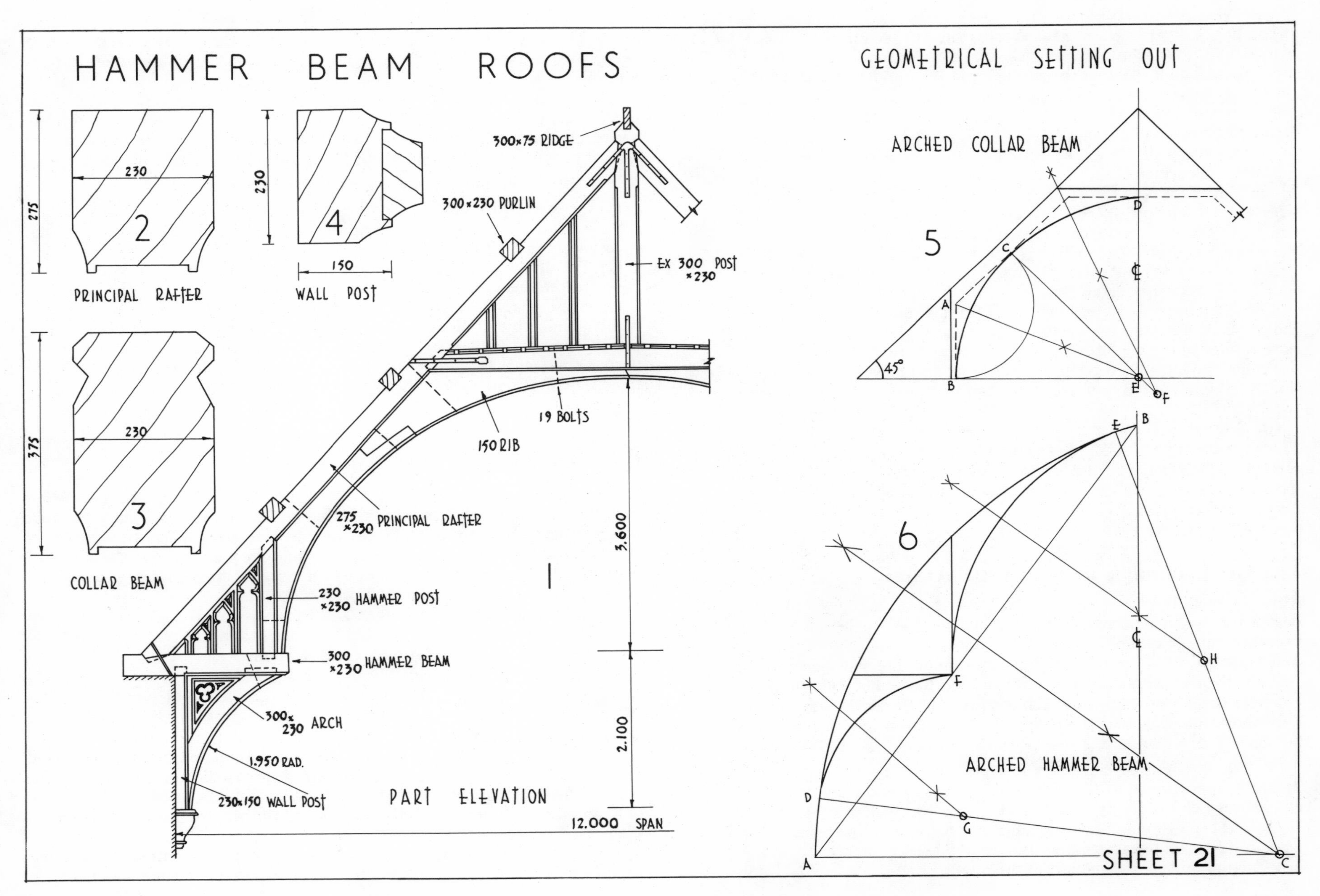
HAMMER BEAM ROOFS
GEOMETRICAL SETTING OUT
275
230
2
PRINCIPAL RAFTER
230
4
150
WALL POST
375
230
3
COLLAR BEAM
300×75 RIDGE
300×230 PURLIN
EX 300 POST ×230
19 BOLTS
150 RIB
275 ×230 PRINCIPAL RAFTER
3.600
1
230 ×230 HAMMER POST
300 ×230 HAMMER BEAM
300× 230 ARCH
1.950 RAD.
2.100
230×150 WALL POST
PART ELEVATION
12.000 SPAN
ARCHED COLLAR BEAM
5
45°
A
B
C
D
E
F
6
ARCHED HAMMER BEAM
G
H
SHEET 21

strengthened by buttresses placed opposite the position occupied by the trusses. This is shown in the part elevation of the roof truss, Sheet 22.8. The arch rests upon a stone corbel usually placed well below the eaves level to throw the weight as low as possible on the wall.

The arrangement of the roof members is similar to the previous example except that there is no hammer-beam and post. The principal rafter is bridle jointed to the wall plate, and secured with a heel strap and the wall post jointed to the principal rafter with an oblique tenon and mortise. The tapered collar-beam is jointed to the principal rafter with a similar joint and secured with a strap. The crown post is secured at the bottom by bolting and at the top by a three-way metal strap. The 100 mm thick arched ribs are housed into the lower edges of the principal rafter and collar-beam, and the face of the wall post as shown in Sheet 22, figs. 9 and 10. Bolts used at the positions indicated secure the curved ribs.

The 125 mm by 125 mm purlins are framed into the principal rafters and the backs of the latter are on the same plane as the 100 mm by 50 mm common rafters which are placed at 600 mm centres, boarded and slated.

Vertical boarding is used at the top of the inside wall to complete the wall line to the pitch of the roof and is finished with a cornice mould between each truss.

An alternative form of truss giving the same traditional outline as Sheet 22.8 is shown in Sheet 22.11. Here a glued laminated arch rib is used. The section of the rib and wall post are shown in Sheet 22.14 with the details of the jointing at the head of the wall post shown in Sheet 22.12 and the collar beam in Sheet 22.13.

The Fire Endurance of Timber Structures

This is generally very considerable. Nevertheless, wherever timber is suggested for use in a structure, the fact which tends to be considered to the exclusion of all others is that wood burns whereas alternative structural materials do not. Yet timber structures under fire will generally fail less quickly than comparable structures of unprotected metals or pre-stressed concrete structures, both of which are rapidly affected by heat.

In short, fire resistance is a property distinct from combustibility and must be considered separately. These points should be noted in this connection:

(1) The strength of timber is not affected by heat: loss of strength is directly proportional to the reduction of cross-section due to the effect of charring which occurs at a steady rate regardless of temperature.

(2) The expansion of timber in a fire is very low and side pressure by timber roofs or floors on the external walls does not develop: hence the building does not fracture or collapse in case of fire.

(3) The high fire endurance of timber structures allows evacuation of occupants, a long period of salvage and comparatively safe access for fire fighting.

(4) The risk of fire breaking out and spreading is largely determined by the use and contents of a building and not by the materials it is constructed from. There are other considerations in relation to the fire hazard—for instance, the risk of outbreak of fire and that of the rapid spread of flame in a building.

It is useful to remember these points in this connection:

(1) Since the shell of the building is often undamaged, repairs and replacement of burnt-out timber members are easier and less costly than total rebuilding required in the case of collapsed buildings.

(2) The amount of burnable materials used in construction is small compared to the fuel content introduced by occupancy.

(3) Rapid spread of flame is associated not so much with the structure as with inflammable contents or with wall and ceiling linings of combustible materials having a high rate of spread of flame. Linings of timber or timber derivatives (such as fibre boards) can, however, easily be treated to bring their spread of flame classification up to the best class.

Glued Laminated Timber Structures

Lamination is the process of building up comparatively thin pieces of timber into larger members which may be either straight or curved. The pieces are bonded together with glue under pressure to produce a structural member of such size, length, shape and strength as may be required for the purpose. The grain of the laminates is parallel, in contrast to plywood in which the grain alternates with each lamination.

The laminates may be arranged in one of two ways, vertically or horizontally. Vertical lamination is seldom used, and the great majority of structural components are laminated horizontally.

The most important advantage of horizontal lamination is that members may readily be curved. It allows the design of segmental or parabolic arches which are in themselves economical structural forms. It also simplifies the design of fixed joints in such members as portal frames, thereby gaining economy through continuity in the structure. Laminated timber arches and portal frames have an immediate aesthetic appeal which is enhanced by the fine finish which can be given to the material.

Examples of a number of glued laminated timber structures are shown on Plate 4.

Other advantages of horizontal lamination may be summarized briefly. Members may be tapered or moulded to give the greatest strength where it is

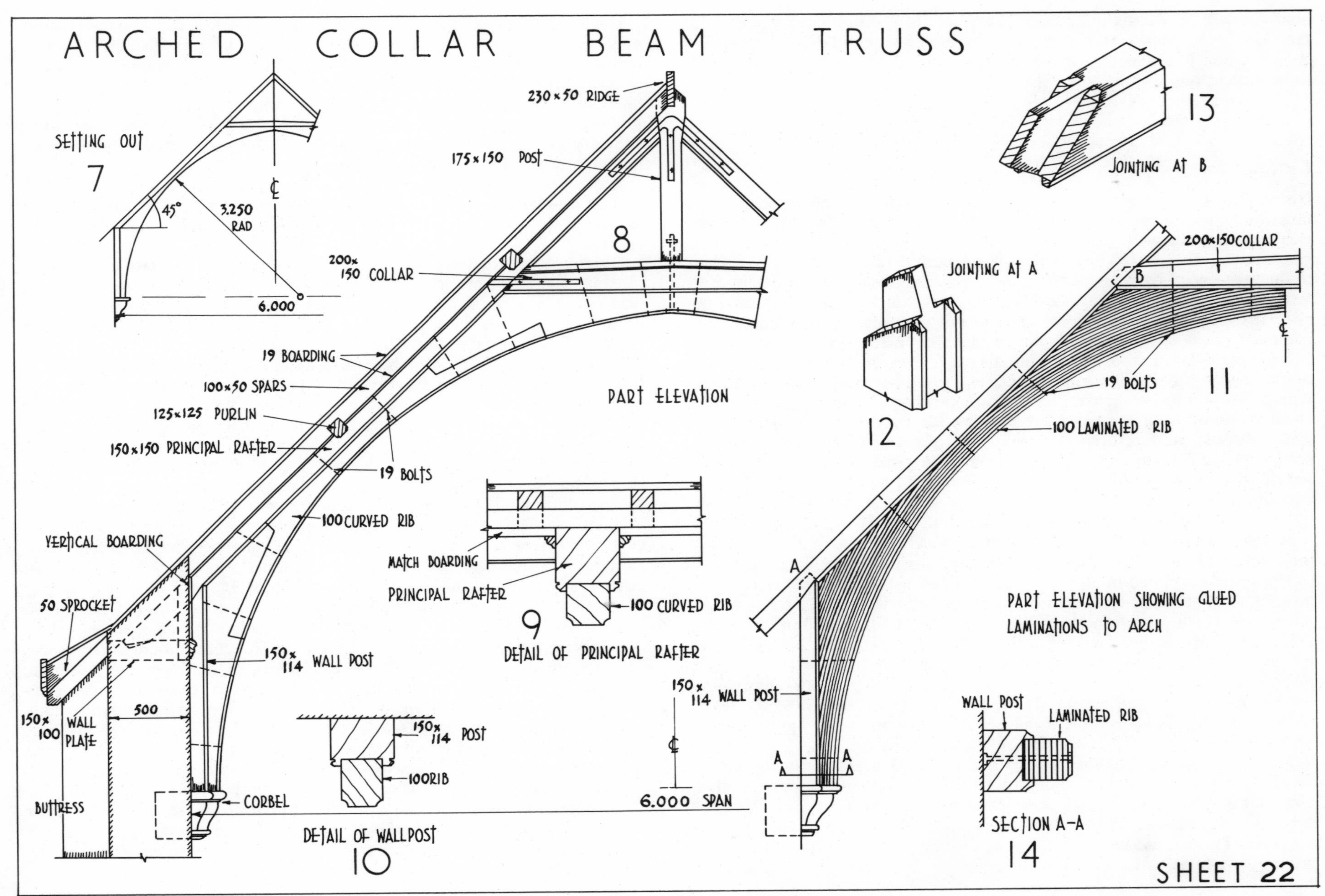
ARCHED COLLAR BEAM TRUSS
SETTING OUT
7
45°
3.250 RAD
6.000
230×50 RIDGE
175×150 POST
8
200×150 COLLAR
19 BOARDING
100×50 SPARS
125×125 PURLIN
150×150 PRINCIPAL RAFTER
19 BOLTS
100 CURVED RIB
PART ELEVATION
VERTICAL BOARDING
50 SPROCKET
150×100 WALL PLATE
150×114 WALL POST
500
BUTTRESS
CORBEL
MATCH BOARDING
PRINCIPAL RAFTER
100 CURVED RIB
9
DETAIL OF PRINCIPAL RAFTER
150×114 POST
100RIB
DETAIL OF WALLPOST
10
6.000 SPAN
13
JOINTING AT B
JOINTING AT A
12
200×150 COLLAR
B
19 BOLTS
11
100 LAMINATED RIB
A
150×114 WALL POST
A A
PART ELEVATION SHOWING GLUED LAMINATIONS TO ARCH
WALL POST
LAMINATED RIB
SECTION A–A
14
SHEET 22

most needed. Seasoning is relatively easy, due to the small cross-section of individual laminates. Components may be built up to any desired length or cross-section. Material may be chosen which would otherwise be too small to be structurally useful.

Adhesives

One of the basic requirements of lamination is that the bonding material must be as strong and durable as the timber itself. In timber structures the weakness has always been in the joints. With the developments in modern glues, joints, when tested to destruction, break in the timber itself rather than in the glue-line. The types of glue most commonly used are the caseins and synthetic resins. Generally speaking casein glues, which are derived from sour milk, are easy to mix and apply and are not over-sensitive to damp. They make extremely strong joints, but lose much of their strength when wetted and are liable to attack by micro-organisms. They should not be used if the timber is likely to attain a moisture content of 20 per cent. If moisture contents above this figure are expected, a synthetic resin adhesive should be used.

All synthetic resin adhesives are plastics, the best-known are those with urea, phenolic and resorcinol bases. They are manufactured as liquids, powders or in dry impregnated film, and may be either thermo-plastic, these soften with the application of heat, or thermo-setting, which do not soften when set, using either hot or cold. Synthetic resins are immune from attack by moulds and bacteria and are resistant to moisture. There are two principal types: the urea-formaldehyde for interior work, and the phenol-formaldehyde for exterior work with which can be included the resorcinol group.

The urea is supplied in two parts, the resin in the form of either syrup or powder and the hardener. They may be purchased together as a ready-mixed powder to which water is added to make the liquid adhesive. When they are used separately, the resin is applied to one face of the joint and the hardener to the other, the setting takes place when the two are brought together and pressure applied. Both casein and urea glues can be set at room temperature but resorcinol resin glues generally require additional heat, although many resin glues are supplied in forms that set quickly without the application of heat.

To promote setting of the resins within a reasonable time when gluing normally in the workshop, one of two methods or a combination of both is used. Heat greatly accelerates setting, as has already been mentioned. A hardener, often called an 'accelerator' or 'catalyst', is added or applied to the resin. This causes the resin to become thicker, then rubbery and finally promotes setting.

Heat is used in conjunction with hardeners when it can be conveniently applied, and produces setting within a very short time. In the furniture industry heat can be economically and conveniently applied by radio-frequency, low voltage heating and heated hydraulic presses to raise the glue-line temperatures, enabling joints to be made in minutes.

Casein glues in powder form have a storage life of about a year if kept in a cool dry place, and resin will remain usable for three to six months after manufacture under normal temperature conditions.

A quick-acting hardener shortens the length of time for which resin glues remain usable, this is known as the 'pot life'. This is usually quoted by the makers at various temperatures. As the temperature rises the pot life becomes shorter.

Glues used for building structures must be gap-filling. That is to say they must be capable of bridging gaps up to 1.25 mm in joints where the surfaces may not be in close contact owing to the impossibility of adequate pressure or any inaccuracies. These gap-filling properties are obtained by the addition to the resins of suitable fillers by the manufacturers.

Stress grading of timber is becoming more and more an essential part of timber engineering where the strength of a laminated member is governed by the minimum cross-section of clear timber in it. With solid timber, defects such as knots, shakes, splits, etc. reduce the strength of the particular member, whereas in laminated work these defects may be present in individual laminates, but they do not occur upon one another and therefore only affect the strength of that particular lamina. While lower-grade timber may be used for the centre laminates, the outer laminates should be selected clear grades.

In straight laminated members the thickness of the laminates is governed only by the need to obtain close contact at the glue-lines, and 50 mm nominal material is commonly used. With curved members, however, the timber must be bent without breaking in a dry state, unheated and with glue applied. The thickness of the laminates will depend on the radius of curvature. This should be taken as 25 mm thick to 3.750 m radius for practical purposes.

In fabricating glued laminated members the moisture content of the timber is controlled at between 8 and 18 per cent, depending on the type of glue used, and where the member is to be used. The surfaces to be glued are machined to fairly accurate dimensions, and scarf joints are cut for the jointing of laminations in the same layer. Plain scarfs with a slope of not more than 1 in 12 are made on a circular sawbench adapted for the purpose. Some form of jig is necessary for curved members which may be arranged vertically in the form of a centre, or horizontally where wood or metal posts in the floor are used. The jig must be correctly shaped to the profile of the concave side of the member to be fabricated. The first laminate is fastened to the jig and thereafter the remaining laminates are passed through a mechanical glue-spreader and assembled one on top of the other.

The pressure required is normally 690 to 1380 kN/m^2 for softwoods, and the best method of obtaining this is by using closely spaced clamps tightened with powered nut-runners. A final squeeze-out of glue is obtained with manual ratchet wrenches. Cramps may be spaced as far as 375 mm apart, but on curved surfaces they may be as close as 100 mm. Pressure is maintained from six to twelve hours, depending on the type of adhesive used and the shop conditions. When the member is removed from the jig the faces are planed to final dimensions and sanded. A preservative seal is applied, and, for protection, components are wrapped in waterproof paper or other suitable covering which is left in place until after erection.

Bowstring Truss

This is a modern development of the traditional Belfast truss and provides a far more efficient structural unit. Basically a bowstring truss consists of a curved laminated top chord and main tie, and bracing members of solid timber. The top chord and main tie or bottom chord are generally double, the bracing members being placed between them and the joints made with timber connectors.

No single factor was more responsible for revolutionizing timber design than the introduction and development of the timber connector, so widely used today. Previously, joints in large structures had been made with bolts and straps which required the use of wider members due to the low efficiency of the bolt as a mechanical connector.

Various types of connectors have been developed and these are dealt with more fully on page 57.

The shape of the bowstring truss is such that a very large proportion of the forces is carried by the top chord and the main tie, and unless the truss is asymmetrically loaded the forces in the bracing members are small. The thrust of the top chord at the eaves is taken by a metal shoe or strap which passes round the end of the main tie.

Sheet 23.1 shows the small key-line diagram of a laminated bowstring truss suitable for a span of 21.000 m.

Bowstring trusses are the most economical form of roof construction for warehouses, timber-yards, workshops, etc., where clear spans of 13.500 to 60.000 m and more are possible.

Since the radius of curvature of the top chord is large, 50 mm laminations are used. Double top chords built up with four laminations each 75 mm by 50 mm. Douglas fir are glued with Recorcinol resin adhesive and nailed together as shown in Sheet 23.4. The tie or bottom chord, Sheet 23.5, is built up in a similar way of three laminates each 150 mm by 25 mm laminated vertically and made double with a camber of 75 mm at the centre.

The struts arranged as shown in Sheet 23.2 are placed between the double sets of chords with the joints made using 75 mm diameter timber connectors with 13 mm diameter bolts and square washers.

The 200 mm by 50 mm purlins are spaced at 600 mm centres with purlin blocks placed between the double top chords to position and support intermediate purlins as shown.

The detail of the finish at the eaves is shown in Sheet 23.3 where a 6 mm mild steel heel strap is used and secured with bolts and shear plate connectors to take the thrust of the top chord. A 230 mm by 50 mm eaves purlin with the top edge bevelled is fixed across the bottom chords and braced by short angle struts.

Trusses can be erected on timber or steel columns, brick piers or load-bearing walls. 200 mm by 200 mm timber stanchions are shown in the diagram, tenoned to a laminated head.

The trusses are spaced at 4.200 m centres, braced diagonally with 100 mm by 50 mm members and horizontally with 150 mm by 50 mm at the crown as shown in Sheet 23.6. This bracing is used at the positions marked A and B in Sheet 23.2 and repeated for the other half of the truss.

Roofs may be covered in any of the usual materials such as corrugated asbestos, galvanized iron or close-boarded timber and roofing felt.

Glazing and ventilation may be arranged by incorporating centre lanterns or glazing runs.

Trusses for very large spans are generally built in two sections for ease of handling and transporting and bolted together on the site, the chords being spliced at the centre.

Plate 4.8 shows laminated bowstring trusses being erected by a mechanical crane.

Timber Shell Roofs

These are now being used quite frequently in modern building as a result of technological advances made in obtaining and formulating data on the new built-up structural components in timber.

A shell roof consists of a thin sheet of material which is light in weight, and rigid as a result of curvature. Where timber is used the shell of the roof will generally consist of three or more layers of boarding placed in different directions and glued and nailed together.

Shell roofs are, by their size, restricted to site construction where centring is used to form the necessary curves. A number of the different constructions of shell roofs are illustrated on Sheet 24.

The hyperbolic paraboloid timber shell roof shown in the sketch in Sheet 24.1 consists of a covering made up of four panels in which each panel is a

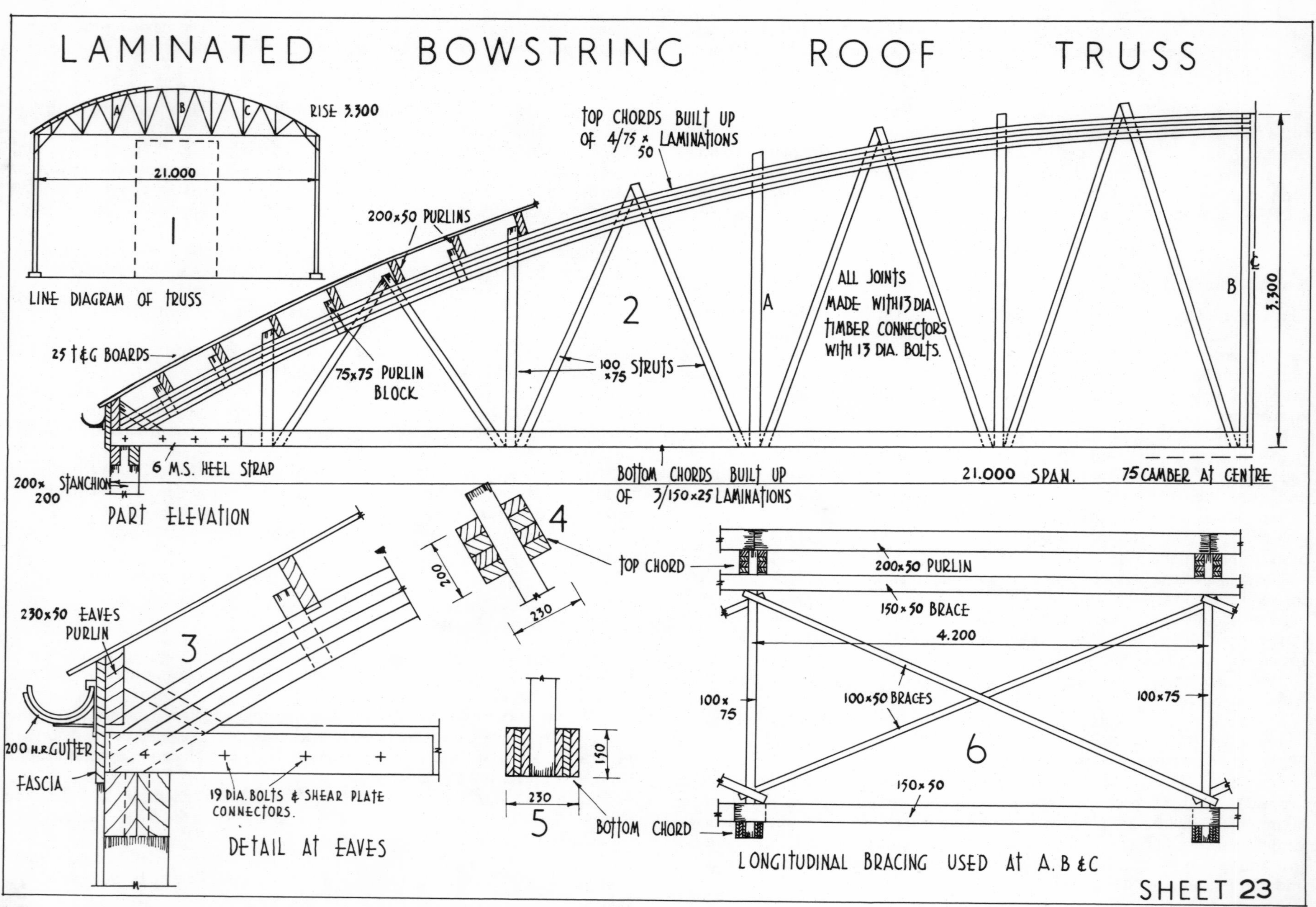
LAMINATED BOWSTRING ROOF TRUSS
A
B
C
RISE 3.300
21.000
1
LINE DIAGRAM OF TRUSS
TOP CHORDS BUILT UP OF 4/75 x 50 LAMINATIONS
200x50 PURLINS
25 T&G BOARDS
75x75 PURLIN BLOCK
100 x75 STRUTS
2
ALL JOINTS MADE WITH 13 DIA. TIMBER CONNECTORS WITH 13 DIA. BOLTS.
3.300
6 M.S. HEEL STRAP
200x 200 STANCHION
BOTTOM CHORDS BUILT UP OF 3/150x25 LAMINATIONS
21.000 SPAN.
75 CAMBER AT CENTRE
PART ELEVATION
4
200
230
TOP CHORD
230x50 EAVES PURLIN
3
200 H.R. GUTTER
FASCIA
19 DIA. BOLTS & SHEAR PLATE CONNECTORS.
DETAIL AT EAVES
150
230
5
BOTTOM CHORD
200x50 PURLIN
150x50 BRACE
4.200
100x 75
100x50 BRACES
100x75
6
150x50
LONGITUDINAL BRACING USED AT A.B&C
SHEET 23

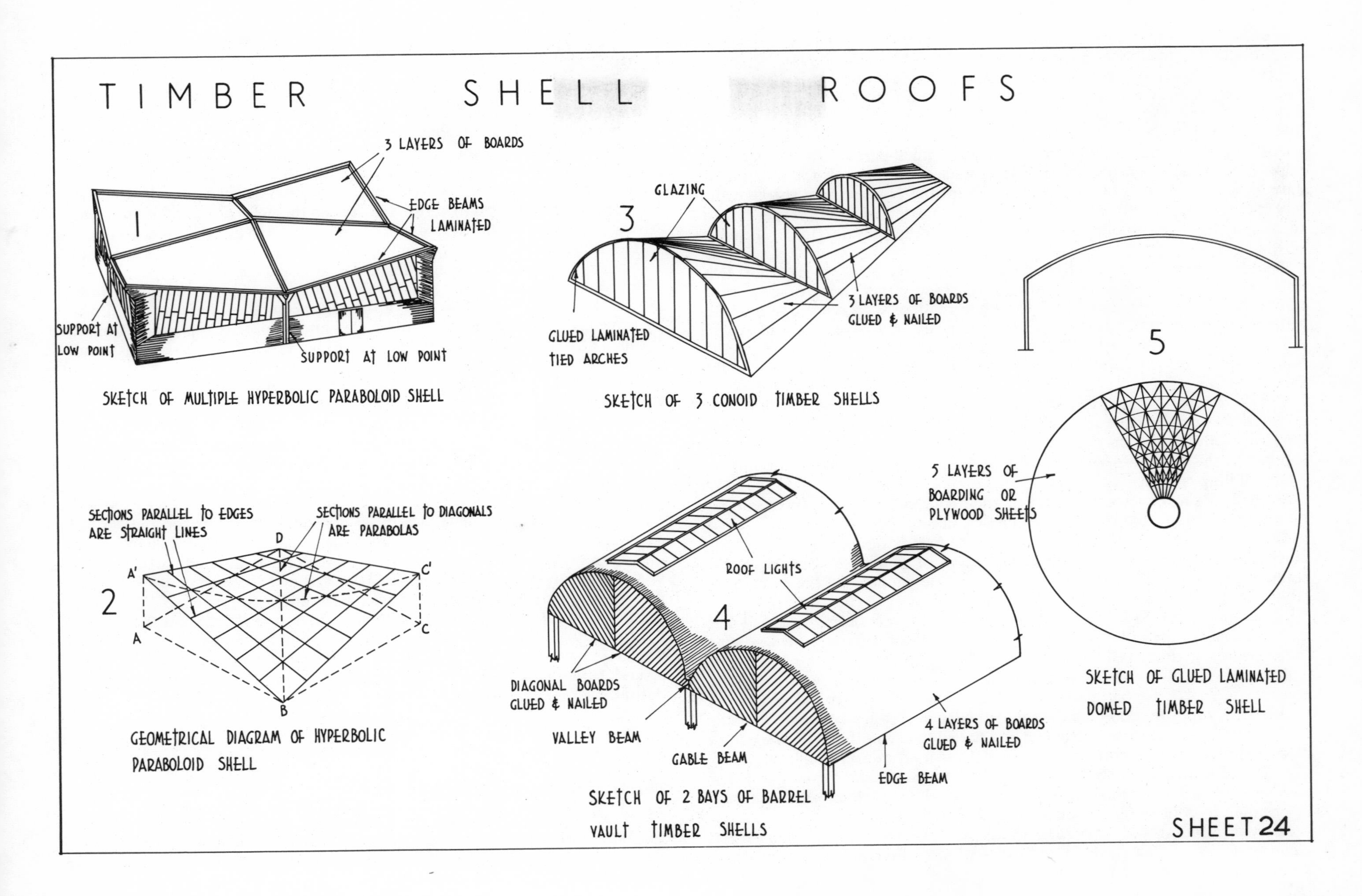
TIMBER SHELL ROOFS
3 LAYERS OF BOARDS
EDGE BEAMS
LAMINATED
1
SUPPORT AT
LOW POINT
SUPPORT AT LOW POINT
SKETCH OF MULTIPLE HYPERBOLIC PARABOLOID SHELL
GLAZING
3
3 LAYERS OF BOARDS
GLUED & NAILED
GLUED LAMINATED
TIED ARCHES
SKETCH OF 3 CONOID TIMBER SHELLS
5
5 LAYERS OF
BOARDING OR
PLYWOOD SHEETS
SKETCH OF GLUED LAMINATED
DOMED TIMBER SHELL
SECTIONS PARALLEL TO EDGES
ARE STRAIGHT LINES
SECTIONS PARALLEL TO DIAGONALS
ARE PARABOLAS
D
A'
C'
2
A
C
B
GEOMETRICAL DIAGRAM OF HYPERBOLIC
PARABOLOID SHELL
ROOF LIGHTS
4
DIAGONAL BOARDS
GLUED & NAILED
VALLEY BEAM
GABLE BEAM
4 LAYERS OF BOARDS
GLUED & NAILED
EDGE BEAM
SKETCH OF 2 BAYS OF BARREL
VAULT TIMBER SHELLS
SHEET 24

4*

three dimensional surface. Sheet 24.2 shows one of the panels constructed in the following manner: four straight lines are arranged to form a square with the two diagonally opposite corners raised. The sides are divided into an equal number of parts and the corresponding points on each pair of opposite sides joined by straight lines.

Although the surface is formed of straight lines parallel to the edges in plan, it is in fact doubly curved. The cross-section through the two raised corners is a parabola (concave upwards) and the cross-section through the low corners is a parabola (concave downwards). All cross-sections parallel to the diagonals are parabolic, and have the same form, only varying in the level of the apex.

The shell need not have all four sides with the same slope. It may have two adjacent sides horizontal and two sides inclined either upwards or downwards. Alternatively it may have two adjacent sides with one slope and the other two sides a different slope. Finally all sides may have different slopes. All these shells are parts of the same geometrical surface (the hyperbolic paraboloid) provided that they are square on plan and have straight sides.

The method of support for the roof will depend on whether single panels are used or whether panels are grouped together as in Sheet 24.1. Although the load of the roof is carried at the low corners it is advisable to give some support at the high corners to limit deflection. Normally curtain walling or panel infilling is used.

Sheet 24.3 shows a second form of shell roof consisting of three conoid timber shells supported on glued laminated tied timber arches.

The Timber Research and Development Association have carried out much research in the design and construction of timber shell roofs in this country and have erected and tested full-scale models at their research laboratories.

The conoids formed in position consist of three layers of boards glued and nailed to each other with vertical glazing used in the spaces as shown, to give the necessary lighting below.

A third form of timber shell roof is shown in the sketch, Sheet 24.4. The construction is of timber barrel-vault shells carried on vertical laminated timber ribs, supported on glued-laminated edge and valley beams. The shells consist of four layers of boards glued and nailed to each other with the gables diagonally boarded. A continuous roof light is used in each shell.

A domed timber shell shown in Sheet 24.5 gives a further example of the wide possible application of this type of roof, the actual shape of the roof determined by a need for an uninterrupted floor space.

As with all timber engineering, design is the concern of the structural engineer whilst the carpenter's concern is that of fabrication, construction and erection of the various members.

Hyperbolic Paraboloid

The plan and elevation of a hyperbolic shell roof consisting of four panels are shown in Sheet 25, figs. 1 and 2. In this example the spans are not great for this form of roof, being 4,800 m square with a rise of 1.500 m. Support for the roof is by timber columns at the low points with 22 mm diameter tie rods to resist the outward thrust as shown. To carry the forces around the edges of the shell, edge beams are required, these may either be cut from solid timber or laminated. Solid timber is only used for small shells up to 6.000 m square. Most edge beams are fabricated in the workshop from glued laminated timber.

The slope of the shell varies along the edge beam, this requires the beams to be laminated with a twist in them during construction.

For erection on the site, tubular steel scaffolding is provided to the underside of the shell. The bottom half of the edge beam is put in position followed by three layers of boards glued and nailed and finally the top half of the edge beam. The purpose of gluing the boards in timber shell construction is mainly to increase the stiffness of the membrane.

Sections taken at the high, mid and low points in the edge beams are shown in Sheet 25, figs. 3, 4 and 5.

Sheet 25.6 shows the tie rod and mild steel anchor fixing at the low points to the timber columns. In order to throw the rain-water clear over the edge of the beam the area shown in the plan at the low points is boarded as shown in Sheet 25.6.

Now that timber shell roofs have become established, new improved nails have been introduced for securing the boarded laminations in the roofing. They are of two types—square twisted shank nails and ringed shank nails.

Square twisted shank nails are used between the laminated membrane and edge beams and are available up to 200 mm long. They are made from high-carbon-content steel and can be driven without bending.

Ringed shank nails are wire nails up to 56 mm in length and are used for board nailing. They do not withdraw easily and as the wood fibres enclose the ringed serrations of the nail shank, the final pressure imparted to the layers of boards is retained. This pressure between the layers when gluing is desirable.

Laminated Roof Beams

The illustrations on Sheet 26 show examples of glued laminated beams for varying spans with alternative fixings to timber stanchions and brickwork.

In Sheet 26.1 a 6 mm mild steel plate is housed into the beam and stanchion on each side and fixed by screwing.

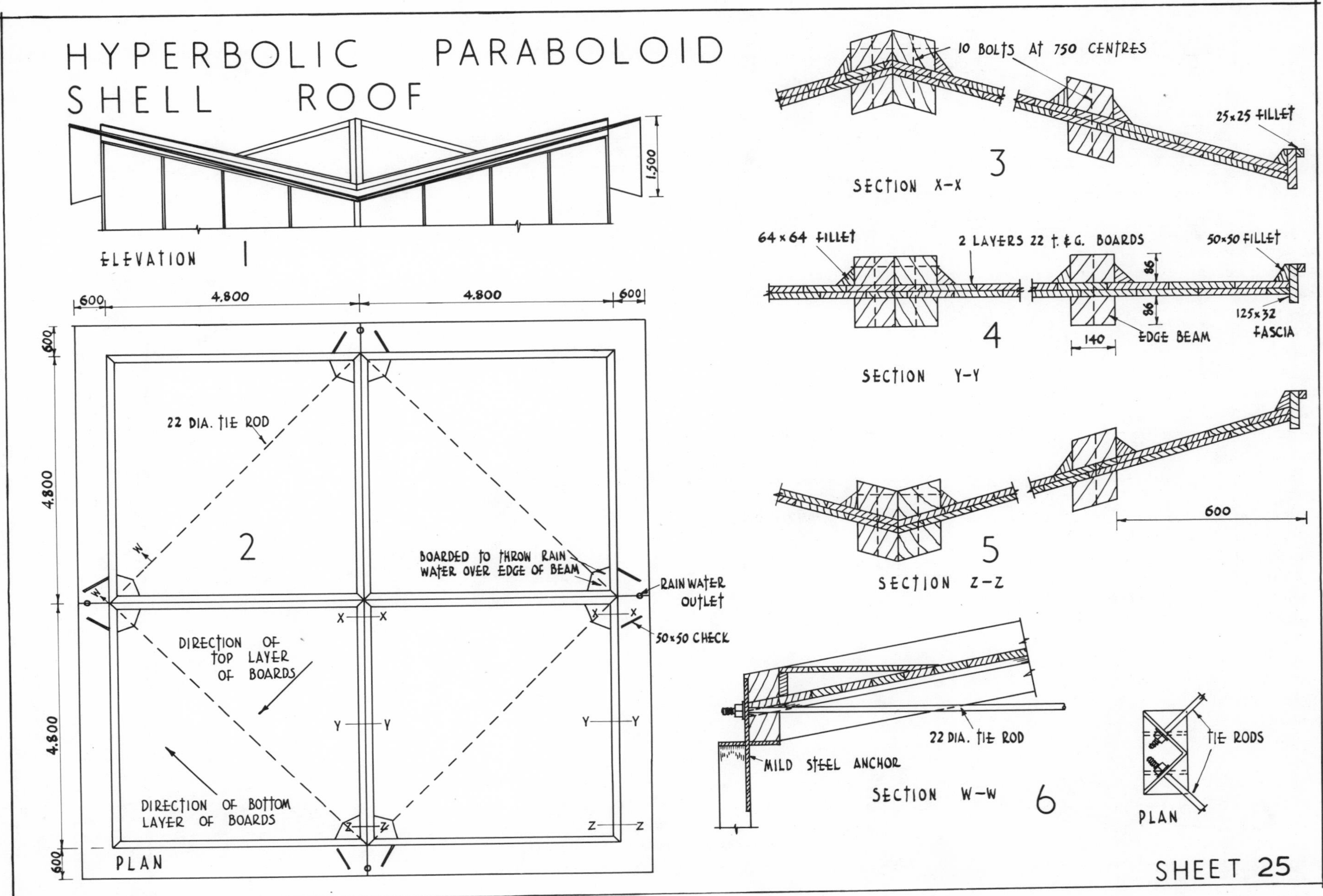
HYPERBOLIC PARABOLOID SHELL ROOF
1,500
ELEVATION 1
600
4.800
4.800
600
22 DIA. TIE ROD
2
BOARDED TO THROW RAIN WATER OVER EDGE OF BEAM
RAIN WATER OUTLET
50×50 CHECK
DIRECTION OF TOP LAYER OF BOARDS
DIRECTION OF BOTTOM LAYER OF BOARDS
PLAN
10 BOLTS AT 750 CENTRES
25×25 FILLET
3
SECTION X–X
64×64 FILLET
2 LAYERS 22 T. & G. BOARDS
50×50 FILLET
86
125×32 FASCIA
4
140
EDGE BEAM
SECTION Y–Y
600
5
SECTION Z–Z
22 DIA. TIE ROD
MILD STEEL ANCHOR
SECTION W–W
6
TIE RODS
PLAN
SHEET 25

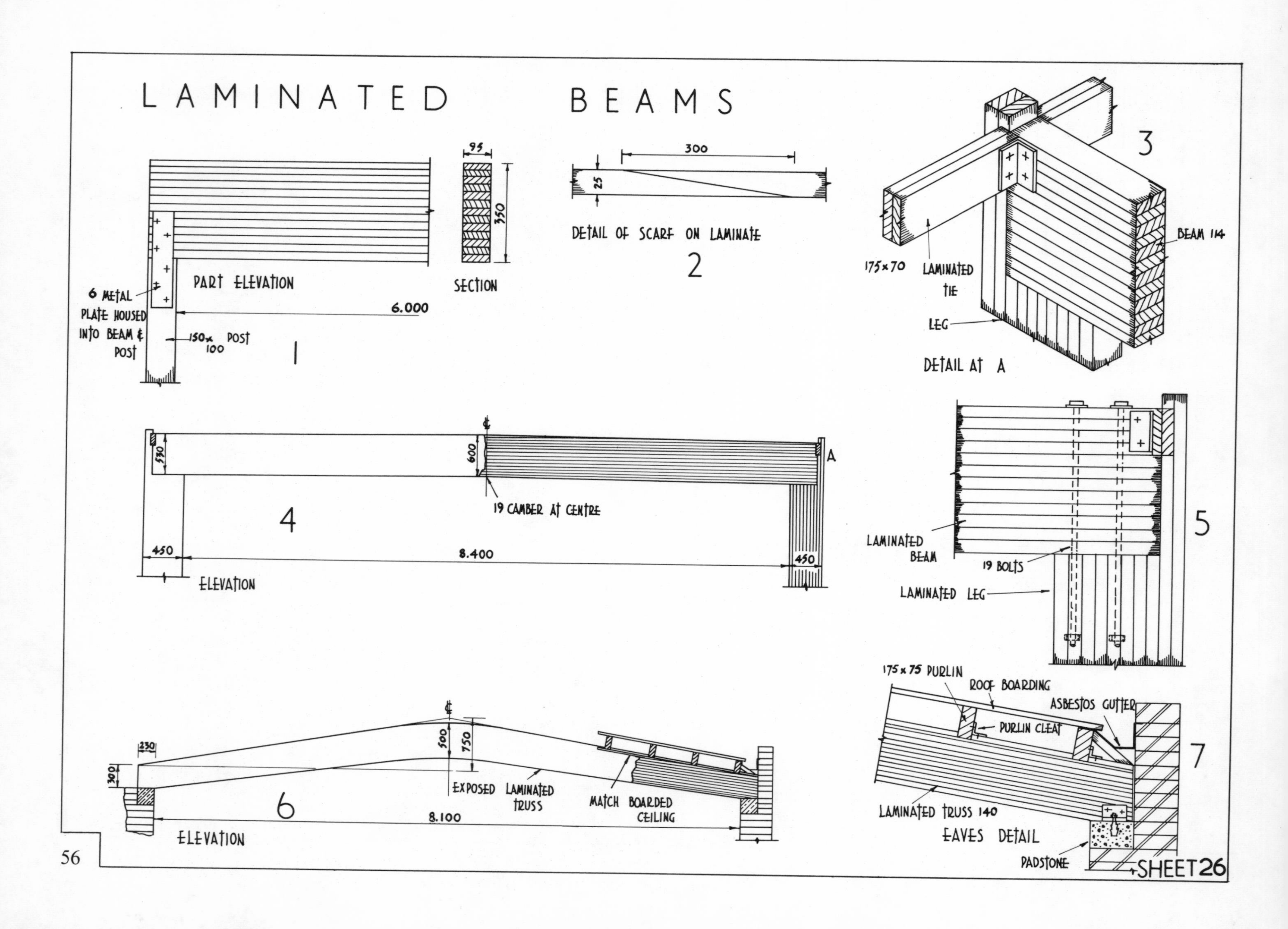
LAMINATED BEAMS
95
350
PART ELEVATION
SECTION
6 METAL PLATE HOUSED INTO BEAM & POST
6.000
150× 100 POST
1
300
25
DETAIL OF SCARF ON LAMINATE
2
3
BEAM 114
175×70
LAMINATED TIE
LEG
DETAIL AT A
℄
530
600
A
19 CAMBER AT CENTRE
4
450
8.400
450
ELEVATION
LAMINATED BEAM
19 BOLTS
LAMINATED LEG
5
175×75 PURLIN
ROOF BOARDING
ASBESTOS GUTTER
PURLIN CLEAT
7
LAMINATED TRUSS 140
EAVES DETAIL
PADSTONE
230
300
500
750
EXPOSED LAMINATED TRUSS
MATCH BOARDED CEILING
6
8.100
ELEVATION
SHEET 26

Sheet 26.2 shows a scarf for the jointing of laminates.

A tapered laminated beam suitable for a span of 8.400 m is shown in Sheet 26.4. This has a camber of 19 mm at the centre and is fixed at the eaves to glued laminated legs. The fixing is by two 19 mm bolts bored as shown in Sheet 26.5. Sheet 26.3 shows the fixing of the laminated ties between the beams where mild steel angles are coach screwed to the ties and the beam on each face.

A cranked glued laminated beam is shown in Sheet 26.6. This is carried on padstones where mild steel angles are used with coach screws into the beam and rag-bolts into the padstone. Purlins, 175 mm by 75 mm in section, are spaced at 600 mm centres and notched over the truss and cleated. The roof is boarded, with an asbestos eaves gutter as shown in Sheet 26.7. Matchboarding is used to finish the ceiling with the exposed trusses varnished.

Boxbeams

Sheet 27.1 shows the part elevation of a boxbeam faced with 13 mm Douglas fir plywood on each face. This type of beam has a high degree of stiffness combined with a low weight factor and is cheaper than the equivalent laminated beam.

The beam consists of a softwood glued core built up of 100 mm by 100 mm top and bottom chords with 100 mm by 50 mm stiffeners at 600 mm centres. A wedge-shaped fillet glued and nailed to the top chord may be used to give the required fall in the finished roof covering. The face ply is glued and nailed to the core with the face grain of the plywood running along the direction of the span.

Sheet 27.2 shows the detail of the jointing between the beam and laminated post. This is forked over the beam on each face and screwed as shown.

A section through the beam is shown in Sheet 27.3.

'Kaybeam' Roof Truss—Pat. No. 754739

This, shown in Sheet 27.4, which combines strength with attractive appearance, is ideally suited for use in assembly halls, libraries and other public buildings. The basic structure follows the lines of the familiar Pratt girder, with a relatively flat pitch, the internal frame being formed by the top, bottom and vertical members. The latter are spaced more closely together as the loading increases, with the diagonals applied externally to form the outer stressed skin.

Sheet 27.5 shows the construction of the beam which can be designed for spans of up to 15.000 m in either softwood or hardwood and finished to tone with any desired decorative scheme. The trusses are illustrated on Plate 4.6.

A section through the beam is shown in Sheet 27.6.

Timber Connectors

These have been developed in numerous forms, each possessing certain advantages suitable for particular situations in which they are employed.

Toothed Connectors

This type is a round or square metal plate having projecting teeth around the edges. They are supplied in either single- or double-sided forms. The single-sided connector is used in cases where prefabricated or easily demountable structures are projected, each piece of timber having a connector embedded, these connectors coming back-to-back when the members are assembled.

The double-sided connector is used single where joints are permanent, and is embedded in the contact faces simultaneously when the members are drawn together by the bolt passing through the connector and the timbers. This form of connector is used with the Timber Research and Development Association design truss on Sheet 28 and the trusses on Sheets 23 and 29. No pre-cut grooves are required for these toothed connectors, the teeth being embedded by the drawing together of the two members in contact; for this purpose a special long bolt is used, as it will be appreciated that the normal bolt is not long enough to take the cut before members are drawn up, due to the thickness of the not-yet-embedded connectors.

Shear Plate Connectors

These are somewhat similar to single-sided toothed connectors but have a solid rim in place of teeth and therefore require pre-cut circular grooves to house the rims. They have a higher load-carrying capacity than the toothed rings and are therefore preferred in making timber-to-steel connections, or in providing demountable joints. As they have no teeth to be embedded, the special bolt for drawing-up is not required, the connector rims being lightly tapped into position in the grooves. This type is illustrated in Sheet 31.5.

Split Ring Connectors

These are simple split circular steel bands, which are embedded in pre-cut circular grooves cut in the members being jointed, half the depth of the ring being embedded in each timber. They have a higher load-carrying capacity than toothed or shear plate connectors and are employed where necessary to maintain joints at a reasonable size. This type is shown in Sheet 31.3.

Spike Grid Connectors

This type is specially designed for the purpose of connecting poles or baulks together, such as are used in the construction of piers, jetties, etc. They take several forms such as:

(1) The flat grid for joining two flat surfaces.
(2) The single curve for joining a flat and curved surface.
(3) The double curve for joining two curved surfaces.

As they are, in effect, spiked types of connectors, somewhat similar in application to the toothed plate connectors previously described, they are installed by inserting between the surfaces to be joined and embedded by applying pressure by means of a through bolt.

The round type of toothed connectors are available in 50 mm, 64 mm, and 75 mm diameters and the square type of toothed connector is available in 38 mm, 44 mm, 50 mm, 64 mm, 75 mm and 88 mm diameters. Shear plate connectors are available in 68 mm and 100 mm diameters, while split rings are manufactured in 64 mm and 100 mm diameters.

Stressed-Skin Plywood Panels

A stressed-skin panel is formed by combining one or two skins of a suitable thin material with a system of suitably spaced beams and headers which not only form a stiff frame but also contribute towards the load-carrying cross-section. Suitable materials for this form of structure would be plywood (capable of very high permissible stressing) for the skin or skins and solid carcassing timber for the beams or studding, the skins being glued to the latter members.

This type of panel undoubtedly makes great use of the inherent high strength of plywood and in the case of the type employing upper and lower skins can, by careful selection of facing veneers, eliminate the need for separate materials for the ceiling or internal cladding, depending on whether the unit is used for flooring or for walling.

Being extremely light in weight and thin in cross-section, the panels are ideal for use in prefabricated housing for walls, floors and roofs, being simple to erect. They also present possibilities for use as bridge decks, for shuttering for concrete, advertisement hoardings, etc., and are capable of acting as slabs spanning simultaneously in two directions, with consequent shallower depth of thickness.

For a panel to qualify as a stressed-skin panel, the following criterions must be satisfied:

(1) Skins must be glued to the framing.
(2) Skins and frames must be continuous or adequately spliced longitudinally.
(3) Headers must be provided if thin, deep framing members are used.
(4) The effective width of the plywood flanges depends upon certain relations between stress and strain functions and the geometry of its construction, i.e., the number of parallel and perpendicular plies respectively.

The bonding of the plywood to the framing could also be effected by nailing or by a combination of nailing and gluing, but in this case the nailing sizes and spacings require a considerable amount of careful calculation and gluing is recommended, the panels being prefabricated in controlled workshops.

If the panel length exceeds the available stock length of the plywood, suitable joints will be required in the skins and framing members in order to develop their full strength at the joints. These can be simple butt joints with internal glued cover plates which are simple to fabricate. Glued scarf joints of bevel 1 in 12 can also be provided, although with more difficulty, and these scarf joints have a higher efficiency rating than the spliced butt joints. With regard to the types of plywood skins to be used for particular situations or requirements, the possible variations of species and combinations of plywoods with other materials are considerable. For insulation purposes, thermal and sound, plywood with cores of cork, asbestos, rubber, etc. are available and special requirements as to facing veneers for decorative purposes are easily met. Particular requirements calling for metal or plastic facing can also be complied with and the plywood can be bonded at the edges and rendered vermin- and water-proof.

The T.R.A.D.A have recently prepared designs of various methods of cross-wall roof constructions for different types of roof coverings and roof pitches. One design is of light-weight prefabricated sloping roof panels which are entirely self-supporting and capable of carrying all imposed live loading without the necessity for providing trusses and purlins. Plywood girders used with common rafters and plywood girders with scissors trusses are other forms of roof designs for cross-wall construction where the T.R.A.D.A. domestic type roofs and traditional purlin and rafter roofs are not suitable.

The Timber Research and Development Association Roof

The T.R.A.D.A roof design is not a single design for one particular span or pitch but a system of construction covering a number of spans for domestic roofs up to 7.800 m between walls and for two main pitches, 40 degrees for plain tiling, and 35 degrees for interlocking tiles and slates.

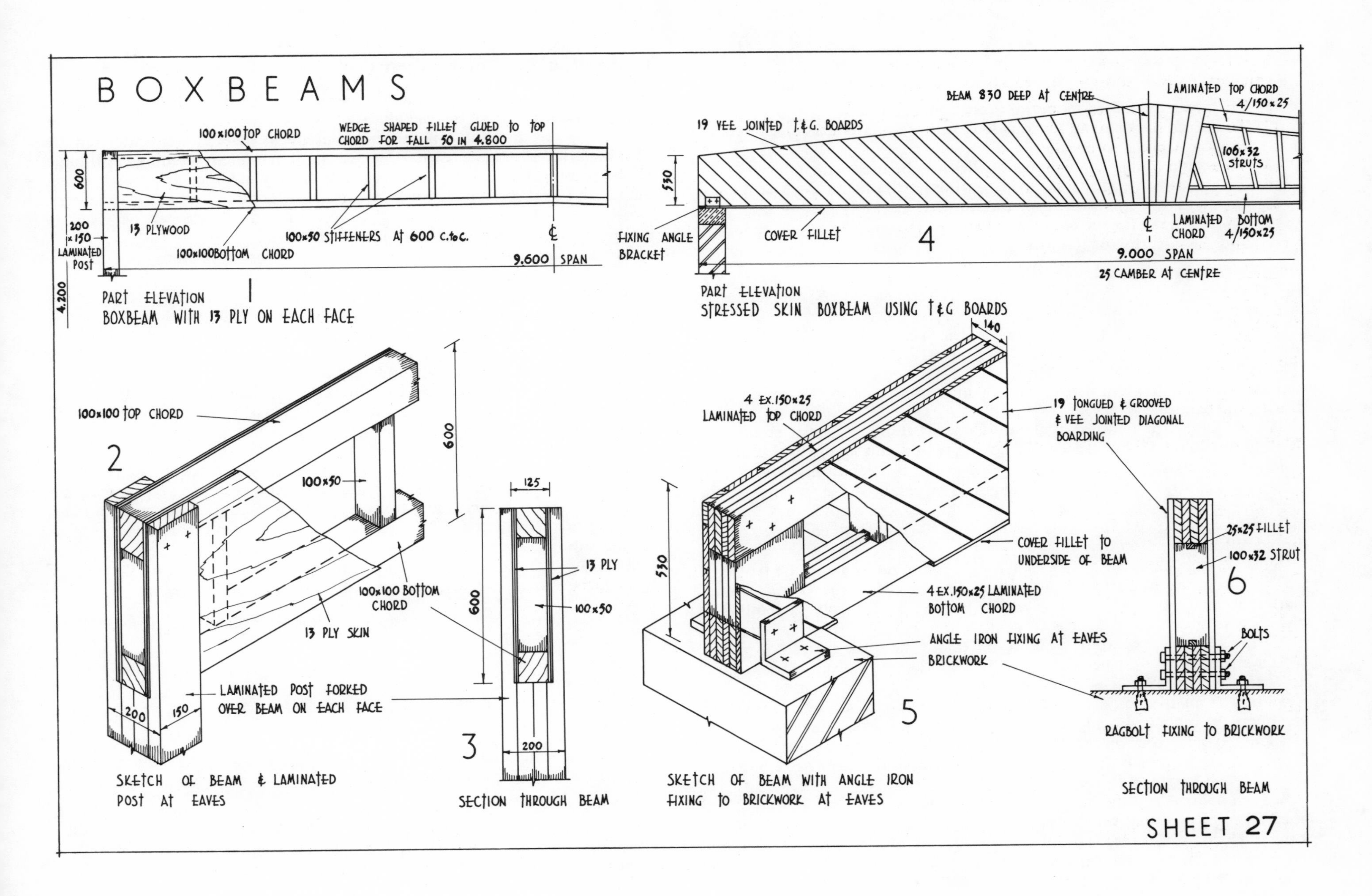

BOXBEAMS
100x100 TOP CHORD
WEDGE SHAPED FILLET GLUED TO TOP CHORD FOR FALL 50 IN 4.800
600
200 x150
LAMINATED POST
13 PLYWOOD
100x100 BOTTOM CHORD
100x50 STIFFENERS AT 600 C. to C.
9.600 SPAN
4.200
PART ELEVATION
1
BOXBEAM WITH 13 PLY ON EACH FACE
BEAM 830 DEEP AT CENTRE
LAMINATED TOP CHORD 4/150x25
19 VEE JOINTED T.&G. BOARDS
106x32 STRUTS
530
FIXING ANGLE BRACKET
COVER FILLET
4
LAMINATED BOTTOM CHORD 4/150x25
9.000 SPAN
25 CAMBER AT CENTRE
PART ELEVATION
STRESSED SKIN BOXBEAM USING T&G BOARDS
140
100x100 TOP CHORD
2
100x50
600
100x100 BOTTOM CHORD
13 PLY SKIN
LAMINATED POST FORKED OVER BEAM ON EACH FACE
200
150
SKETCH OF BEAM & LAMINATED POST AT EAVES
125
13 PLY
600
100x50
200
3
SECTION THROUGH BEAM
4 EX.150x25 LAMINATED TOP CHORD
19 TONGUED & GROOVED & VEE JOINTED DIAGONAL BOARDING
530
COVER FILLET TO UNDERSIDE OF BEAM
4 EX.150x25 LAMINATED BOTTOM CHORD
ANGLE IRON FIXING AT EAVES
BRICKWORK
5
SKETCH OF BEAM WITH ANGLE IRON FIXING TO BRICKWORK AT EAVES
25x25 FILLET
100x32 STRUT
6
BOLTS
RAGBOLT FIXING TO BRICKWORK
SECTION THROUGH BEAM
SHEET 27

A series of design sheets is available issued by the Association.
This type of roof was evolved to satisfy the following requirements:

(1) To secure overall economy in timber consumption.
(2) To reduce the demand for long lengths in standard sizes and to enable random lengths to be used.
(3) To enable timber in available qualities to be used with safety.
(4) To avoid the necessity for load-bearing partitions on the first floor.
(5) To encourage prefabrication and modern assembly methods.
(6) To reduce the time taken on site between the completion of the walling and the covering of the roof.

Sheet 28.1 shows the small-scale key elevation of a truss for a span of 6.200 m. The trusses are fixed at a maximum of 1.800 m centres with common rafters at 450 mm centres between.

A little more than half of a truss is shown in Sheet 28.2. Where a central partition is available to support the ceiling joints, the central hanger is omitted. The 100 mm by 50 mm ceiling binders are fixed to the ceiling joists by skew nailing with two 100 mm wire nails. It is important to strut the ceiling joists during the fixing.

All the joints are made with 50 mm diameter 'Bulldog' timber connectors with 13 mm diameter bolts, each bolt having two 38 mm square by 3 mm mild steel washers as shown in Sheet 28.3. One purlin is used in this type of roof positioned centrally on the rafter length. The detail of the jointing of the purlins in length is shown in Sheet 28.9. A 50 mm lap is cut with 13 mm diameter holes bored as shown to prevent splitting, and 100 mm skew nails used in pre-bored holes.

Sheet 28, figs, 3, 4, 5, 6, 7 and 8 show the six main joints in the truss.

The two halves of a trussed rafter should be identical, but where a central hanger is required it will be necessary to use two longer bolts in the half of the truss to which the hanger is to be attached. Fabrication of the trusses is therefore a simple operation as both halves are the same hand, not paired. When erecting it is only necessary to take two identical halves, reversing one to form the complete trussed rafter unit. It is important to note that any deviation from the design which results in a different arrangement of the members or joints will complicate the assembly and defeat the object of the design. A typical case where this can lead to complication is the use of a single length ceiling joist or bottom tie, and it is far better to make the ceiling joist or tie in two lengths as shown.

It is emphasized on the design sheets that efficient joints are essential, and it is recommended that where undue shrinkage may occur through the use of timber which has not received sufficient drying, inspection of the joints should be made and bolts tightened where necessary before the end of the contract maintenance period.

Protecting Timber Against Fire

The increasing use of timber as a structural material has brought with it a growing recognition of the usefulness of preservative treatment as a means of improving the durability of the material. Local by-laws frequently require that materials should be rendered fire resistant in modern buildings. Although timber cannot be rendered incombustible by any methods at present in use, steps can be taken to increase its resistance to ignition and to delay or even prevent its active participation in a fire. Such resistance can be given to timber by the application of fire-retardants.

Modern fire-retardants are of two types: (1) those applied to the timber by pressure impregnation; (2) those for surface application.

The first method is normally used when there is a risk of intense fire in other adjoining materials such as structural timbers in warehouses, garages, hangers. Surface coatings are generally used for lightweight and decorative wood, plywood and wallboards where the risk is of ignition from a fairly small source and may be applied either by brush, spray or dipping.

The protection of timber against fire is distinct from its preservation from attack by destructive insects and fungi, which will have been dealt with in earlier work, but the idea of a combined treatment should be possible.

Among the large number of chemicals which have been suggested for the protection of timber against fire are monoammonium dihydrogen orthophosphate, ammonium sulphate, boric acid, zinc chloride and ammonium chloride. These substances and compositions containing one or more of them are the principal water-soluble fire-retardants used in pressure impregnation today, although some are also used in surface coatings.

A fire-retardant surface coating will protect the surface of the timber against the spread of flame and ignition in one or more ways:

(1) The coating may swell to form an insulating layer which retards the access of heat.

(2) The fire-retardant may change the thermal decomposition of the timber.

(3) The coating may liquefy by the heat to produce glassy substances which coat the surface and prevent access of oxygen.

(4) The coating may be thick, heavy and inactive, keeping out oxygen and absorbing or dissipating the heat so that the timber remains cool.

Timber which has been treated and is to be used in the open should be protected from the weather by painting it with a paint of low combustibility and, if timber which has been treated with a fire-retardant is to be glued, the makers of the particular adhesive used and the manufacturers of the fire-retardant should be consulted.

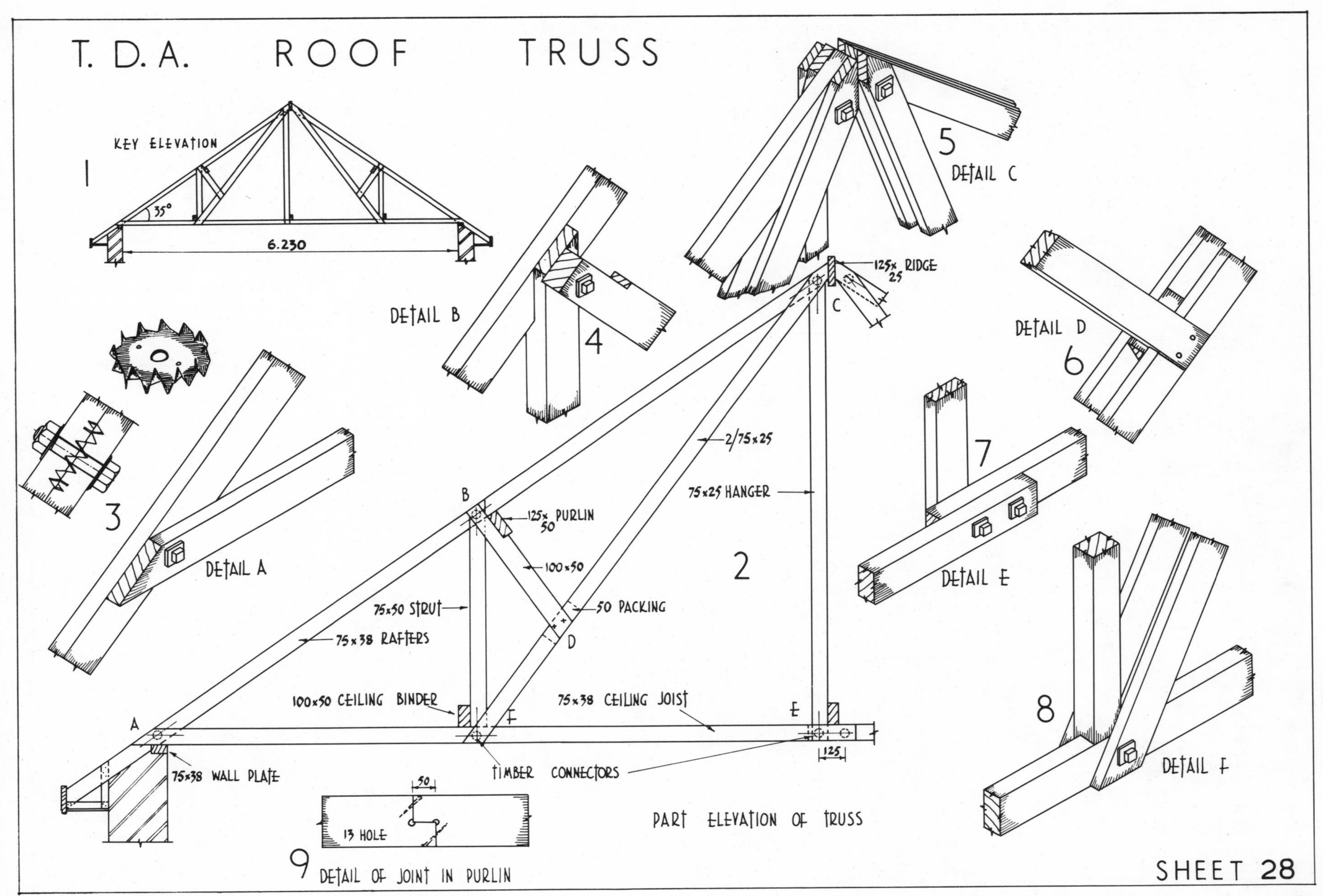
T. D. A. ROOF TRUSS
KEY ELEVATION
1
35°
6.230
DETAIL B
4
5
DETAIL C
125x RIDGE 25
C
DETAIL D
6
DETAIL A
3
2/75x25
75x25 HANGER
7
DETAIL E
B
125x PURLIN 50
100x50
2
75x50 STRUT
50 PACKING
D
75x38 RAFTERS
8
DETAIL F
100x50 CEILING BINDER
75x38 CEILING JOIST
A
F
E
125
75x38 WALL PLATE
TIMBER CONNECTORS
50
13 HOLE
PART ELEVATION OF TRUSS
9 DETAIL OF JOINT IN PURLIN
SHEET 28

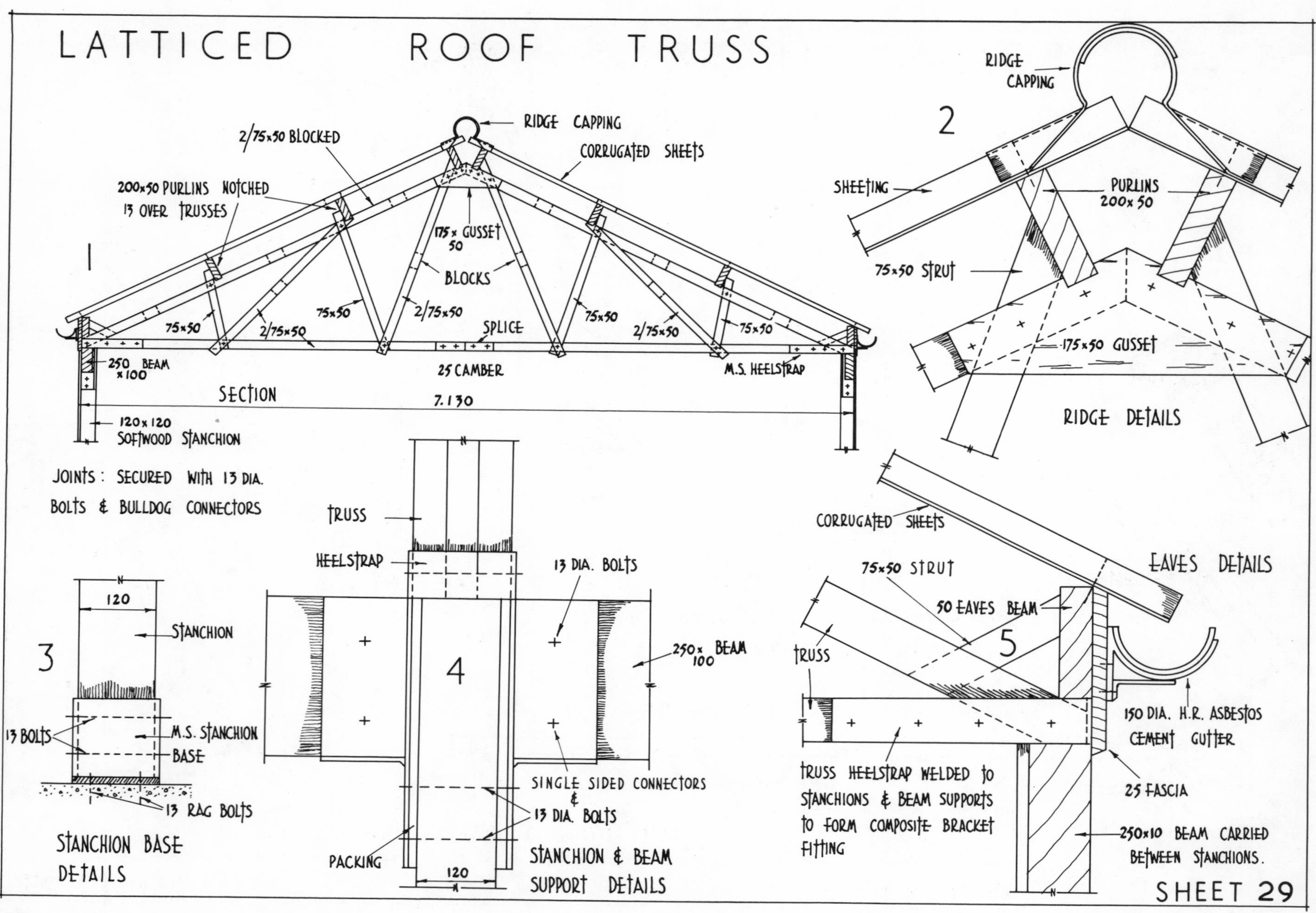
LATTICED ROOF TRUSS
RIDGE CAPPING
2/75×50 BLOCKED
CORRUGATED SHEETS
200×50 PURLINS NOTCHED
13 OVER TRUSSES
175×50 GUSSET
BLOCKS
1
75×50
2/75×50
75×50
2/75×50
SPLICE
75×50
2/75×50
75×50
250×100 BEAM
25 CAMBER
M.S. HEELSTRAP
SECTION
7.130
120×120
SOFTWOOD STANCHION
JOINTS: SECURED WITH 13 DIA.
BOLTS & BULLDOG CONNECTORS
2
RIDGE CAPPING
SHEETING
PURLINS
200×50
75×50 STRUT
175×50 GUSSET
RIDGE DETAILS
TRUSS
HEELSTRAP
13 DIA. BOLTS
250×100 BEAM
4
120
STANCHION
3
13 BOLTS
M.S. STANCHION
BASE
13 RAG BOLTS
STANCHION BASE
DETAILS
SINGLE SIDED CONNECTORS
&
13 DIA. BOLTS
PACKING
120
STANCHION & BEAM
SUPPORT DETAILS
CORRUGATED SHEETS
EAVES DETAILS
75×50 STRUT
50 EAVES BEAM
5
TRUSS
150 DIA. H.R. ASBESTOS
CEMENT GUTTER
TRUSS HEELSTRAP WELDED TO
STANCHIONS & BEAM SUPPORTS
TO FORM COMPOSITE BRACKET
FITTING
25 FASCIA
250×10 BEAM CARRIED
BETWEEN STANCHIONS.
SHEET 29

The effectiveness of fire-retardant treatment is determined by testing treated timbers in controlled conditions and comparing the results with similar untreated timbers in the same conditions. The accepted tests in this country are those to British Standard 476/1953.

Latticed Roof Truss

Sheet 29.1 shows a section through the pitched roof of a timber storage shed. The design of the truss is based on the T.R.A.D.A. industrial roof truss and is supported on timber stanchions having a span of 7.130 m with no ceiling. It will be seen that as with the domestic design of truss, the members are small compared with traditional forms of roof trusses. The same methods of fabrication of the truss are used whatever the span. As the span increases, the greater the sections of members required. With wider spans rigidity is afforded to the columns by knee braces from the trusses and by bracing to the eaves beam. Such trusses are commonly called knee-braced trusses.

The rafters and tie-beams comprise two 75 mm by 50 mm members with 75 mm by 50 mm struts placed between. The dotted lines on the drawing show the positions of 50 mm blocking pieces nailed between each pair of members.

The purlins are 200 mm by 50 mm and are notched over the trusses and supported by the ties as shown.

The joint at the end of the tie-beam and rafter is made with a truss heel-strap welded to the stanchion and beam supports to form a composite bracket shown in Sheet 29, figs. 4 and 5.

Sheet 29.2 shows a detail at the ridge where a gusset plate is used. The roof covering is of corrugated asbestos sheets finished with a ridge capping as shown.

Sheet 29.5 shows the finish at the eaves with the fixing at the base of the stanchions where mild steel stanchion bases are used, shown in Sheet 29.3. The trusses are spaced at 3.000 m centres and all joints are secured with 13 mm bolts and 50 mm diameter double-sided connectors.

A more recent development in the U.S.A. in the assembly of trussed rafters is the use of sheet-metal gussets and splice plates. The plate connectors include flat plates nailed to both sides or inserted into saw kerfs of nominal 50 mm timber. They include H-shaped sheet metal connectors which are fastened to the members with notched nails for automatic clinching during driving. Toothed plates with protrusions and flat plates with a gang of nails punched from the plate forming an integral part of the connector to penetrate into the timber members are used.

Nailed trussed rafters can be assembled on the site in jigs by nailing the members together, or in the shop for mass production using special roller type presses, and where flat metal plates are nailed to both sides of the members, large diameter hardened high carbon steel helically threaded nails are used.

Portal Frames

Laminated timber portal frames are coming into wider use in the construction of churches, halls, gymnasiums and similar buildings. They may be designed as plywood-faced portals using a softwood core, or as glued laminated arches where a certain dignified architectural appeal is realized and timber's versatility is exploited to the full.

Sheet 30.1 shows the elevation of a plywood-faced portal frame where it will be noted that alternative designs are used in the construction. To the left, the face plies to the boom or rafter are allowed to run through to the outside edge of the leg at the knee, with the plies to the leg jointing to them. To the right, an alternative finish at the knee is shown.

A larger detail of a portal frame with the face ply removed to show the arrangement of the members forming the core is shown in Sheet 30.3. This portal is suitable for a span of 9.000 m having a rise of 1.500 m. The frames are made in two sections and are jointed at the crown. Each section or frame is made up of a top boom and a leg, constructed of 125 mm by 38 mm chords in machined redwood, with ribs as stiffeners arranged as shown. These are butt jointed to the chords being glued and nailed. The stiffeners in the top boom may be arranged vertically or at right-angles to the top chord. The face plies have to be spliced every 2.400 m. This may be done on a stiffener or by gluing a splice plate on the inner face of each of the plies to be joined.

Additional bracing is necessary at the knee position. This is in the form of diagonal knee bracing as shown.

A method of jointing the frames at the crown is shown in Sheet 30.4. A solid core is used with a handrail type bolt. An alternative is to use a split-ring connector between the cores with a mild steel bolt.

The face plies forming the skin are of 6 mm exterior grade Douglas fir plywood which may be glued or glued and nailed, or glued and screwed to the softwood framing.

At the base, a softwood core is used. Here the portals are housed in a mild steel shoe and fixed by rag-bolting to the concrete base. Coach bolts are used at the positions shown in Sheet 30.5 to secure the leg of the frame in the metal shoe.

A section through the leg of the frame is shown in Sheet 30.6.

Sheet 30.7 shows an alternative jointing of the two frames at the crown where a plywood gusset plate is glued and nailed to each face.

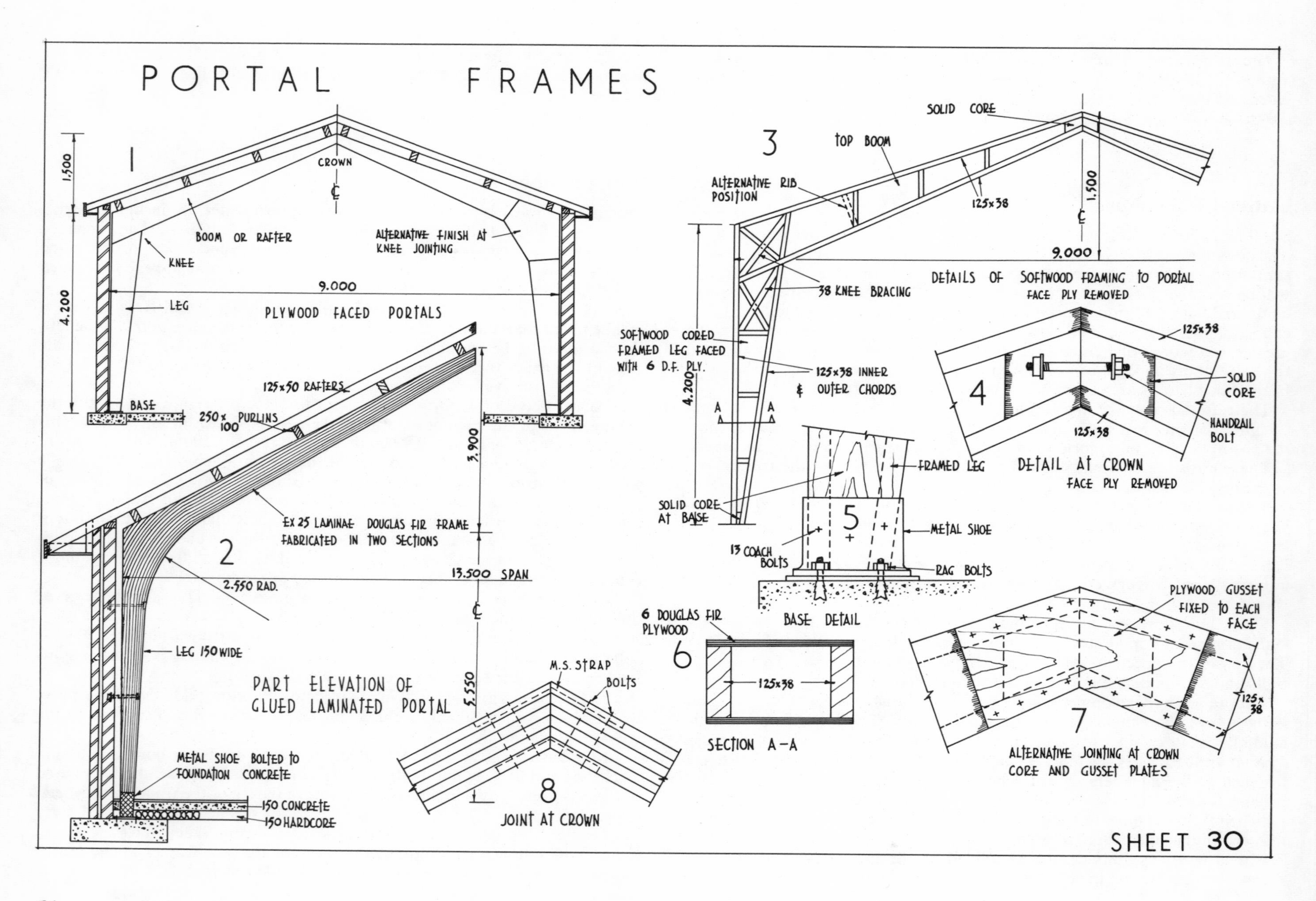

PORTAL FRAMES
1
1.500
4.200
CROWN
BOOM OR RAFTER
KNEE
ALTERNATIVE FINISH AT KNEE JOINTING
9.000
LEG
PLYWOOD FACED PORTALS
BASE
125×50 RAFTERS
250×100 PURLINS
3.900
EX 25 LAMINAE DOUGLAS FIR FRAME FABRICATED IN TWO SECTIONS
2
2.550 RAD.
13.500 SPAN
LEG 150 WIDE
PART ELEVATION OF GLUED LAMINATED PORTAL
5.550
METAL SHOE BOLTED TO FOUNDATION CONCRETE
150 CONCRETE
150 HARDCORE
M.S. STRAP
BOLTS
8
JOINT AT CROWN
3
SOLID CORE
TOP BOOM
ALTERNATIVE RIB POSITION
125×38
1.500
9.000
38 KNEE BRACING
DETAILS OF SOFTWOOD FRAMING TO PORTAL
FACE PLY REMOVED
SOFTWOOD CORED FRAMED LEG FACED WITH 6 D.F. PLY.
4.200
125×38 INNER & OUTER CHORDS
A A
SOLID CORE AT BASE
125×38
4
SOLID CORE
HANDRAIL BOLT
125×38
DETAIL AT CROWN
FACE PLY REMOVED
FRAMED LEG
5
METAL SHOE
13 COACH BOLTS
RAG BOLTS
BASE DETAIL
6 DOUGLAS FIR PLYWOOD
6
125×38
SECTION A–A
PLYWOOD GUSSET FIXED TO EACH FACE
125×38
7
ALTERNATIVE JOINTING AT CROWN CORE AND GUSSET PLATES
SHEET 30

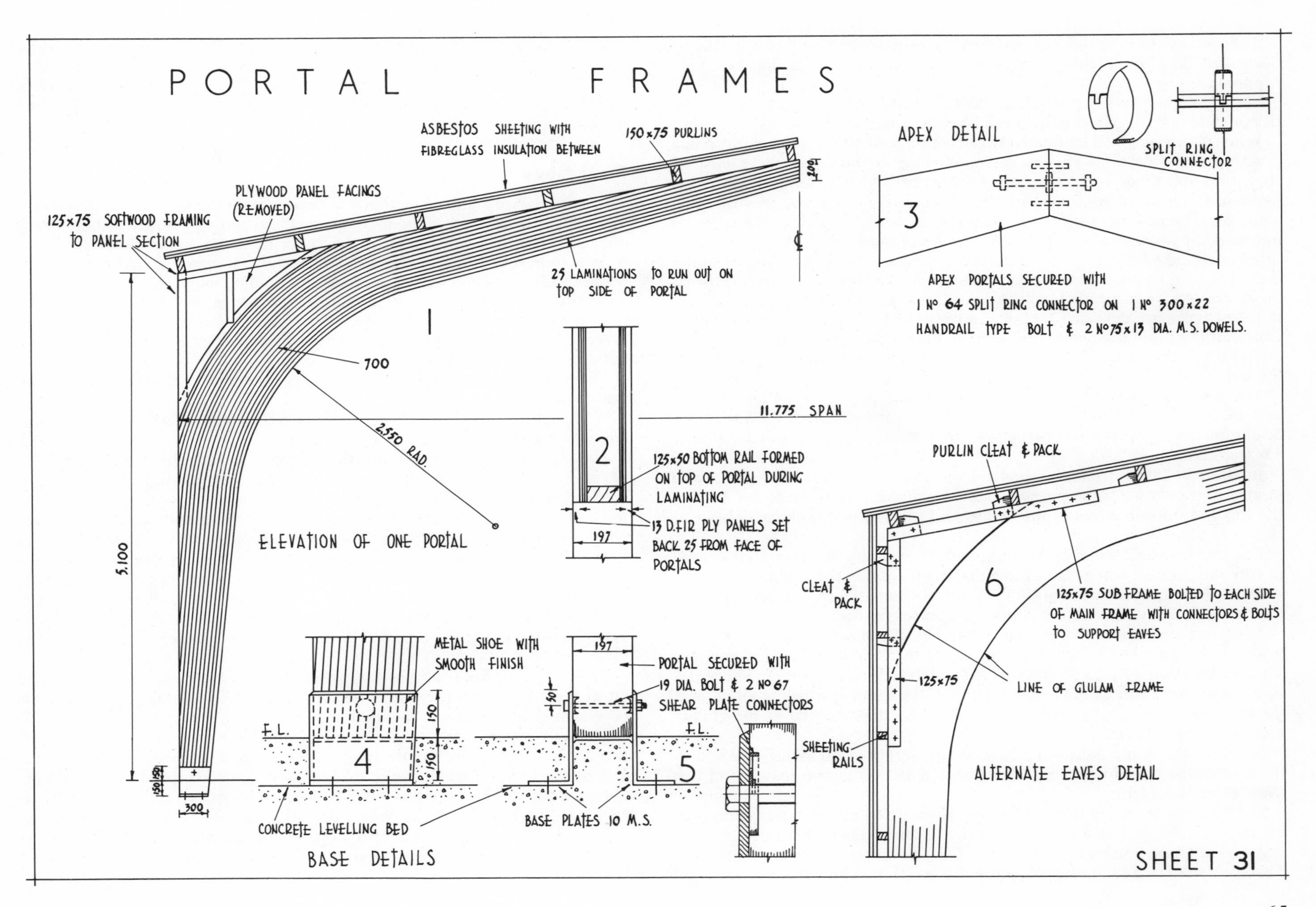
PORTAL FRAMES
ASBESTOS SHEETING WITH FIBREGLASS INSULATION BETWEEN
150×75 PURLINS
PLYWOOD PANEL FACINGS (REMOVED)
125×75 SOFTWOOD FRAMING TO PANEL SECTION
25 LAMINATIONS TO RUN OUT ON TOP SIDE OF PORTAL
200
1
700
2.550 RAD.
11.775 SPAN
5.100
ELEVATION OF ONE PORTAL
300
2
125×50 BOTTOM RAIL FORMED ON TOP OF PORTAL DURING LAMINATING
13 D.FIR PLY PANELS SET BACK 25 FROM FACE OF PORTALS
197
APEX DETAIL
SPLIT RING CONNECTOR
3
APEX PORTALS SECURED WITH 1 Nº 64 SPLIT RING CONNECTOR ON 1 Nº 300×22 HANDRAIL TYPE BOLT & 2 Nº75×13 DIA. M.S. DOWELS.
PURLIN CLEAT & PACK
CLEAT & PACK
6
125×75 SUB FRAME BOLTED TO EACH SIDE OF MAIN FRAME WITH CONNECTORS & BOLTS TO SUPPORT EAVES
125×75
LINE OF GLULAM FRAME
SHEETING RAILS
ALTERNATE EAVES DETAIL
METAL SHOE WITH SMOOTH FINISH
F.L.
4
150
CONCRETE LEVELLING BED
BASE DETAILS
PORTAL SECURED WITH 19 DIA. BOLT & 2 Nº 67 SHEAR PLATE CONNECTORS
50
5
BASE PLATES 10 M.S.
SHEET 31

Purlins notched over the truss support the rafters which are usually clad on the underside with matchboarding or insulation board, the purlins being left exposed.

Sheet 30.2 shows the part elevation of a glued laminated timber portal frame of 13.500 m span suitable for a church or hall (see Plate 4.1).

When choosing glued laminated construction for buildings there are certain practical considerations to bear in mind. One is the question of cost. A structural unit made by lamination is more costly than a solid member, but lamination serves many functions which cannot as readily be met by other methods. The cost may therefore be justified either by greater efficiency or improved appearance. The attractive profiles of laminated components have often encouraged designers to leave them exposed to the interiors of buildings. This can result in a saving of the total cost of construction since it has made the use of suspended ceilings unnecessary. Laminated arch forms of different types are designed to suit a variety of requirements. These include two- and three-pin segmental and semicircular arches, parabolic and Gothic arches for spans up to 75.000 m. Examples of laminated work are shown on Plate 4.

The portal frames illustrated on Sheets 30 and 31 are in the latter group and are three-pinned arches, being hinged at the crown and at the springings or abutments. Arches in roof structures are normally spaced from 3.000 m to 6.000 m centres.

In Sheet 30.2 the portal is laminated from 25 mm Douglas fir laminates and jointed at the crown using purpose-made mild steel straps which are let into the top and bottom edges of the arches and secured with mild steel bolts as shown in Sheet 30.8.

The purlins are 250 mm by 100 mm with 125 mm by 50 mm rafters spaced at 600 mm centres, being clad on the underside leaving the purlins exposed.

Sheet 31.1 shows the elevation of a portal frame which is built up at the haunch with a panel section to support the eaves purlin.

The panel section is built up using 125 mm by 75 mm softwood framing. The lower curved member is formed on the top edge of the portal during laminating and the 13 mm plywood panels glued and nailed to the framing. Sheet 31.2 shows a section through the lower portion of the framing which is set in from the edge of the laminated arch on each face, so that the line of the glulam arch is clearly defined.

Sheet 31, figs. 4 and 5 show the base details. A fabricated metal shoe houses the foot of the portal which is secured with a 19 mm diameter bolt and two shear plate connectors.

Sheet 31.3 shows the jointing at the crown using a 64 mm split ring connector with a handrail type bolt and two 75 mm by 13 mm diameter mild steel dowels.

An alternative finish at the haunch of the portal is shown in Sheet 31.6. This consists of two 125 mm by 75 mm softwood frames bolted one to each side of the main arch with timber connectors to support the two lower purlins. Cleats bolted between the frames support the purlins and act as spacers.

The roof covering is asbestos sheeting with fibreglass insulation between, nailed to the purlins, and the sides are sheeted with asbestos fixed to sheeting rails as shown.

Roof Bevels

There are two recognized methods of determining the lengths and bevels for members in roof work. One is by geometrical setting out, the other setting out using the steel square. There are other methods, under the headings of rule of thumb and trial and error, which the craftsman should ignore if he desires to become efficient in all classes of roofing work.

A further method is by calculation, a knowledge of the properties of the right-angled triangle or of trigonometry being necessary. Calculations are useful for checking lengths and taking off quantities of materials, but are seldom used on the site.

The steel square method is based on geometrical principles, and will be dealt with later.

When setting out bevels and obtaining lengths geometrically, line diagrams only are used.

The most simple form of hipped roof over a rectangular plan is shown in Sheet 32.1. A hipped roof square at one end and splayed at the other is shown in Sheet 33.9. These examples have been chosen to illustrate that the method of determining the bevels and lengths of the members is the same for whatever shape the building may be, irrespective of the roof pitch.

In the section Sheet 32.2, the span and the rise determine the pitch and also the common rafter bevels at A and B. As single lines only are being used, care must be taken in determining the actual lengths of roof members. Sheet 32.3 shows the marking out for the common rafter. If the three roof surfaces have the same pitch, the hips will bisect the angles between the wall plates in plan.

The hip bevels are found by taking the rise H from the section and setting off at right-angles to the hip A′–B′ in plan. A line joining A′–D gives the bevels and the length and inclination of the hip.

The jack rafter bevels are found by rotating the roof surface to the horizontal plane and projecting the rafter L on to this developed surface from the plan at M. The dihedral angle or backing bevel for the hip is found by taking a square cut across the hip in plan at E. A line is drawn through E at right-angles to the hip cutting the wall plates at F and G. With E as centre and radius EJ, an arc of circle is drawn tangential to the hip A′–D, to cut the plan of the hip at H. Lines joining points F–H and H–G give the required backing bevel.

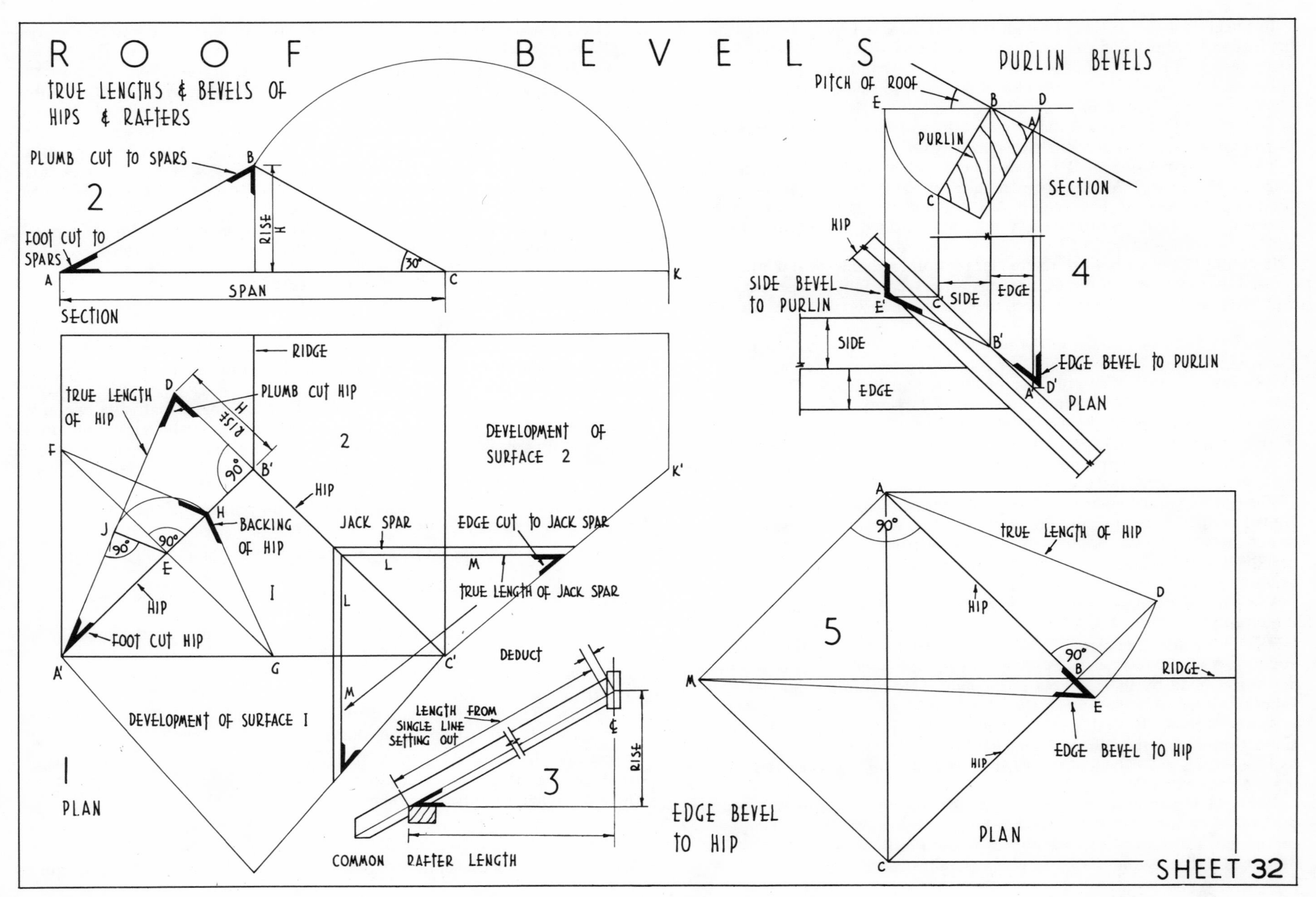
ROOF BEVELS
TRUE LENGTHS & BEVELS OF HIPS & RAFTERS
PLUMB CUT TO SPARS
2
FOOT CUT TO SPARS
A
B
C
K
RISE H
30°
SPAN
SECTION
RIDGE
TRUE LENGTH OF HIP
PLUMB CUT HIP
D
H
RISE
90°
B'
F
2
DEVELOPMENT OF SURFACE 2
K'
HIP
J
90°
90°
E
H
BACKING OF HIP
JACK SPAR
EDGE CUT TO JACK SPAR
L
M
TRUE LENGTH OF JACK SPAR
HIP
I
L
FOOT CUT HIP
A'
G
C'
DEDUCT
M
DEVELOPMENT OF SURFACE I
LENGTH FROM SINGLE LINE SETTING OUT
RISE
3
I
PLAN
COMMON RAFTER LENGTH
EDGE BEVEL TO HIP
PURLIN BEVELS
PITCH OF ROOF
E
B
D
A
PURLIN
C
SECTION
HIP
SIDE BEVEL TO PURLIN
E'
C'
SIDE
EDGE
4
SIDE
B'
EDGE BEVEL TO PURLIN
EDGE
A'
D'
PLAN
A
90°
TRUE LENGTH OF HIP
D
HIP
5
90°
B
RIDGE
M
E
EDGE BEVEL TO HIP
HIP
PLAN
C
SHEET 32

In Sheet 32.4 the section through the roof, together with the plan line of the hip and the plan of the purlin, are drawn to a larger scale to show the purlin bevels. The edge and side of the purlin A–B and B–C are rotated to a horizontal line, giving points B–D and B–E. These points are projected into plan to give the true widths of the edge and side and if the original intersection points A′ and C′ are moved out horizontally to E′ and D′, lines drawn from B′ to these points will give the two purlin bevels.

The two hips A–B and B–C drawn in the plan Sheet 32.5 fit against the sides of the ridge, and therefore require an edge bevel. The hip A–B is produced to show its true length A–E. A line at right-angles to A–E, to intersect with the ridge produced at point M, is drawn; this point joined to E will give the edge bevel.

It should be pointed out to the reader that other geometrical methods may be used to obtain certain bevels, i.e., jack rafters, edge-cut to hips and purlins, and that the method which he understands best is the one he should adopt.

Sheet 33.6 shows a sketch of the two purlins, cut so that the lower edges or lips which project below the hip, are mitred together on the underside. The bevel to cut the projecting portions is called the purlin lip cut. The method of determining this bevel is shown in Sheet 33.7. The plan of the hip A–B is drawn with one half of the elevation directly above it. The horizontal trace of the underside of the hip is drawn at right-angles to the hip in plan, to intersect the X–Y line at point E as shown. A line drawn from point E to the ridge gives the vertical trace.

The purlin, shown enlarged in section, is drawn with the side produced to cut the vertical trace at C and the X–Y line at B. With B as centre rotate B–C to cut the X–Y line at point F. A line drawn from F to D will give the lip bevel. The application of this bevel to the purlin is shown in Sheet 33.6 and again in Sheet 33.8.

The plan of a hipped roof, square at one end and splayed at the other, is shown in Sheet 33.9.

When drawing the plan of the roof, as in the case of the rectangular building, if all the roof surfaces are inclined at the same pitch, the hips bisect the angle made by the wall plates in plan. This condition applies also to valleys at internal angles, when they are used in a roof.

As the hips A′–B′ and B′–C′ are of different lengths and have different inclinations, then the dihedral angles also will be different, and each will require setting out. The method used is similar to that already described in the previous example.

Two sets of purlin bevels are required for cutting the purlins. One set for the purlins fitting against the shorter hip, the other for those fitting against the longer hip. Again, the method of obtaining these has previously been described and only one set is shown in the illustration.

Two lip bevels are required for the purlins at the splayed end of the roof and are dealt with as before.

An alternative method to that used in the last example for obtaining the lengths and bevels of the jack rafters is shown. The longer side of the roof is developed by rotating the roof surface A–B in the section, to the horizontal plane at point K. A vertical line drawn from this point to intersect with a line drawn out at right-angles from A′ in plan, will give K′. A line drawn from K′ to the point B′ gives the development of the roof surface. The jack spar drawn in the correct position on this developed roof surface will be its true length and will contain the correct edge bevel.

It should be realized that the jack spars fitting against the longer hip will require a different edge bevel to those fitting against the shorter hip.

Steel Square

Among all the tools used by the carpenter there is, perhaps, none so useful, simple and indispensable as the modern steel square. There is not a tool that may be so readily applied to the quick solution of the many difficult problems of laying out work as the square.

The steel square is made in the form of a right angle, that is its two arms (the body and the tongue) make an angle of 90 degrees, which is a right angle. A line joining points A and C Sheet 34.1 completes the triangle which is a right triangle. Therefore the steel square as well as all roof framing is based on the principle of a right triangle.

The following terms identifying the different portions of the square should be noted and remembered:

(1) *Blade*. The blade is the longer and wider part 24 in. (600 mm) long and 2 in. (50 mm) wide.

(2) *Tongue*. The tongue is the shorter and narrower part and usually 16 in. (400 mm) long and 1½ in. (38 mm) wide.

(3) *Heel*. The point at which the blade and tongue meet on the outside edge of the square is called the heel.

(4) *Face*. The face of the square is the side on which the name 'STANLEY' is stamped, shown in Sheet 34.2.

(5) *Back*. The back is the side opposite to the face, Sheet 34.3. The modern square usually has two kinds of markings, scales and tables.

It should be pointed out that, this craftsman's tool has not, as yet, been produced by the makers with metric graduations. The uses and applications therefore relate to imperial dimensions for the present.

Scales

These are the inch divisions found on the outer and inner edges of the square

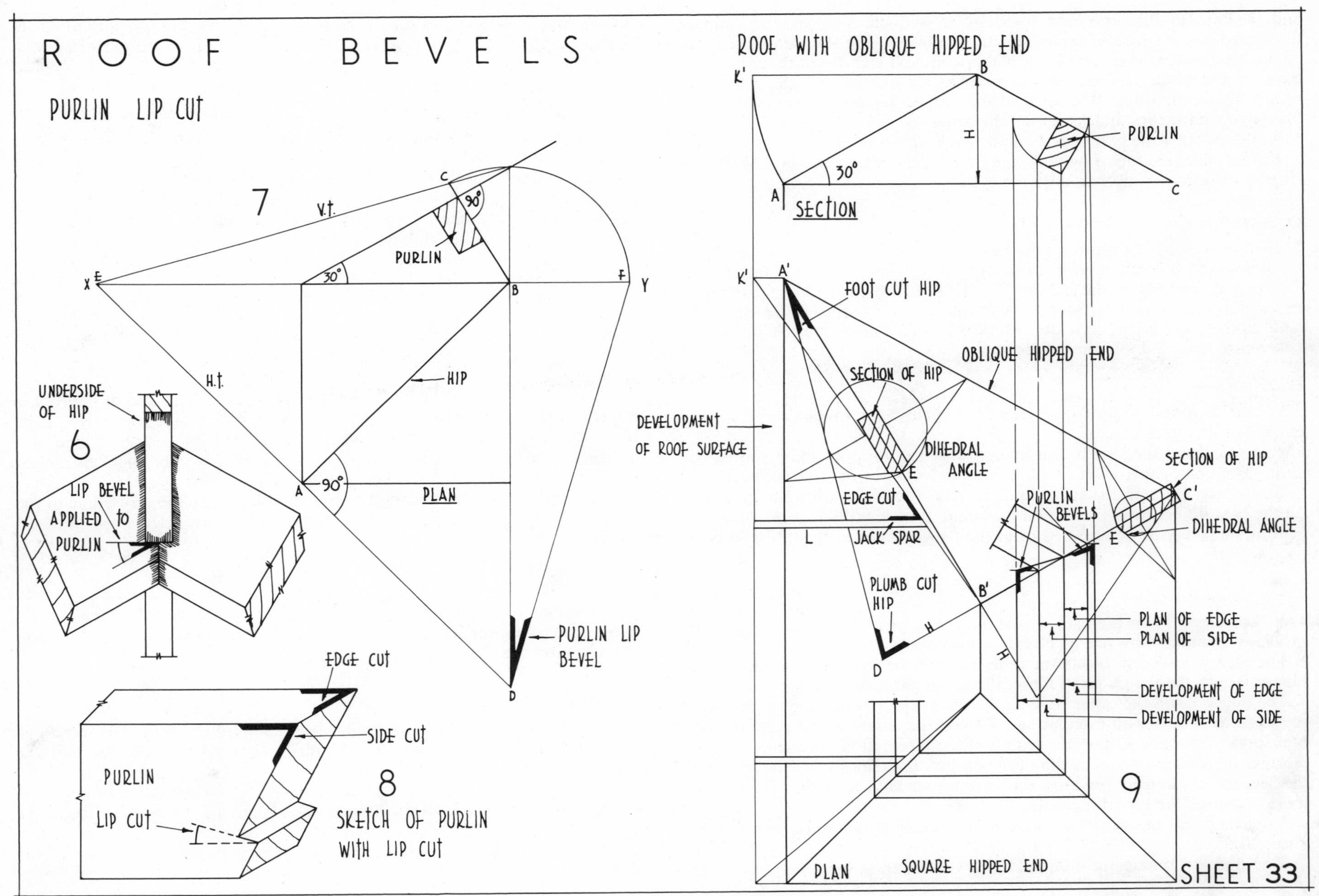
ROOF BEVELS
PURLIN LIP CUT
7
V.T.
C
90°
PURLIN
30°
X E
B
F Y
H.T.
HIP
A
90°
PLAN
D
PURLIN LIP BEVEL
UNDERSIDE OF HIP
6
LIP BEVEL APPLIED to PURLIN
EDGE CUT
SIDE CUT
PURLIN
LIP CUT
8
SKETCH OF PURLIN WITH LIP CUT
ROOF WITH OBLIQUE HIPPED END
K'
B
H
PURLIN
30°
A
C
SECTION
K'
A'
FOOT CUT HIP
OBLIQUE HIPPED END
SECTION OF HIP
DEVELOPMENT OF ROOF SURFACE
DIHEDRAL ANGLE
E
EDGE CUT
L
JACK SPAR
PURLIN BEVELS
SECTION OF HIP
C'
DIHEDRAL ANGLE
E
PLUMB CUT HIP
B'
H
D
PLAN OF EDGE
PLAN OF SIDE
DEVELOPMENT OF EDGE
DEVELOPMENT OF SIDE
9
PLAN
SQUARE HIPPED END
SHEET 33

and the inch graduations into fractions of an inch. The 'Stanley' square illustrated has the face of the blade and the face of the tongue divided into inches and sixteenths on one edge, with inches and eighths on the other. The back of the blade has inches and twelfths on one edge with inches and sixteenths on the other. The back of the tongue has inches and twelfths on one edge with inches and tenths on the other.

A hundredth scale is stamped on the back of the tongue.

Rafter tables are found on the face of the blade for determining rapidly the lengths of rafters as well as their cuts.

Octagon Square

This scale is along the centre of the face of the tongue, and is used for laying off lines to cut an 'eight-square' or octagon piece of timber from a square one. To mark an octagon on the end of a 9 in. (230 mm) by 9 in. (230 mm) timber, proceed as shown in Sheet 34.4. Step off the same number of divisions on the scale as there are inches in width. Mark off this distance on each side of the two centre lines and join the points obtained.

Brace Measure

This table is found along the centre of the back of the tongue and gives the lengths of common braces.

Example. Find the length of a brace whose run on post and beam equals 39 in. The brace table gives 55·15, being the length of the brace, Sheet 34.5.

Braces may be regarded as common rafters. Therefore, when the brace run on the post differs from the run on the beam, their lengths as well as top and bottom cuts may be determined from the figures given in the tables of common rafters.

Essex Board Measure

This table appears on the back of the blade and gives the contents in board measure of almost any size of board or timber.

The inch graduations along the outer edge of the square are used in combination with the values given along the seven parallel lines.

The figure 12 on the outer edge represents a 1 in. board 12 in. wide and is the starting point for all calculations. To find the contents of a piece of lumber: under the mark 12 find the length of the piece; along the same scale of inch graduations locate the width of the timber; then follow the line on which the length is stamped towards the column under the width. The figure indicated gives the board measure.

Example. Find the board measure of a board 8 ft. long and 11 in. wide, Sheet 34.6.

First find 8 ft. under the column 12 for the length in feet, then find 11 in. on the edge of the square for the width in inches. Follow the line to where they come together at 7–4 or seven and four-twelfths is found to be the number of feet in the board.

Rafter Tables

The rafter tables on the 'Stanley' squares are based on the 'rise per foot run', which means that the figures in the tables indicate the length of rafters per one foot of common rafters for any rise of roof, Sheet 35.7.

The reason for using this 'per foot run' method is that the length of any rafter may be easily determined for any width of building. The length per foot run will be different for different pitches, therefore, before the length of the rafter can be established, the rise of the roof in inches or the 'rise per foot run' determined.

The rise per foot run of a rafter is found by taking the rise of the roof in inches and dividing it by the number of feet in the run. Using the table on the square, under the figure for the rise per foot run, read the appropriate length which is the length of the rafter in inches per foot run. The length given must be multiplied by the number of feet run of the rafter.

Example. Find the length of a common rafter where the rise of the roof is 8 inches per foot run. From the square, Sheet 35.8, on the inch edge of the blade under the figure 8 will be found 14·42 which is the length of the rafter in 'inches per foot run'. The total length of the rafter will be 14·42 multiplied by 10 which equals 144·20 inches or 12·01 feet.

The lengths of rafters obtained from the tables are to the centre line of the ridge. The thickness of half of the ridge should be deducted from the total length.

Bevels

To obtain the top and bottom bevels take 12 in. on the blade and the rise per foot run on the tongue. 12 in. on the blade will give the horizontal cut and the figure on the tongue the vertical cut.

Applying the Square

After the total length of the rafter has been established, both ends should be marked and allowance made for the tail or eaves finish and for half the thickness of the ridge.

Both cuts are obtained by applying the square so that the 12 in. mark on the blade and the mark on the tongue that represents the rise shall be at the edge of the stock.

All cuts for common rafters are made at right-angles to the side of the rafter.

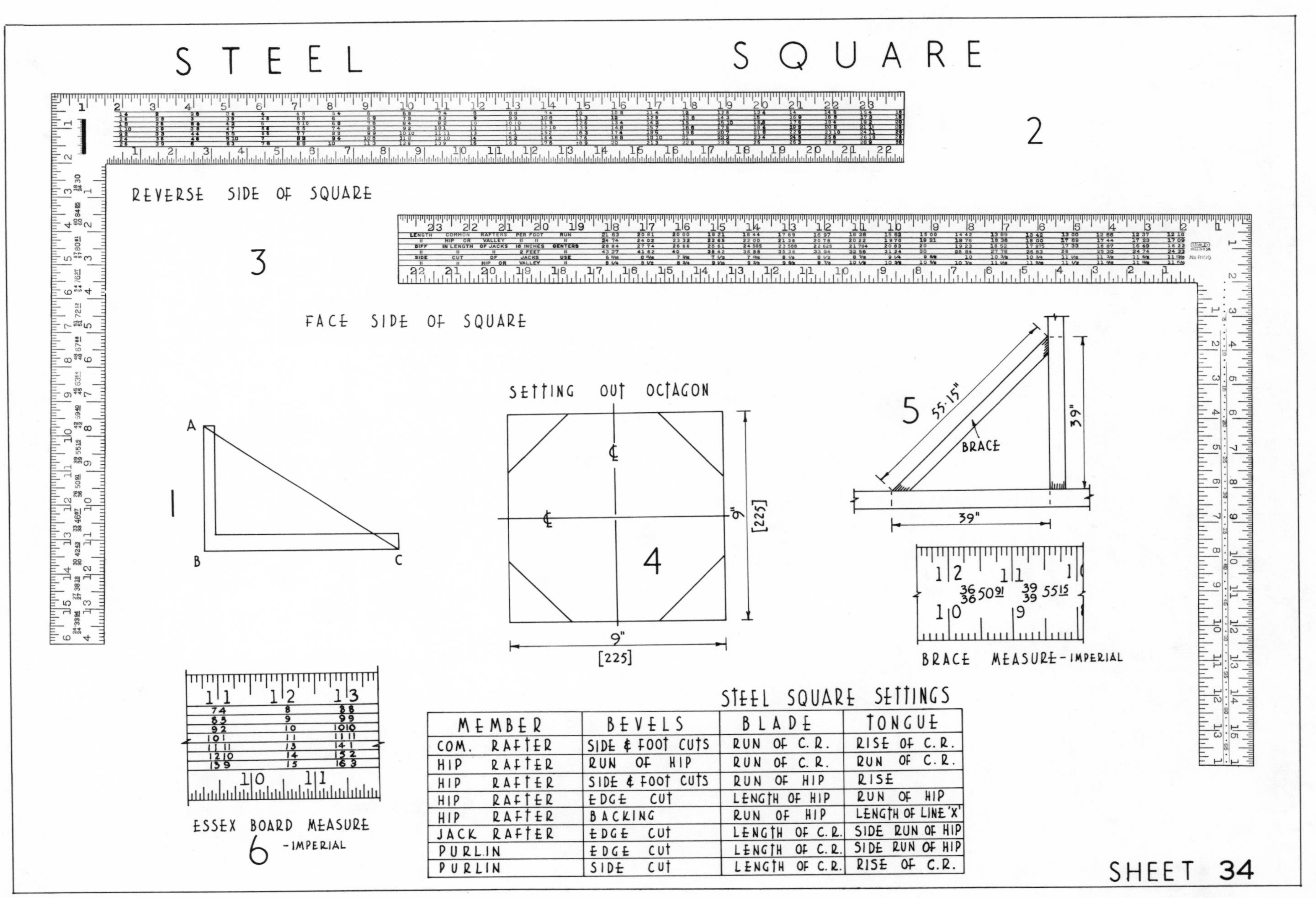

STEEL SQUARE SETTINGS

MEMBER	BEVELS	BLADE	TONGUE
COM. RAFTER	SIDE & FOOT CUTS	RUN OF C.R.	RISE OF C.R.
HIP RAFTER	RUN OF HIP	RUN OF C.R.	RUN OF C.R.
HIP RAFTER	SIDE & FOOT CUTS	RUN OF HIP	RISE
HIP RAFTER	EDGE CUT	LENGTH OF HIP	RUN OF HIP
HIP RAFTER	BACKING	RUN OF HIP	LENGTH OF LINE 'X'
JACK RAFTER	EDGE CUT	LENGTH OF C.R.	SIDE RUN OF HIP
PURLIN	EDGE CUT	LENGTH OF C.R.	SIDE RUN OF HIP
PURLIN	SIDE CUT	LENGTH OF C.R.	RISE OF C.R.

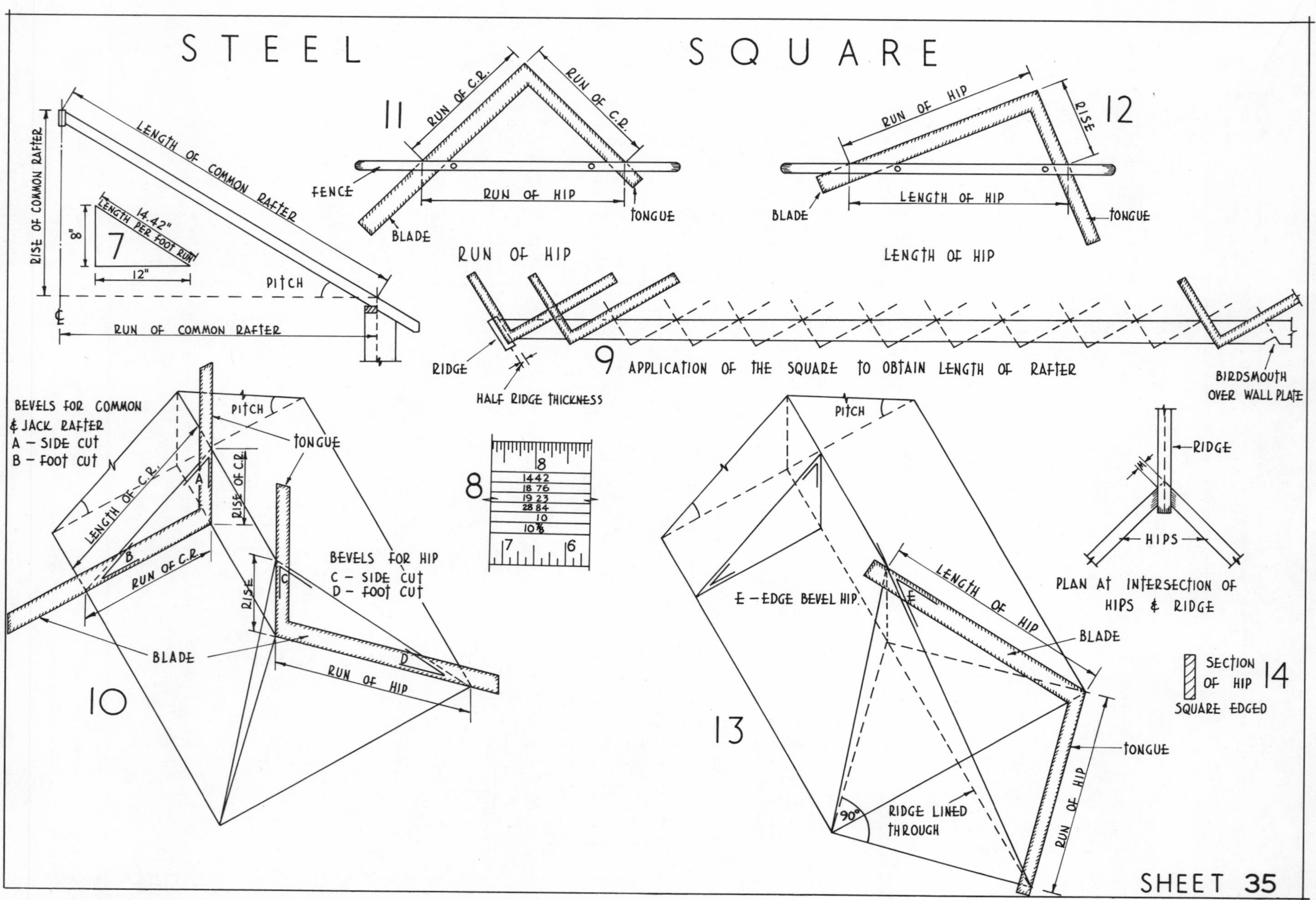
STEEL SQUARE
RISE OF COMMON RAFTER
LENGTH OF COMMON RAFTER
14.42"
LENGTH PER FOOT RUN
8"
7
12"
PITCH
RUN OF COMMON RAFTER
11
RUN OF C.R.
RUN OF C.R.
FENCE
RUN OF HIP
TONGUE
BLADE
RUN OF HIP
12
RUN OF HIP
RISE
BLADE
LENGTH OF HIP
TONGUE
LENGTH OF HIP
RIDGE
HALF RIDGE THICKNESS
9 APPLICATION OF THE SQUARE TO OBTAIN LENGTH OF RAFTER
BIRDSMOUTH OVER WALL PLATE
BEVELS FOR COMMON & JACK RAFTER
A – SIDE CUT
B – FOOT CUT
PITCH
TONGUE
LENGTH OF C.R.
RISE OF C.R.
RUN OF C.R.
BEVELS FOR HIP
C – SIDE CUT
D – FOOT CUT
RISE
BLADE
RUN OF HIP
10
8
1442
18 76
19 23
28 84
10
10⅜
PITCH
E – EDGE BEVEL HIP
LENGTH OF HIP
BLADE
TONGUE
RUN OF HIP
RIDGE LINED THROUGH
90°
13
RIDGE
HIPS
PLAN AT INTERSECTION OF HIPS & RIDGE
SECTION OF HIP 14
SQUARE EDGED
SHEET 35

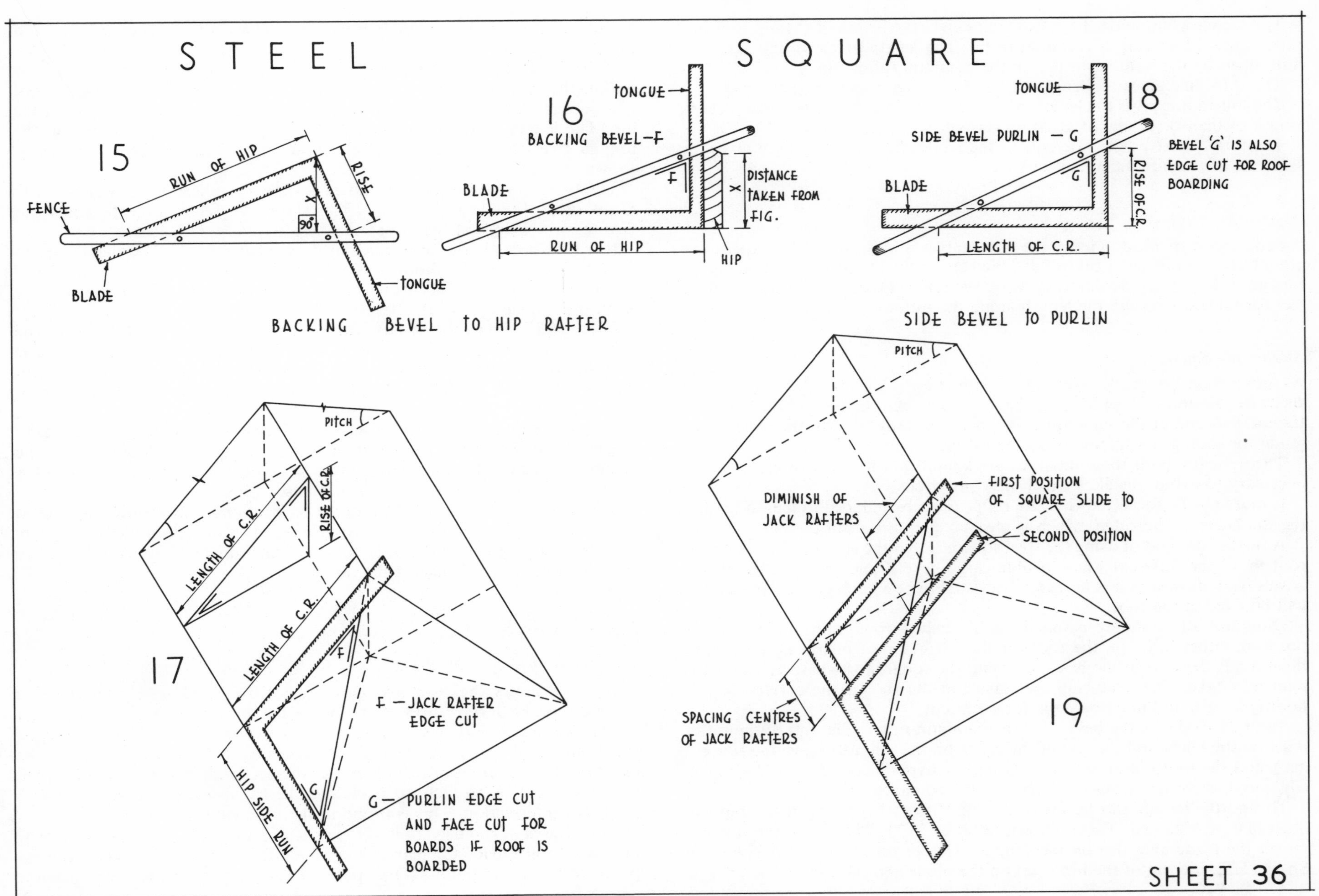
STEEL SQUARE
15
RUN OF HIP
RISE
FENCE
90°
X
BLADE
TONGUE
16
TONGUE
BACKING BEVEL–F
BLADE
F
X
DISTANCE TAKEN FROM FIG.
RUN OF HIP
HIP
18
TONGUE
SIDE BEVEL PURLIN – G
BEVEL 'G' IS ALSO EDGE CUT FOR ROOF BOARDING
BLADE
G
RISE OF C.R.
LENGTH OF C.R.
BACKING BEVEL TO HIP RAFTER
SIDE BEVEL TO PURLIN
PITCH
LENGTH OF C.R.
RISE OF C.R.
LENGTH OF C.R.
17
F
F – JACK RAFTER EDGE CUT
G
HIP SIDE RUN
G – PURLIN EDGE CUT AND FACE CUT FOR BOARDS IF ROOF IS BOARDED
PITCH
DIMINISH OF JACK RAFTERS
FIRST POSITION OF SQUARE SLIDE TO
SECOND POSITION
SPACING CENTRES OF JACK RAFTERS
19

SHEET 36

The second column on the rafter tables gives the length of the hip or valley rafter for a 12 in. run of common rafter. The length given on the square is multiplied by the number of feet in the common rafter run.

The third line gives the length of hip and valley rafters per foot run.

The fourth line gives the length of the first jack rafter and the differences in length of the others spaced at 24 in. centres.

The fifth line gives the side cuts of the jack rafters and the sixth gives the side cuts of the hip and valley rafters.

Alternative Method

An alternative method to using the rafter tables is by scaling. The dimensions are set up to a scale of 1 in. to 1 ft., the run on the blade and the rise on the tongue. A rule is laid across to measure the rafter directly and then, counting feet for inches, measure the length along the rafter.

Sliding the Square

As inches have been called feet in the last method, by sliding the square along the rafter 12 times, Sheet 35.9, the bevels can be marked off and the length of the rafter found at the same time. A fence is usually fixed to the square as a guide for sliding the square along the rafter.

Theoretically both these methods are sound, but in practice great care is necessary to avoid mistakes which may prove costly.

It must not be forgotten that the rafter must be reduced in length by half the thickness of the ridge, which is measured square from the cut.

A further method of using the steel square is that the square will fit into the roof in all the different ways to obtain the bevels, and in working out the bevels from drawings, it is necessary to visualize how each particular member will be fixed in the roof.

Diagrams of a square-planned equal-pitched roof with the square in position, either in or on the roof for the various roof members, have been illustrated; these form the basis for using the square for whatever shape the plan may take. This picturing the square in the roof is the correct way of finding bevels, and must be done for every cut.

Sheet 35.10 shows the bevels for the common rafter. The run of the rafter is set on the blade and the rise of the rafter on the tongue to give the side cut at A and the foot cut at B. The diagonal or hypotenuse of the right-angled triangle thus formed gives the length of the common rafter.

To find the length and bevels of the hip rafter, the length of the hip run must first be obtained. This is shown in Sheet 35.11. The run of the rafter is set on the blade and also on the tongue. The length of the hip is shown in Sheet 35.12, the run of the hip is set on the blade and the rise on the tongue. These settings give the side cut for the hip at C and the foot cut at D, as seen in Sheet 35.10.

The edge bevel for the hip rafter to fit against the ridge is determined by finding the traces of a plane which contains the edge of the hip. When the roof is square in plan, the horizontal trace of this plane is equal to the run of the hip. The length of the hip set on the blade and the run of the hip on the tongue gives the hip edge bevel E, Sheet 35.13. The section of the hip is shown in Sheet 35.14.

The backing bevel for the hip rafter is shown in Sheet 36.16. Reference to Sheet 36.15 should first be made. This shows the settings on the square to give the true length of the hip already dealt with.

It will be seen that a perpendicular drawn from the heel of the square to the fence line gives the distance X which is necessary for the backing bevel setting on the square. This is the run of the hip on the blade and the distance X on the tongue to give the bevel F.

Sheet 36.17 shows the square setting for the jack rafter edge bevel. The length of the common rafter on the blade and the hip side run on the tongue give the edge bevel H, as shown. The same square setting gives the edge cut for the purlin and the face cut for boards, if the roof is to be boarded, shown at J.

The side bevel for the purlin is shown in Sheet 36.18. The square setting for this bevel is the length of the common rafter on the blade and the rise of the common rafter on the tongue to give the bevel G. This is also the edge cut for cutting the roof boards if they are to be used.

Sheet 36.19 shows the diminish of the jack rafters. The square is moved as shown to the second position, a distance equal to the spacing of the rafters and the difference in length on the tongue of the square, from the ridge to the hip line, gives the diminish of the rafters.

Intersecting Roofs

Conical Roof with Pitched Roof

Sheet 37.1 shows the plan and elevation of a conical roof intersecting a pitched roof. Both roof surfaces are inclined at 45 degrees. In this example, it is necessary to develop the shape of the curb to be fixed on the main roof which later will receive the feet of the rafters from the conical roof. As the two roof surfaces are the same pitch, the intersection between them is as when the cone is cut by an inclined plane parallel to any one generator, giving a parabolic intersection. This is found in the following manner:

Draw the plan and elevation of the conical roof and in the elevation, draw the main roof. Divide the plan of the cone into 12 equal parts and join these points to the centre point 8. Only points 1 to 7 on the right of the centre line

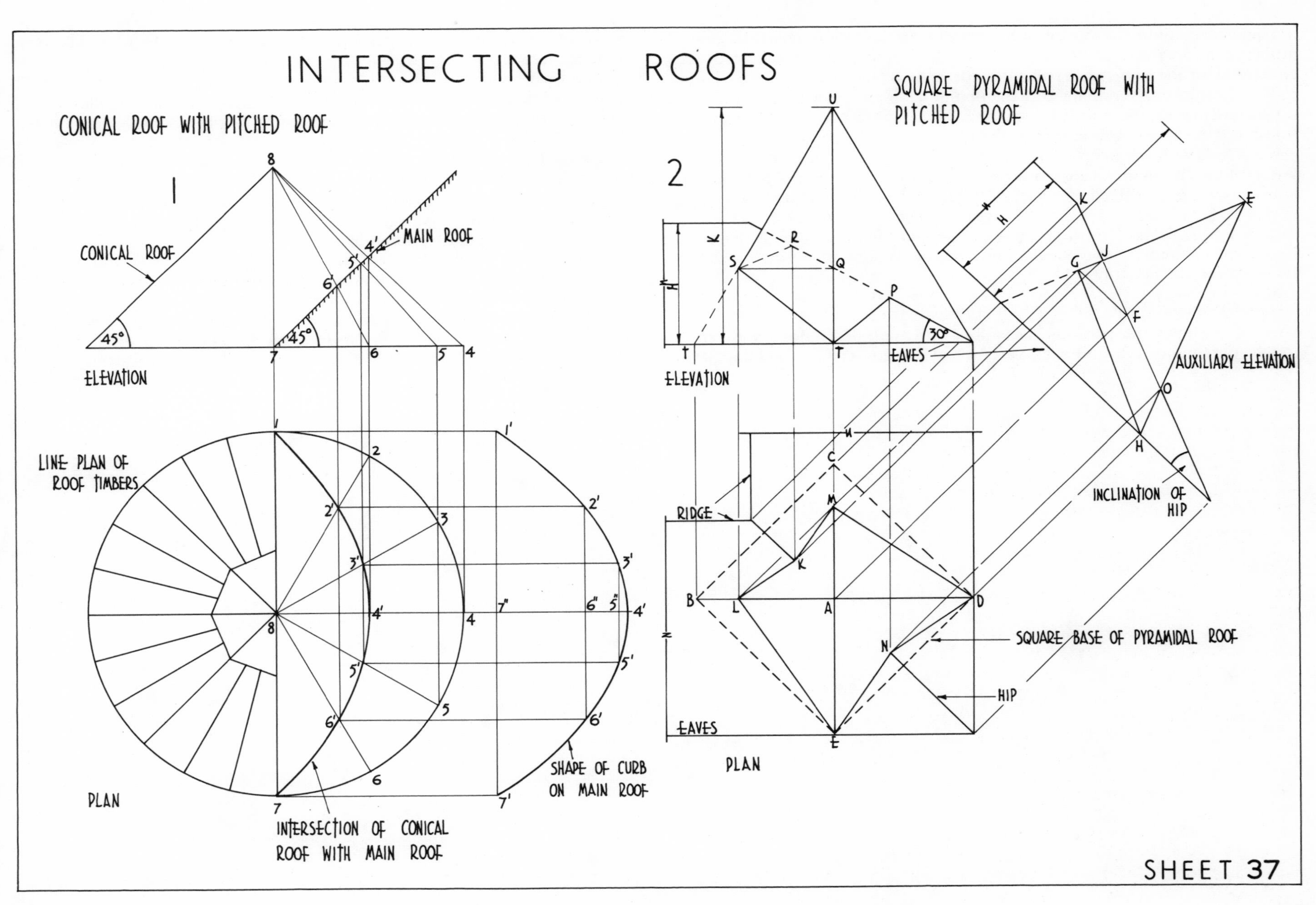
INTERSECTING ROOFS
CONICAL ROOF WITH PITCHED ROOF
1
CONICAL ROOF
MAIN ROOF
45°
45°
ELEVATION
LINE PLAN OF ROOF TIMBERS
PLAN
SHAPE OF CURB ON MAIN ROOF
INTERSECTION OF CONICAL ROOF WITH MAIN ROOF
SQUARE PYRAMIDAL ROOF WITH PITCHED ROOF
2
30°
EAVES
ELEVATION
AUXILIARY ELEVATION
INCLINATION OF HIP
RIDGE
SQUARE BASE OF PYRAMIDAL ROOF
HIP
EAVES
PLAN
SHEET 37

in plan are required. To the left of the centre line, the roof timbers are shown in line form only. Project these points upwards vertically to the base of the cone in elevation and radiate them to the apex of the cone, point 8. These radiating lines cut the main roof in points 6′, 5′, 4′, which are projected downwards to the plan to give points 2′, 3′, 4′, 5′, 6′. A fair curve drawn through these points will complete the plan and give the intersection of the conical roof with the main roof.

To obtain the shape of the curb on the main roof, the points on the plan intersection are projected clear of the plan to the right. On the centre line points 4′, 5″, 6″ and 7″ are struck from the points 4′, 5′, 6′ and 7 in the elevation. Perpendiculars through these points and intersecting with points 1, 2′, 3′, 4′, 5′, 6′ and 7 in the plan will give points through which the parabolic shape of the curb is drawn.

Square Pyramidal Roof Intersecting a Hipped Roof

Sheet 37.2 shows the plan and elevation with the intersections between a square pyramidal roof and a hipped roof at the corner of a building. The pyramidal roof is placed diagonally across the corner of the roof. The plan of the main roof is first drawn with the hip bisecting the corner as shown. The elevation of the main roof is then projected and the height of the ridge H obtained. The auxiliary elevation is drawn at right-angles to the plan of the hip, transferring the height H from elevation to give the inclination of the hip rafter. The sloping sides to the square pyramidal roof are now drawn. A perpendicular from E is drawn cutting the hip in the auxiliary elevation at F. Draw FG parallel to the eaves line cutting the sloping side at G. A line drawn from G to H completes this elevation. Perpendiculars drawn from points J and O in the auxiliary elevation cut the hip in plan at N and K. Project from G to intersect the plans of the hips of the pyramidal roof in M and L. Join MKL and DNE to complete the plan, showing all the lines of intersection.

The height of the square pyramidal roof is positioned in the elevation at U, having been taken from the auxiliary elevation. Points K and N are projected up to points R and P in elevation. A line is drawn from point Q parallel to the eaves line to give point S in the elevation. Lines drawn from P and S to point T complete the elevation with all lines of intersection shown.

Chapter Four: Carpentry

Dormers

Sheet 38.1 shows the key front and side elevations of a segmental headed dormer window in a pitched roof, inclined at 45 degrees.

In this problem the development of the dormer roof and the shape of the opening in the main roof are required to be found.

First the front and side elevations are drawn in Sheet 38.2. The segmental roof line is divided into 12 equal parts and numbered as shown. These points are projected onto the main roof line and numbered 6′, 7′, 8′, 9′, 10′, 11′ and 12′. The plan is next projected below the side elevation with the same ordinates used in the elevation transferred to the plan. These are numbered points 0 to 6. Horizontal projections from these points are drawn to intersect with vertical lines dropped from 6′, 7′, 8′, 9′, 10′, 11′, 12′ in the side elevation. Through these intersections in points 0′, 1′, 2′, 3′, 4′, 5′, 6′, etc., a fair curve is drawn to complete the plan.

The development of the dormer roof is carried out as follows:

Points 0 to 12 from the segmental roof line in elevation are stepped out on the vertical line shown in the development. These points are projected horizontally and the distance 0–0′ taken from the plan, stepped off on the outside edge. The distances marked a, b, c, d, e, f and g in the plan are transferred to the development and lines drawn vertically from these points to intersect with the horizontals, give the points 0′, 1′, 2′, 3′ etc. through which a curve is drawn to complete the development of the roof surface.

In obtaining the true shape of the opening in the roof, only the shaped portion is shown projected at right-angles to the main roof line. The points projected intersect with the ordinates taken from the plan to give the shape on the main roof.

Turrets

Sheet 38.3 shows the plan and part section of a small turret roof suitable for an out-building. The building is octagonal in plan with 50 mm hips secured at the top to an octagonal finial. 100 mm by 50 mm jack spars are used between the hips. Joists secured to the wall plates serve as ties and carry the lining to the ceiling. The roof is boarded as shown with the hip rafters backed.

To the right of the centre line, the necessary bevels for the backing of the hips and the mitring of the boards are shown.

The dihedral angle has been dealt with under roof bevels and will not be repeated here. It will be seen that one-half of the dihedral angle gives the required edge bevel for cutting the boards. The side bevel requires the development of one of the roof surfaces to be done as follows:

A–B is the length of one of the sides in plan, and C′ the height of the hip. With C′ as centre and radius C′A, describe an arc to A′ in C′ produced in the section. Draw horizontal lines from the side A–B in plan to intersect with a vertical line drawn from A′ in elevation giving points A′, B′. Join these points to C to give the development of the roof surface ABC which gives the required side bevel for cutting the boards.

Timber Spires

Spires may be described as pyramidal roofs having a base outline square, octagonal or circular in plan. Being subject to considerable wind pressure, special care must be taken with the construction of the spire in bracing and anchoring to the main structure.

Sheet 39.1 shows the elevation of a timber spire 6.000 m high, octagonal at the base and covered in copper. The spire is mounted on a stone tower 2.700 m diameter.

A section through the tower and spire is shown in Sheet 39.2. A 275 mm by 75 mm wall plate or curb rests upon the top of the tower and is rag-bolted down flush with its outer face. The angles are secured by halving and bolting. Two 230 mm by 75 mm diagonal ties cross at right-angles and are halved together at their intersection, notched and bolted to the wall plates, thus acting as a tie to the walls. From the centre of these a 125 mm by 125 mm octagonal mast or finial rises to a conical end piece at the top of the spire. The finial is housed into the conical end piece at the top, and the two bored to receive the cross. The foot of the finial is stub tenoned and bolted to the ties. The faces of the octagonal pyramid are inclined at an angle of 80 degrees. The 125 mm by 50 mm hips are birds-mouthed into the angles of the octagonal base and secured with iron angle brackets coach bolted to the rafters and wall plates. The upper ends of the hips are nailed to the finial and end piece. Two

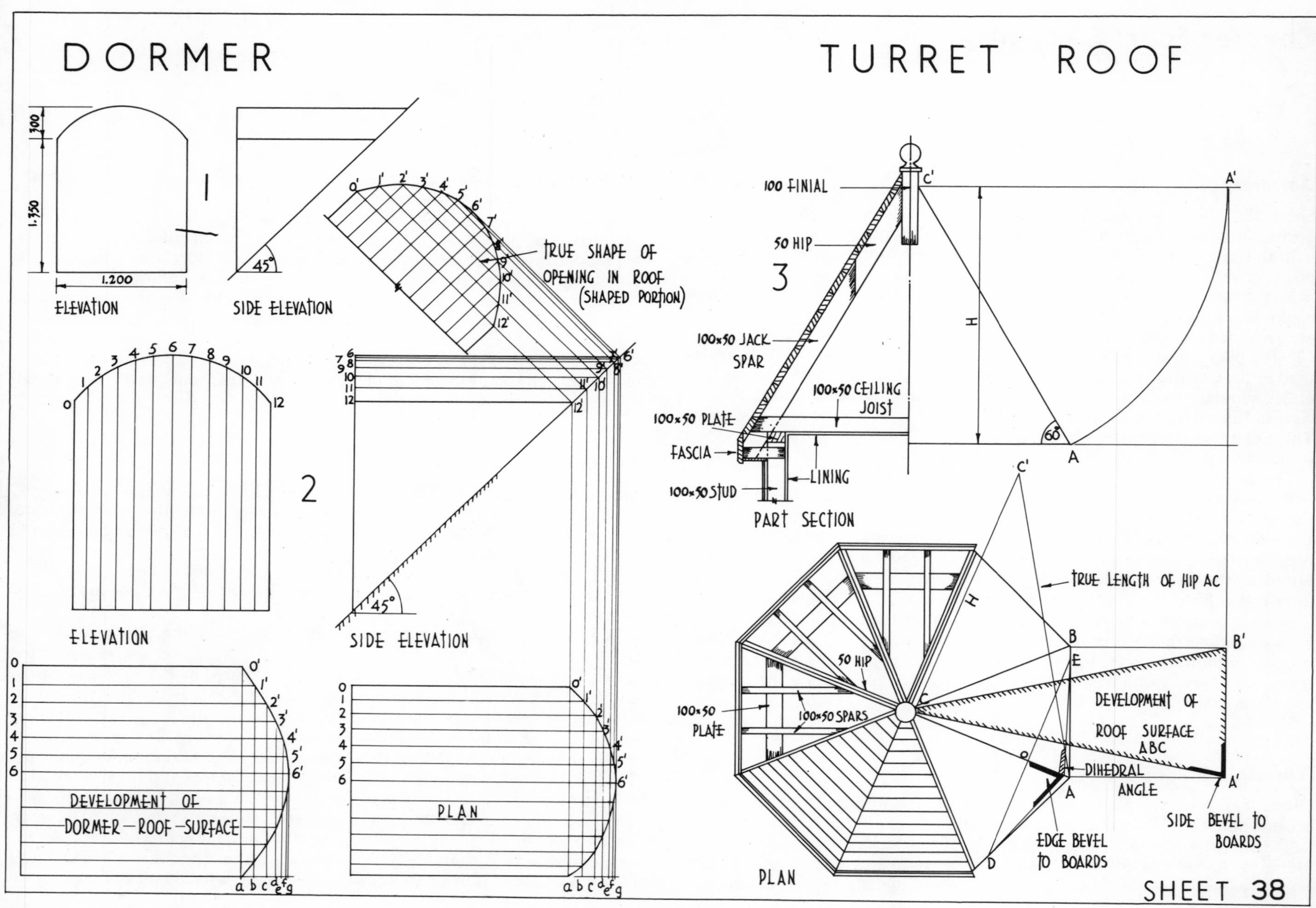
DORMER
TURRET ROOF
300
1.350
1.200
ELEVATION
SIDE ELEVATION
45°
1
TRUE SHAPE OF OPENING IN ROOF (SHAPED PORTION)
2
ELEVATION
SIDE ELEVATION
DEVELOPMENT OF DORMER-ROOF-SURFACE
PLAN
a b c d e f g
100 FINIAL
50 HIP
3
100×50 JACK SPAR
100×50 CEILING JOIST
100×50 PLATE
FASCIA
LINING
100×50 STUD
PART SECTION
H
60°
A
A'
C'
TRUE LENGTH OF HIP AC
50 HIP
100×50 PLATE
100×50 SPARS
B
B'
E
C
DEVELOPMENT OF ROOF SURFACE ABC
DIHEDRAL ANGLE
EDGE BEVEL TO BOARDS
SIDE BEVEL TO BOARDS
D
PLAN
SHEET 38

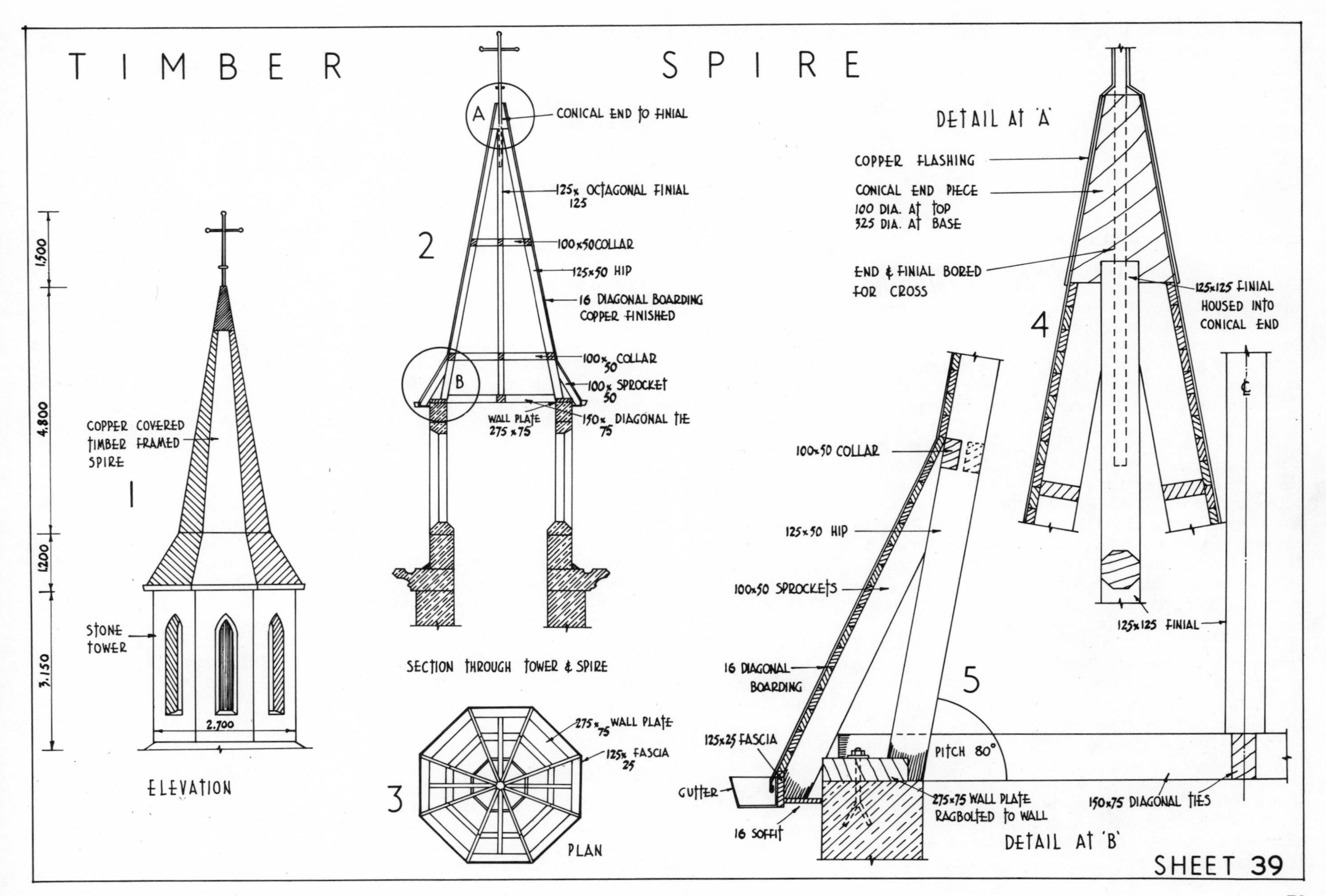
T I M B E R S P I R E
1
COPPER COVERED TIMBER FRAMED SPIRE
STONE TOWER
1,500
4,800
1,200
3,150
2,700
ELEVATION
2
A
CONICAL END TO FINIAL
125x125 OCTAGONAL FINIAL
100x50 COLLAR
125x50 HIP
16 DIAGONAL BOARDING COPPER FINISHED
B
100x50 COLLAR
100x50 SPROCKET
WALL PLATE 275x75
150x75 DIAGONAL TIE
SECTION THROUGH TOWER & SPIRE
3
275x75 WALL PLATE
125x25 FASCIA
PLAN
DETAIL AT 'A'
COPPER FLASHING
CONICAL END PIECE
100 DIA. AT TOP
325 DIA. AT BASE
END & FINIAL BORED FOR CROSS
125x125 FINIAL HOUSED INTO CONICAL END
4
125x125 FINIAL
100x50 COLLAR
125x50 HIP
100x50 SPROCKETS
16 DIAGONAL BOARDING
5
125x25 FASCIA
GUTTER
16 SOFFIT
PITCH 80°
275x75 WALL PLATE RAGBOLTED TO WALL
150x75 DIAGONAL TIES
DETAIL AT 'B'
SHEET 39

series of collars are framed between the hips as shown, and 100 mm by 50 mm sprockets, nailed to the hips and the wall plates, complete the framing. Diagonal boarding 16 mm thick with a 125 mm by 25 mm fascia and soffit complete the spire.

Sheet 39.3 shows the plan of the roof timbers to the spire, with the details at the apex and at the eaves shown in Sheet 39, figs. 4 and 5.

Domes and Pendentives

Sheet 40.1 shows the plan and elevation of a hemisphere, part of which is covering a square plan. The square is exactly contained within the hemisphere. The four vertical walls, which become vertical planes, remove four half-zones leaving the hemisphere as shown in the sketch, Sheet 40.2. This is termed a domical vault.

Sheet 40.3 shows the top zone of the sphere removed from the previous figure, leaving a circular ring on plan. The spherical spandrels above each of the angles in plan are termed pendentives and are shown in the sketch, Sheet 40.4.

Sheet 40.5 shows the part plan and part sectional elevation of a pendentive dome, having a circular opening for a skylight in the centre. The diameter of the dome is equal to the diagonal of the square in the plan.

One half of the plan shows the ribs in position, with the shape of the diagonal rib B developed to the left. This is struck from the centre with a radius equal to that of the dome and is laminated similar to those shown in the Domed Roof on Sheet 41. The spacing of the ribs will depend on the nature of the finish. To obtain the sectional elevation, the springing line of the dome is drawn and from the mid point as centre, and radius equal to one half of the diagonal, describe the inner curve of the rib A′. Four such ribs are required. From the same centre the semicircle, which is the intersection of the hemisphere and the wall, is drawn. The position of the curb for the skylight is projected vertically from the plan. The other ribs are drawn in the elevation by first positioning the ends at the side rib and at the curb. In this view, these ribs will be elliptical curves.

The ribs have different lengths, but the same curvature. The lengths of ribs B, C and D are found by turning them to the centre line in plan at points E, F and G. The diagonal rib B has been redrawn clear of the elevation at B′ as shown. Vertical projections from points E and F on to the inner curve of the developed rib B′, are drawn at E′ and F′. Projections drawn from these points and from the head of the rib B′ give the head and foot cuts of the ribs C′ and D′. Four such ribs as C′ and D′ are required.

Domed Roof

A dome is a vaulted roof having a circular, elliptical or polygonal plan.

Sheet 41.1 shows the plan and elevation of an octagonal dome, the covering being omitted. To the left of the centre, the projections of two hips and one rib are shown, and to the right the method of obtaining the true shape of one of the hips.

The method of determining the elevation of the hips and rib is shown in the next example and fully described.

To obtain the shape of one of the hips, the outer curve of the elevation is divided into any convenient number of parts. Draw vertical projectors from these points on to the centre line of the hip member 0–6 in plan. Projections at right-angles from these points and clear to the right, with corresponding heights from the elevation, enable the true outline of the hip to be determined.

The development of the hip rafter shows the member laminated in two thicknesses, with the joints of each layer falling in the middle of the adjacent layer as in small centring. The number of sections making up the hip depends upon the span and the available width of material. Likewise, the construction of the dome depends upon its size and its location.

Sheet 41.2 is an enlarged plan of the top ends of the hips, showing their connection to the finial. The tenons are each nailed from the side in turn. Purlins, at a convenient position, are housed into the hip rafters and provide a fixing for the head of the ribs.

Sheet 41.3 shows the true shape of the rib AB projected from plan. This is struck with a radius equal to the radius of the dome from the centre O. The construction of the ribs is similar to that for the hips.

When the hip rafter is cut to the true shape and square to the face, it still requires backing so that the roof boarding will seat properly. This is done by placing a templet cut to the developed shape on the hip and sliding it sideways a distance equal to C–D marked in plan. The hip is again marked to the templet and the process repeated on the other side. When cut to the lines the correct bevel will be found.

It should be noted from the enlarged plan of the hip that the amount of slide is dependent on the thickness of the hip.

Hemispherical Dome

The sketch, Sheet 41.4, shows a hemispherical dome illustrating two approximate methods of covering this figure. To the left of the centre, gores are used. In this case the section planes cutting the sphere are all vertical and pass through the centre.

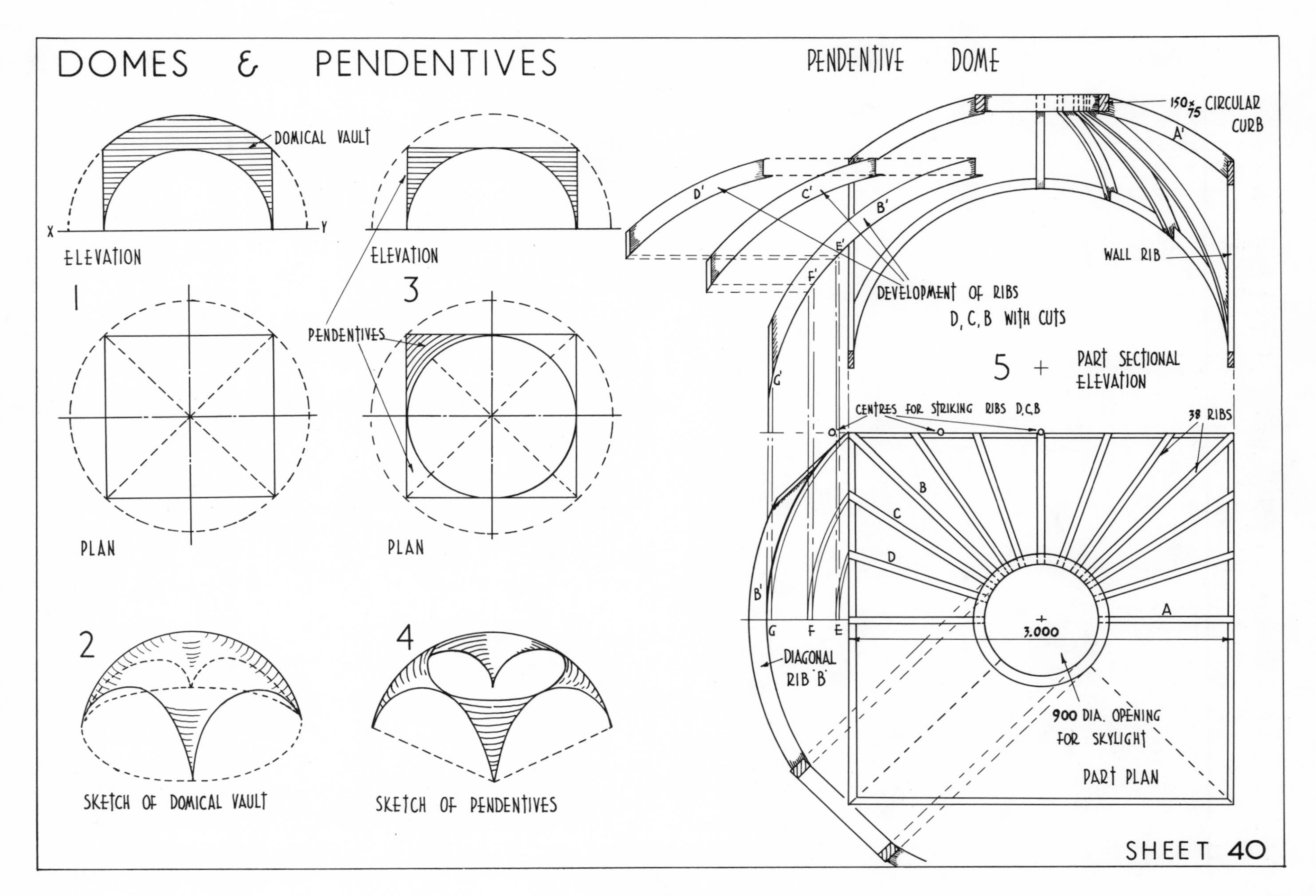
DOMES & PENDENTIVES
DOMICAL VAULT
X
Y
ELEVATION
1
PLAN
2
SKETCH OF DOMICAL VAULT
ELEVATION
3
PENDENTIVES
PLAN
4
SKETCH OF PENDENTIVES
PENDENTIVE DOME
150 x 75 CIRCULAR CURB
A'
D'
C'
B'
E'
F'
G'
WALL RIB
DEVELOPMENT OF RIBS D, C, B WITH CUTS
5 + PART SECTIONAL ELEVATION
CENTRES FOR STRIKING RIBS D,C,B
38 RIBS
B
C
D
A
B'
G F E
3.000
DIAGONAL RIB 'B'
900 DIA. OPENING FOR SKYLIGHT
PART PLAN
SHEET 40

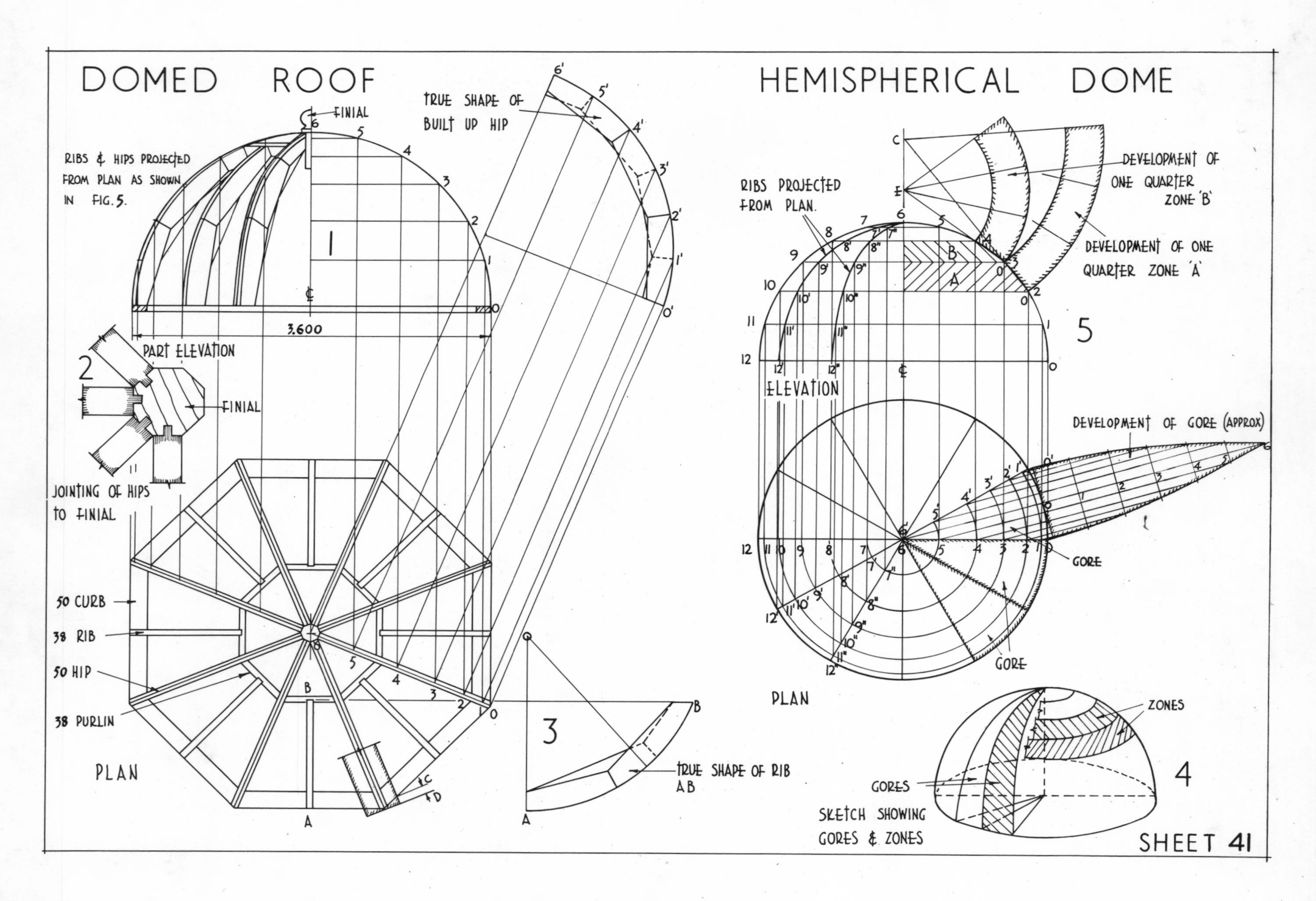
DOMED ROOF
HEMISPHERICAL DOME
FINIAL
TRUE SHAPE OF BUILT UP HIP
RIBS & HIPS PROJECTED FROM PLAN AS SHOWN IN FIG. 5.
3.600
PART ELEVATION
FINIAL
JOINTING OF HIPS TO FINIAL
50 CURB
38 RIB
50 HIP
38 PURLIN
PLAN
TRUE SHAPE OF RIB AB
RIBS PROJECTED FROM PLAN.
DEVELOPMENT OF ONE QUARTER ZONE 'B'
DEVELOPMENT OF ONE QUARTER ZONE 'A'
ELEVATION
DEVELOPMENT OF GORE (APPROX)
GORE
GORE
PLAN
ZONES
GORES
SKETCH SHOWING GORES & ZONES
SHEET 41

To the right of the centre, zones are used. This method of development is based on the assumption that the sphere is made up of a large number of conical surfaces, forming horizontal rings in elevation.

As it is not possible to develop accurately the surface of the sphere, approximations have to be resorted to in order to produce a practically accepted result in covering domes in timber, as a basis for further covering in lead or copper.

Sheet 41.5 shows the plan and elevation of a hemispherical dome. To the left of the centre the elevation of the ribs is determined. Divide the elevation as shown and project the points vertically downwards to the diameter in plan. With point 6 as centre, arcs are drawn from the points on the centre line, to cut the ribs in points 6–12′ and 6–12″. These points are projected back into the elevation and where they intersect with horizontal lines drawn from their opposite numbers will give the points to draw the elevations of the ribs.

To the right of the centre, the development of the sphere is shown. The points 0–6 in elevation are set out as horizontal sections. Draw vertical projectors from these points to the diameter in plan from which the concentric plan circles of six zones can be drawn. A line drawn through points 2 and 3 of zone A in elevation is produced to cut the centre line to give the centre C. With C as centre and C2 and C3 as radii, draw the arcs for 2 and 3. Similarly, from E as centre and radii E3 and E4, draw the smaller zone. The stretch-out for the zones is seen in plan, the distances stepped off and transferred to the development give the stretch-out for one-quarter of the complete zone. Each zone is treated in a similar way.

The approximate development of a gore is shown in the plan. With centre 6 and radii 6–1, 6–2, etc., draw arcs across the section to be developed, to give 1–1′, 2–2′, etc. The centre line is drawn clear of the plan and distances equal to those round the elevation marked off. Projections from the plan points to intersect with the vertical lines through 0, 1, 2, 3, 4, 5, 6 will give points through which to draw the approximate development.

Intersecting Vaults

In vaulting there are three main types: barrel, groined and ribbed.

When a barrel vault is intersected by one or more vaults, as in the corridors of public buildings, the line of intersection is called a groin, and the vault a groined vault.

Sheet 42.1 shows the intersection at right-angles of two vaults of different span, but of the same height. The cross-section of the smaller vault is semicircular and the groins are straight lines on the plan. The shape of the section of the larger vault is found by dividing the semicircle into a number of parts, 1, 2, 3, etc., and projecting these points vertically downwards on to the plan line of the intersection. On projections from these points drawn horizontally, the heights, equal to those in the original section, are marked giving points 1′, 2′, 3′, etc. The curve drawn through these points will give the required cross-section. It is semi-elliptical, its major axis being the span, and its semi-minor axis the rise of the vault.

The points on the plan line of the intersection are again used to obtain the true shape of the groin ribs, the method being similar to that already described. The ribs are built up or laminated, the spacing depending on the nature of the covering. A rebated section shows the method of obtaining the lengths and side cuts of the rib AB. The double lines at A′ and B′ show the bevel at the ends where they meet the groin rib.

Another method of framing for a vault of this type is for the centring for one of the vaults to continue across the intersection using ribs placed and fixed at intervals. The laggings run through to complete the centring for this vault. The ribs for the cross-vault are then fixed with one rib close up to the centre already in position and lagged. The laggings are run over those of the first set and scribed to fit the curve, the overhang being supported by part ribs and fillets. In this method groin ribs are not required.

Niches

A niche is a recess formed in a wall with the head usually semispherical. Sheet 42.2 shows the part elevation and plan of such a niche, framed in timber, it being assumed that the finish is to be plaster. The plan is semicircular and the face of the wall is flat.

The ribs of niches may be placed either in vertical planes as illustrated, or in horizontal planes to form the spherical surface.

In Sheet 42.3 the ribs forming the niche head stand in vertical planes which would pass through the centre if produced. The ribs are of similar curvature because the head of the niche is formed by a quadrant of a circle rotating about the centre. The lengths of the ribs will vary because of the thickness of material. The front or face rib is semicircular, the remaining ribs, all arcs of circles, are struck with the same radii as the front one. Additional length for the trenching of the ribs into the curb must be allowed.

The shapes of ribs A, B and C are shown projected at right-angles to their positions in plan. The joints at the top of the ribs are also positioned from the plan as shown.

Ventilating Turret

Sheet 43.1 shows the key sectional elevation of the base of a ventilator fixed on a roof of half pitch. This is in the form of a frustum of a regular octagonal

VAULTS & NICHES

INTERSECTING VAULTS

1

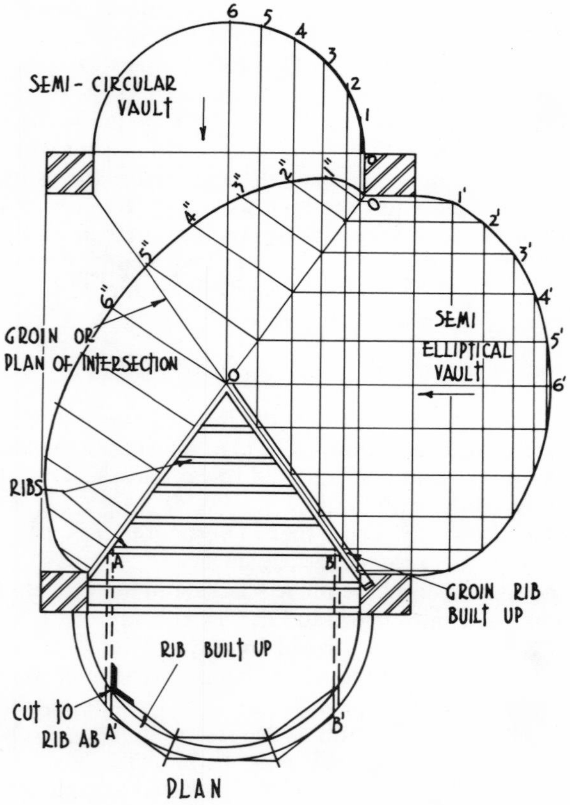

SEMI-CIRCULAR VAULT
6 5 4 3 2 1
0
1" 2" 3" 4" 5" 6"
1' 2' 3' 4' 5' 6'
SEMI ELLIPTICAL VAULT
GROIN OR PLAN OF INTERSECTION
RIBS
A
B
GROIN RIB BUILT UP
RIB BUILT UP
SIDE CUT TO RIB AB
A'
B'
PLAN

NICHE WITH DOMED HEAD

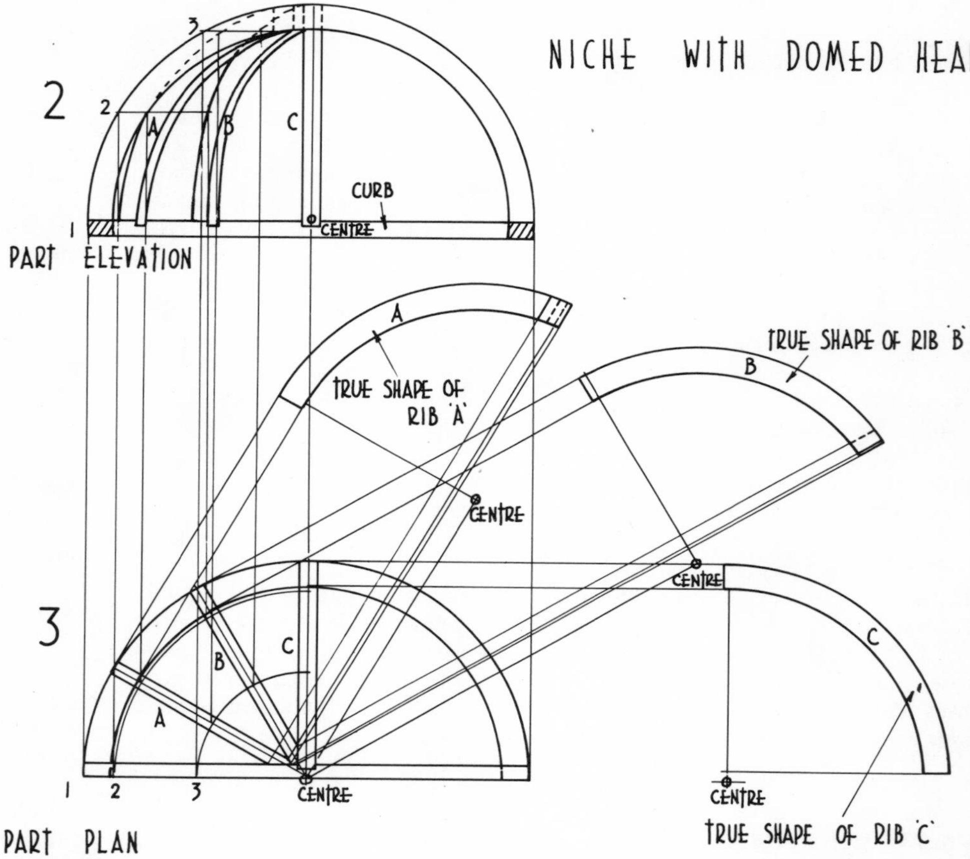

2
3
2
A
B
C
CURB
CENTRE
1
PART ELEVATION
TRUE SHAPE OF RIB 'A'
TRUE SHAPE OF RIB 'B'
CENTRE
3
C
B
A
1 2 3
CENTRE
PART PLAN
C
CENTRE
TRUE SHAPE OF RIB 'C'

SHEET 42

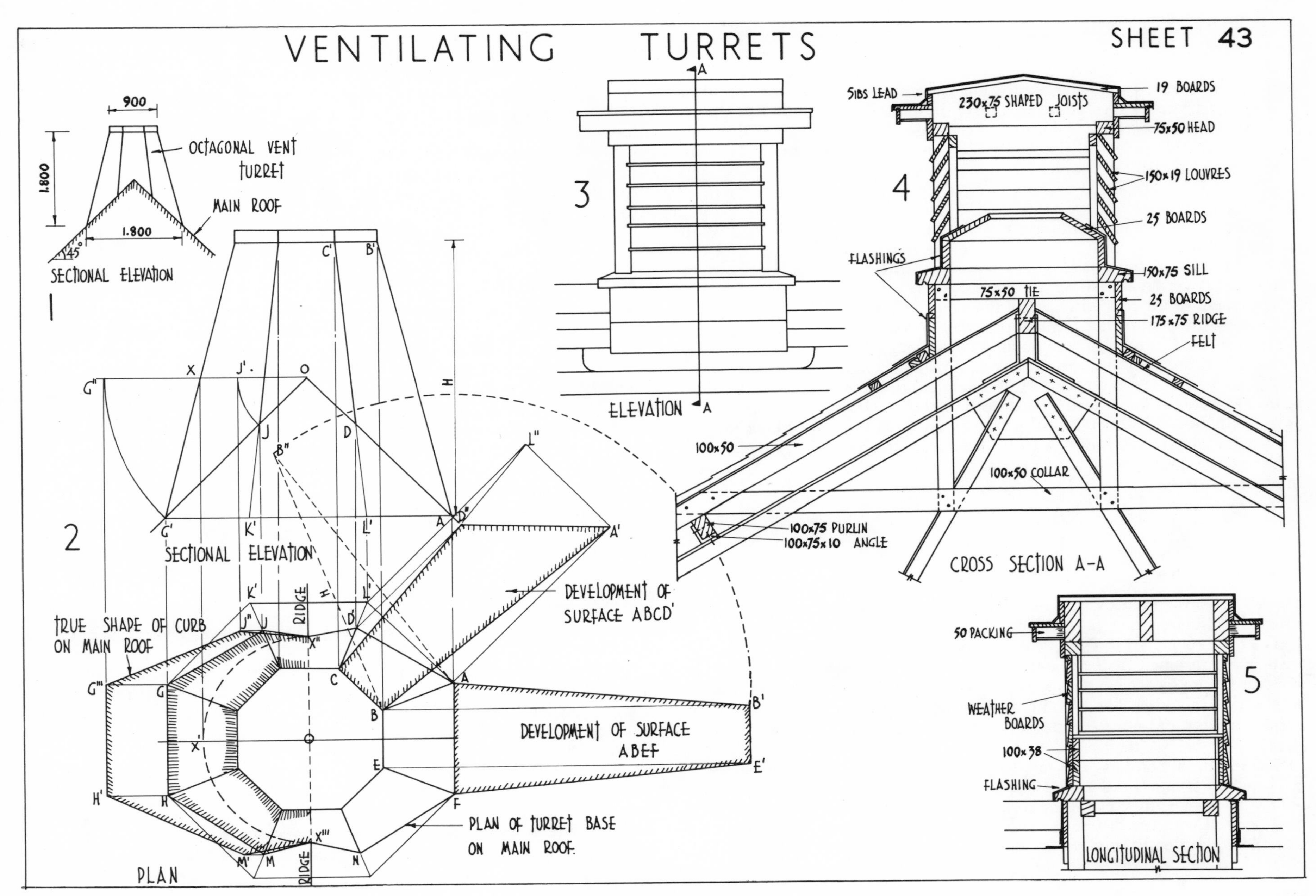
VENTILATING TURRETS
SHEET 43
900
1.800
OCTAGONAL VENT TURRET
MAIN ROOF
1.800
45°
SECTIONAL ELEVATION
1
2
SECTIONAL ELEVATION
TRUE SHAPE OF CURB ON MAIN ROOF
RIDGE
DEVELOPMENT OF SURFACE ABCD'
DEVELOPMENT OF SURFACE ABEF
PLAN OF TURRET BASE ON MAIN ROOF
PLAN
3
ELEVATION
4
5lbs LEAD
230x75 SHAPED JOISTS
19 BOARDS
75x50 HEAD
150x19 LOUVRES
25 BOARDS
FLASHINGS
150x75 SILL
75x50 TIE
25 BOARDS
175x75 RIDGE
FELT
100x50
100x50 COLLAR
100x75 PURLIN
100x75x10 ANGLE
CROSS SECTION A-A
50 PACKING
WEATHER BOARDS
100x38
FLASHING
5
LONGITUDINAL SECTION

pyramid. The development of two surfaces of the base and the plan of the surfaces, where they intersect with the main roof, are shown in Sheet 43.2.

The plan and elevation of the base are set out. To obtain the plan of the surfaces intersecting with the main roof, vertical projections from points J and D are drawn to give points J′ and D′ in plan. A horizontal line drawn from the ridge O cuts the generator of the base at point X. A vertical projection from this point to the centre line in the plan gives point X′. With centre O and radius O–X′ describe an arc to cut the ridges in points X″ and X‴. Points J–X″, D′–X″, M–X‴ and N–X‴ are joined to complete that part of the plan.

The surface of the main roof to the left of the ridge is next developed. With centre O points J and G′ are brought to the horizontal line at J′ and G″. Vertical projectors from these points to intersect with horizontal lines brought out from J, G, H and M in the plan are drawn. X″J″G″H′M′X‴ is the development of the roof surface and the shape of the curb on the main roof.

The development of the surfaces of the base ABEF and ABCD′ is similar for both cases. It is necessary to obtain the true length of one of the sides first. Side A–B in the plan has been chosen and its true length A–B″ determined. B–B″ is drawn at right-angles to A–B and equal to the height of the base H, in the elevation. A line drawn from A to B″ is the true length of the side. With A as centre and A–B″ as radius describe an arc to meet point B brought out horizontally in B′. Join point A to B′. Point E is treated in the same way to complete the development of the surface. The surface A′BCD″ is developed in a similar manner.

Louvred Turret

Sheet 43.3 shows the elevation of a ventilating turret having two of the sides fitted with louvres.

The turret is placed in the centre of a roof constructed with steel trusses. It rests upon a framed curb, the angle posts of which are halved and bolted to collars. 75 mm by 50 mm ties are used between the angle posts at the top and are fixed to the ridge. The 100 mm by 50 mm collars are placed resting on the timber purlins and notched and bolted to the roof spars, as shown in Sheet 43.4.

The ventilator consists of 75 mm by 75 mm jambs or angle posts, framed to a 75 mm by 50 mm head and 150 mm by 75 mm hardwood sill. The sills are mitred and bolted together at the corners, with the posts stub-tenoned to them and secured with coach screws. The curb is boarded below the sill with the sheet lead flashing dressed up the faces of the curb and over the slated roof. Shaped bearers or joists are notched and fixed to the head of the framed sides and covered with 19 mm boards. A small projecting cornice throws the roof water clear of the louvres and is finished with sheet lead or copper. A flashing over the sill is dressed up to the inside of the ventilator on boards nailed to bearers, to throw any driving rain-water clear from the inside. The two ends of the ventilator are finished with horizontal weather-boards, fixed flush with the faces of the angle posts.

Sheet 43.5 shows a longitudinal section through the ventilator.

Where large turrets are used to house bells or a clock, they should be well braced internally and securely anchored to the main roof as they are subject to high wind pressures.

Ventilators

Sheet 44.1 shows the elevation and section through a triangular louvre ventilator. The frame is jointed together using dovetails at the two base angles, these are formed on the rail and a mitred bridle joint at the top. The louvres are inclined at an angle of 45 degrees and are arranged so that the top edge of each is covered by the bottom edge of the one immediately above it. This cover is necessary so that one should not be able to see through to the other side and to prevent driving rain from passing through to the inside. The louvre boards are trenched into the frame as shown, so that it is necessary to develop the shapes of the boards and also one of the sides of the frame to enable the trenchings to be set out, the sides being set out as a pair.

To find the bevel for setting out the trenchings, project from the section the points a, b, c and d on to the inside face of the frame. Rebate the side of the frame AB so as to show its width and project the points a, b, c and d over at right angles to the developed side in a′, b′, c′, d′. Join these points to give the bevel and the setting out of the trenchings.

The shapes of the louvres can be developed in either of two ways. Both methods are shown. The first requires a templet on which the louvres can be marked, Sheet 44.2.

Points A and O in elevation are projected over to the right of the section in A′ and O′. A line is drawn from O′ at the same inclination as the louvre boards, in this case 45 degrees, to give the centre line. A line drawn at right-angles to this through O′ in B′ and C′, equal to BC in elevation, gives the base of the templet.

A louvre board marked E is projected from the section on to the centre line of the templet in points 1′, 2′, 3′, 4′. From the centre line the points are projected at right-angles to the edges of the templet to give the true shape of the top and bottom faces of the louvre E and the side bevel.

The second method of finding the bevels to cut the louvre boards is shown in Sheet 44.3.

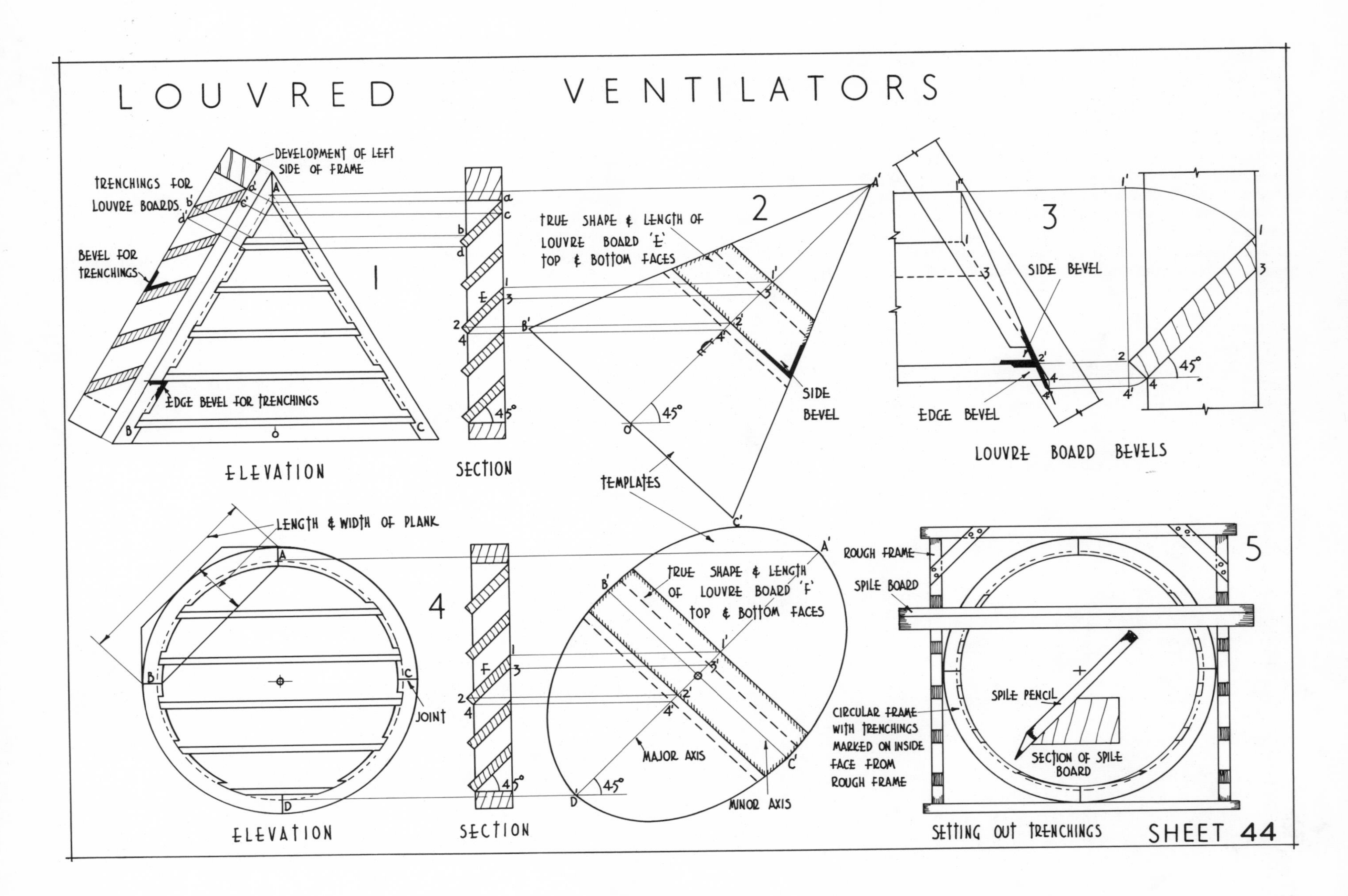
LOUVRED VENTILATORS
DEVELOPMENT OF LEFT SIDE OF FRAME
TRENCHINGS FOR LOUVRE BOARDS.
BEVEL FOR TRENCHINGS
EDGE BEVEL FOR TRENCHINGS
1
ELEVATION
SECTION
TRUE SHAPE & LENGTH OF LOUVRE BOARD 'E' TOP & BOTTOM FACES
2
SIDE BEVEL
45°
TEMPLATES
3
SIDE BEVEL
EDGE BEVEL
LOUVRE BOARD BEVELS
LENGTH & WIDTH OF PLANK
4
JOINT
ELEVATION
SECTION
TRUE SHAPE & LENGTH OF LOUVRE BOARD 'F' TOP & BOTTOM FACES
MAJOR AXIS
MINOR AXIS
ROUGH FRAME
SPILE BOARD
5
SPILE PENCIL
SECTION OF SPILE BOARD
CIRCULAR FRAME WITH TRENCHINGS MARKED ON INSIDE FACE FROM ROUGH FRAME
SETTING OUT TRENCHINGS
SHEET 44

Draw a vertical line through 2 to represent the vertical plane. With centre 2 and radius 2–1 draw an arc cutting the line in 1′. Draw a horizontal line from 1′ to give 1″ on a vertical line from point 1 in elevation. Join 2′–1″ to give the side bevel to the louvre board. The edge bevel is obtained in exactly the same way.

A circular louvre ventilator is shown in Sheet 44.4. The frame in this case is built up in four sections and jointed with either handrail bolts and dowels, or double hammer-headed keys. The method of finding the shapes of the louvre boards is similar to the previous example, and will not be repeated here. The templet in this case is elliptical. The setting out of the trenchings in the frame requires special treatment and a good deal of skill. A rough frame is made to receive the circular one as shown in Sheet 44.5. The positions of the louvres as given by the section are marked on the outer sides of the rough frame shown shaded. A spile board is then used to mark the positions of the trenchings on the front face of the frame. A section through the spile board is shown, having the front edge splayed at the same pitch as the louvre boards. This allows a pencil, sharpened so that the lead is in the same plane as the spile board edge, to be used to mark the housings on the inside face of the circular frame.

The louvre boards are next marked for shaping by first squaring a centre line on all faces of the boards. The templet is placed on each of the boards in turn with the centre line of the board corresponding with that of the templet. The ends of the boards are then marked by pricking through the templet on both sides of the board, making sure the templet is in the correct position.

Chapter Five: Joinery

Sliding and Sliding-folding Doors and Partitions

There are three types of sliding doors:

(1) Doors in one or more sections to slide to one or both sides.
(2) Doors in hinged sections to slide and fold to one or both sides.
(3) Doors in hinged sections on track which turns the doors to slide to one side and round the side wall inside the building.

The doors may be suspended from wheels running in a top track or fitted with swivel rollers running in a floor channel track. There is now a wide range of sliding and folding doors, but it is important to select the appropriate type to suit the plan, and the correct fittings for the size and weight.

Folding partitions work on the same principle as folding doors. Where there is wall space at each side of the opening, doors sliding along the face of the wall or in the cavity are suitable.

The illustrations on the following pages show different types of doors suitable for varying conditions, but the actual construction of the doors has not been dealt with as this is done in the normal way and forms part of earlier work. The details of tracks and door gear are those supplied by P. C. Henderson Ltd.

Sheet 45.1 shows a two-leaf arrangement for garage doors with double top track sliding to one side, as group one. The doors lap the jambs and each other by 50 mm to 75 mm and are suspended by hangers, two to each leaf, each having four wheels and running in a steel track. An enlarged detail of the hanger or trolley is shown in the elevation. These are manufactured in various patterns and may be a single pair of wheels fixed centrally to the top of the hanger, or a double pair with the hanger between as shown. Brackets fixed to the lintel at not more than 900 mm apart carry the track with a closed end-bracket at each end. The jointing of sections of track is made by positioning an open bracket to carry the two lengths at the joint. The top edge of the doors should cover the bottom edge of the lintel by about 25 mm and a 50 mm clearance from the top edge of the doors to the underside of the track allows for vertical adjustment to be carried out if necessary. Corner protection plates are fixed to the outside bottom corners of the doors where they strike ground stops when opening and closing. The latter should be fixed securely with rag bolts to prevent hanger wheels hitting the end brackets.

The bottom of the doors is held in position by side guide rollers running in a steel channel let into the floor, bent fixing lugs bolted under the track at intervals firmly key the channel. Sheet 45.2 shows this detail with a 10 mm clearance between the bottom of the doors and the floor. Alternative designs of guides are shown in Sheet 45.3. To complete the work a filling piece is fixed on the inside of the pier to close the gap caused by the offset of the doors in using double track. The doors are normally secured by means of spring bolts into the floor with an outside fastening on the end leaf, using various types of locking bar used with a padlock or a cylinder lock.

If the doors are to be fixed on the outside face of a building, it is necessary to fix a canopy to protect the whole of the track and door gear from the weather, as shown in Sheet 45.4. One hanger is shown in section and part of the second in elevation in this detail. The steel machined wheels are supplied with ball bearings or roller bearings grease packed, and lubricators which with proper maintenance and regular lubrication will ensure long and satisfactory life.

Sliding-folding Doors

Sheet 46.5 shows the plan and elevation of a system of sliding-folding doors 4.050 m high, which fall in the second group, suitable for transport depots, warehouses or garages. The doors consist of four leaves hinged to a post and folding behind a reveal to give a 3.150 m clear opening in plan.

A recommended thickness for the doors is 54 mm and 800 mm wide, with the joints between the leaves tongued and grooved or rebated. Wide openings may be covered by any number of units hinged to a post, but no unit should exceed six leaves. When an odd number of leaves has to be used, one leaf can be a swing door or 'swinger' as it is termed, and used as a pass door, but high doors operate best without a 'swinger' attached. A wicket for service entrance is shown which simplifies the locking problem and permits the main doors to remain closed in bad weather. This should be 1.500 m minimum height above the bottom rail, and although it has been incorporated in the end leaf, it may be provided in any leaf. It should be noted that the end or outside leaf, not being a swinger and not hinged from posts, must be 38 mm wider than the other leaves to cover the wheels of the end hanger.

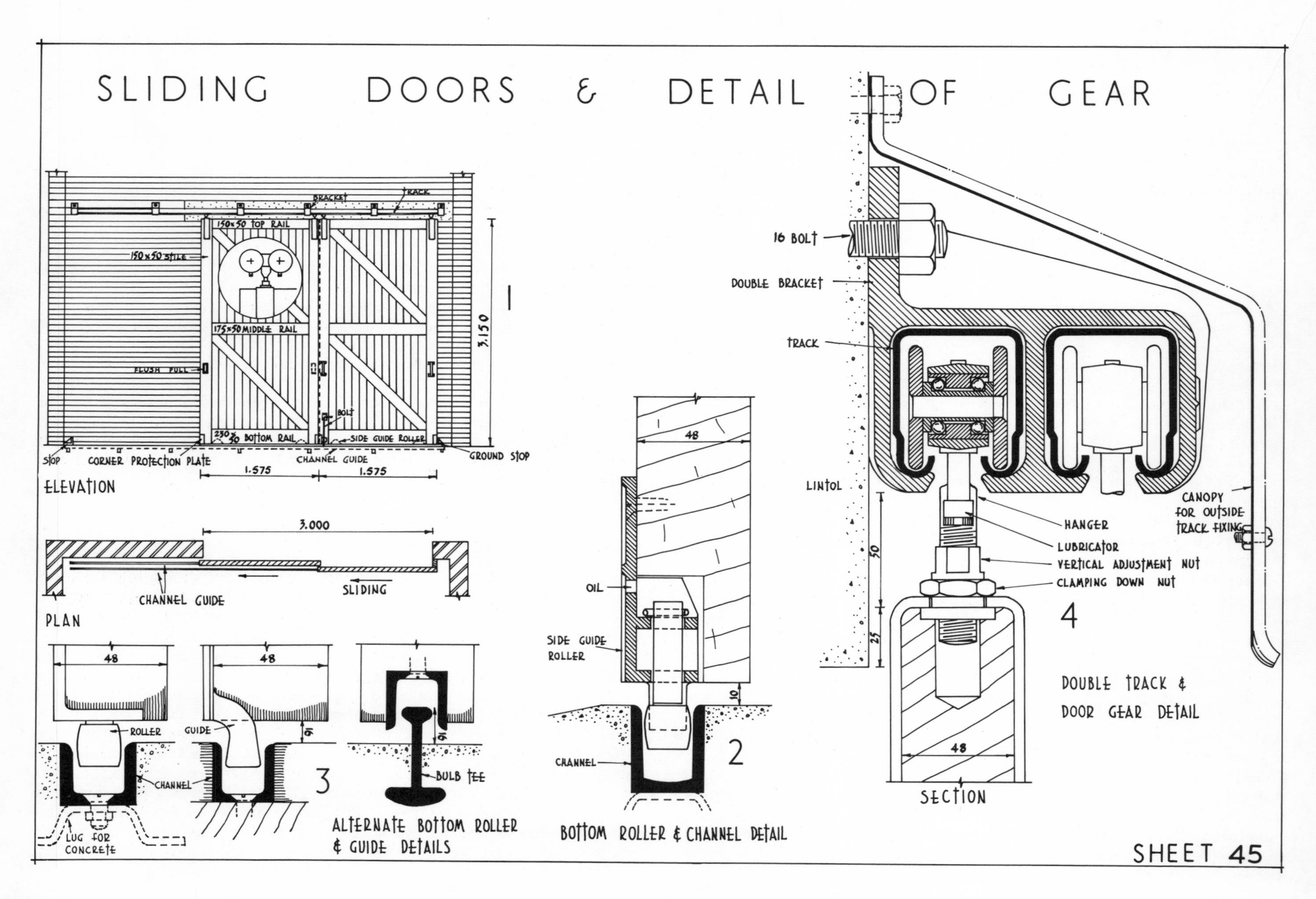
SLIDING DOORS & DETAIL OF GEAR
BRACKET
TRACK
150×50 TOP RAIL
150×50 STILE
175×50 MIDDLE RAIL
FLUSH PULL
BOLT
230×50 BOTTOM RAIL
SIDE GUIDE ROLLER
3.150
1
STOP
CORNER PROTECTION PLATE
CHANNEL GUIDE
GROUND STOP
1.575
1.575
ELEVATION
3.000
CHANNEL GUIDE
SLIDING
PLAN
48
ROLLER
CHANNEL
LUG FOR CONCRETE
GUIDE
16
3
BULB TEE
ALTERNATE BOTTOM ROLLER & GUIDE DETAILS
OIL
SIDE GUIDE ROLLER
10
CHANNEL
2
BOTTOM ROLLER & CHANNEL DETAIL
16 BOLT
DOUBLE BRACKET
TRACK
LINTOL
HANGER
LUBRICATOR
VERTICAL ADJUSTMENT NUT
CLAMPING DOWN NUT
CANOPY FOR OUTSIDE TRACK FIXING
50
25
4
DOUBLE TRACK & DOOR GEAR DETAIL
SECTION
SHEET 45

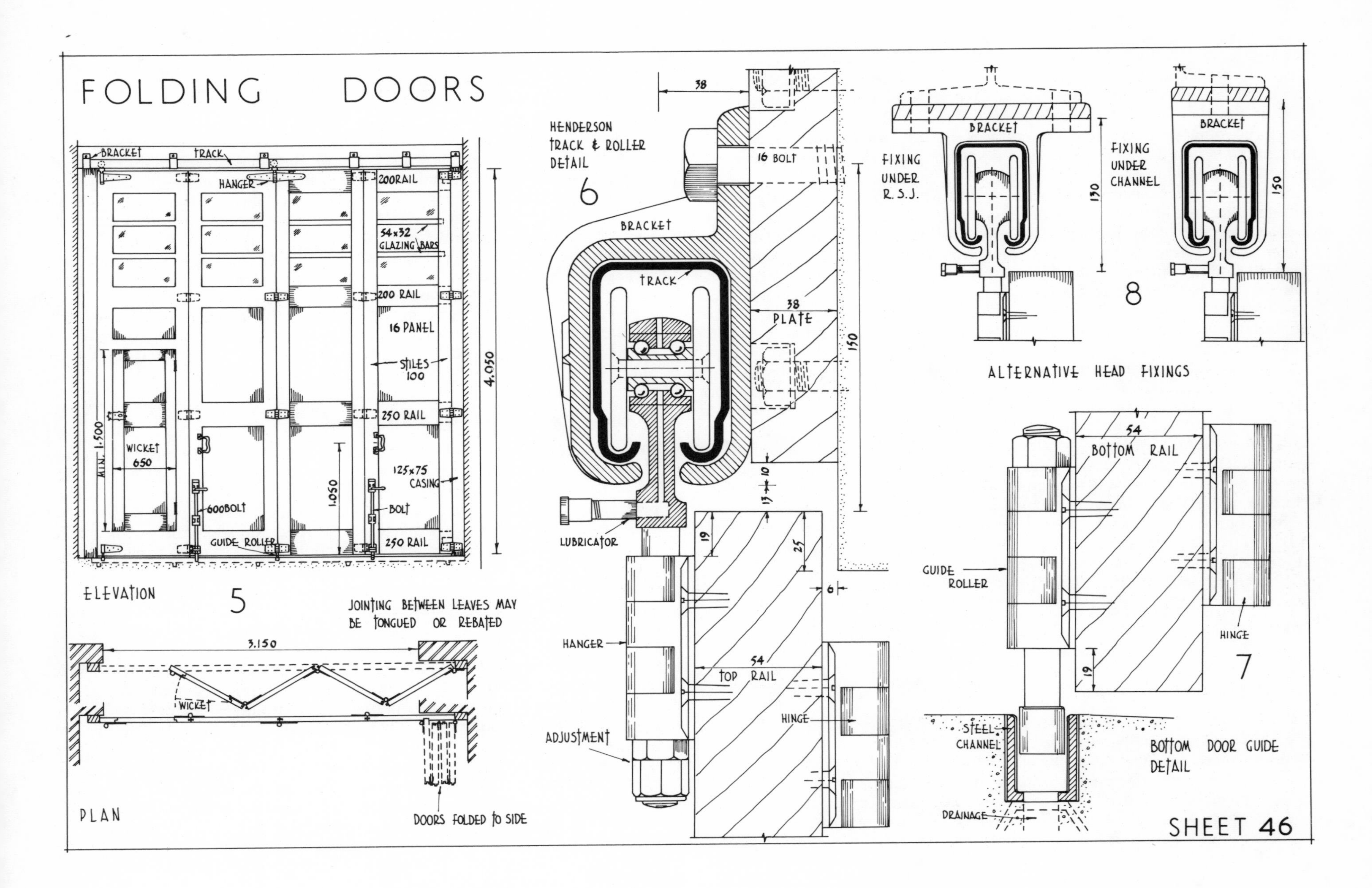
FOLDING DOORS
BRACKET
TRACK
HANGER
200 RAIL
54x32 GLAZING BARS
200 RAIL
16 PANEL
STILES 100
250 RAIL
WICKET 650
MIN. 1.500
125x75 CASING
1.050
600 BOLT
BOLT
GUIDE ROLLER
250 RAIL
4.050
ELEVATION
5
JOINTING BETWEEN LEAVES MAY BE TONGUED OR REBATED
3.150
WICKET
PLAN
DOORS FOLDED TO SIDE
HENDERSON TRACK & ROLLER DETAIL
6
38
16 BOLT
BRACKET
TRACK
38
PLATE
150
10
13
LUBRICATOR
19
25
6
HANGER
54
TOP RAIL
HINGE
ADJUSTMENT
FIXING UNDER R.S.J.
BRACKET
130
FIXING UNDER CHANNEL
BRACKET
150
8
ALTERNATIVE HEAD FIXINGS
54
BOTTOM RAIL
GUIDE ROLLER
HINGE
19
7
STEEL CHANNEL
BOTTOM DOOR GUIDE DETAIL
DRAINAGE
SHEET 46

A timber plate 38 mm thick is bolted to the lintel to carry the track fixed by brackets at 900 mm centres bolted or coach screwed through the plate into the lintel, Sheet 46.6. This of course should be dead level with track joints butted tightly together in the centre of the track bracket. The closed end brackets keep the joints perfectly tight. An extra bracket is used to take the weight of the unit when folded to one side, this is shown in the elevation and accounts for the uneven spacing of the brackets.

The doors are measured and made so that when fixed, the top edge of the doors covers the bottom edge of the lintel by 25 mm and a clearance of 13 mm to the underside of the track brackets allows for vertical adjustment. When the doors are prepared they are laid out flat across joists supported on saw stools in pairs and the hinges set out. These are set out four in height on the outside face of the doors for strength and rigidity, and are of malleable iron with steel spindles. The fittings on the inside of the doors consist of hangers at the top, hinges in the middle and bottom guide rollers. All of these are next screwed on to the inside faces of the doors.

The doors can now be suspended from the track and hinged to the 125 mm by 75 mm casing ready for fixing the bottom channel, where a 10 mm clearance is again necessary, Sheet 46.7. The floor must be level and even for 900 mm or more inside the opening to permit the leaves to fold with proper floor clearance and no scraping. The concrete outside should slope away from the entrance. The bottom channel guide should be cemented in after the doors have been tested when hung for plumb, and the guide is level and parallel with the top track. This is holed for drainage and should be kept greased.

Bow pull handles are screwed on the inside of the doors at a height of 1.050 m from the floor and 600 mm surface bolts let into the floor secure the doors. A cylinder lock is used on the wicket.

Sheet 46.8 shows alternative head fixings for the track, using special brackets where the doors are to be fixed under a rolled steel joist lintel or under a steel channel.

Sliding Doors—Around the Corner

Sheet 47.9 shows the plan and elevation of a system of doors which fall into the third group—sliding to one side and round the side wall inside the building. They are ideal for domestic garages, and as they must of necessity be made up of narrow leaves to allow turning, these may be designed as glazed, panelled, boarded or flush doors. Alternative designs are shown in the elevation 44 mm thick in four leaves. There must not be less than three leaves nor more than four hinged together, with the first leaf under the curved track forming a swinger. The leaves may be rebated together or left square, their width will depend on the actual opening but should not exceed 800 mm.

Rebated or square locking and stopping stiles the full height of the doors are fixed to the brickwork in the reveals with the doors having 25 mm overlap on the brickwork on each side of the opening and across the lintel.

The fixing of the track differs from the previous example only in making provision for the curved portion and the straight return along the inside wall. The corner track detail is shown in Sheet 47.10 with the necessary dimensions given. A 64 mm thick batten for packing out the track along the side wall is fixed to the brickwork giving the doors clearance along the side wall. A short length of this batten is also required for fixing the curved portion of the track and is fixed to the side and front battens.

The fixing of the track and channel is similar to the previous example and will not be repeated here. It will be seen that the patent fittings in this case are fixed on the inside of the doors only.

Sheet 47.11 shows sections through the stiles of the two types of doors and a sketch of the joint at the middle rail is shown in Sheet 47.12. Sheet 47.13 shows the door gear and fittings while a sketch of a bracket is shown in Sheet 47.14.

Sliding Centre-folding Partitions

Internal folding partitions employed in schools, restaurants, churches, assembly halls, etc. may be hung at the top with hangers or provided with swivel action rollers at the bottom. They may be centre-folding or end-folding; both types will be dealt with in the following illustrations.

Manufacturers recommend partitions be top hung when the weight of the gear can be safely carried overhead without any deflection, a guide being provided at floor level. Also that bottom rollers, primarily designed for interior work, are especially suitable when the weight of the partition can be conveniently carried on the floor.

The range of gears allows the widest scope for design in partitions so far as the joiner's work is concerned. The doors may be fully or partly glazed, flush or panelled, permitting individuality of design, and are suitable for any width of opening up to 4.800 m high. Wide screens usually fold in two units hinged to jambs at the right and left of the opening. Extra wide openings should be covered by three or more units, the intermediate units 'floating', that is, not hinged from either jamb. 'Floating' units can slide either way any distance necessary to fold behind piers and clear of the opening.

For end-folding screens the number of leaves per unit should not exceed six. 'Floating' units should always be four or six—never an odd number, because of the tendency to tilt and instability.

For centre-folding screens the maximum number of leaves hinged to a jamb should not exceed seven and a half leaves, while 'floating' units should be in five or seven leaves.

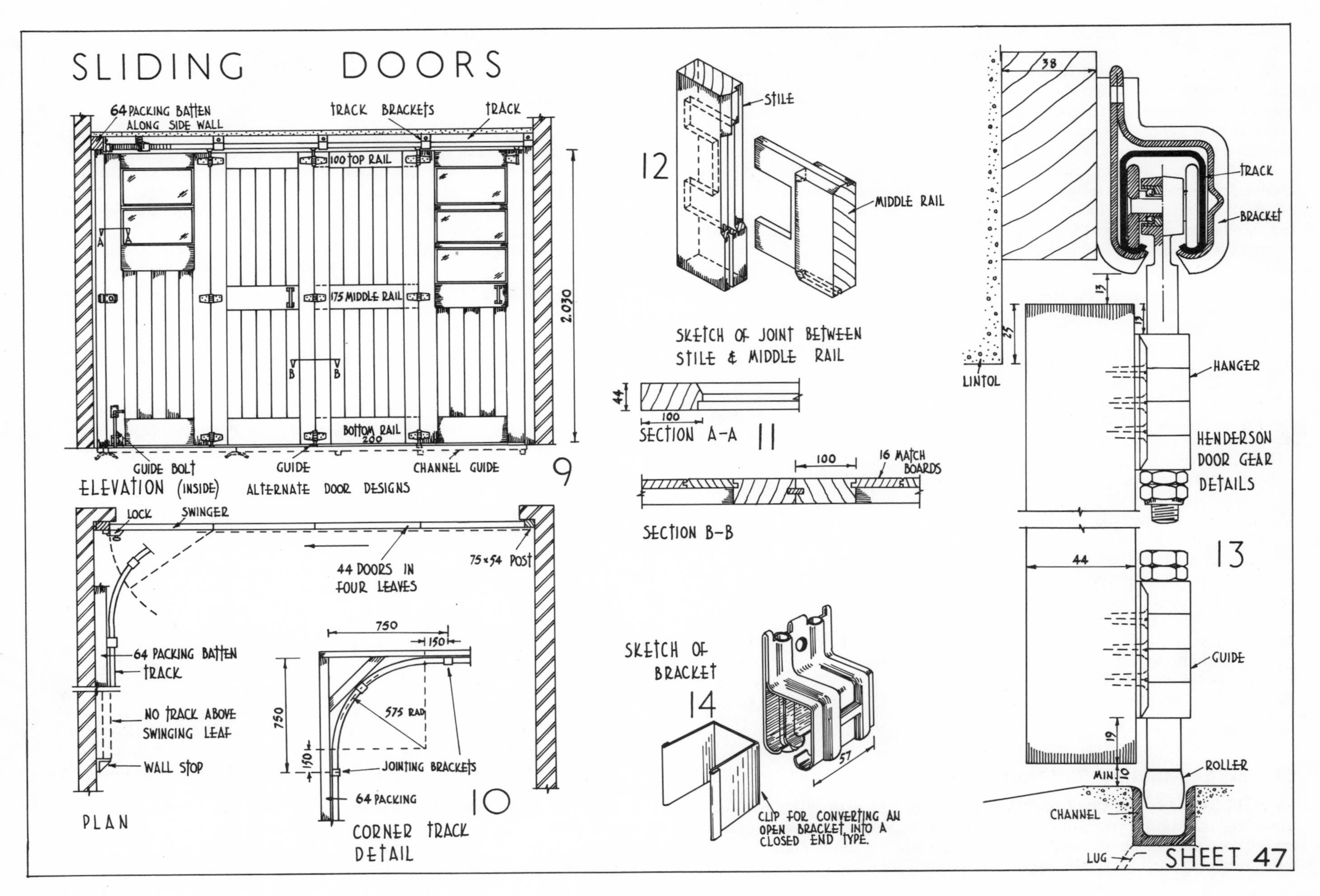
SLIDING DOORS
64 PACKING BATTEN ALONG SIDE WALL
TRACK BRACKETS
TRACK
100 TOP RAIL
175 MIDDLE RAIL
BOTTOM RAIL 200
2.030
GUIDE BOLT
GUIDE
CHANNEL GUIDE
ELEVATION (INSIDE)
ALTERNATE DOOR DESIGNS
9
LOCK
SWINGER
44 DOORS IN FOUR LEAVES
75×54 POST
64 PACKING BATTEN
TRACK
NO TRACK ABOVE SWINGING LEAF
WALL STOP
PLAN
750
150
575 RAD
JOINTING BRACKETS
64 PACKING
10
CORNER TRACK DETAIL
12
STILE
MIDDLE RAIL
SKETCH OF JOINT BETWEEN STILE & MIDDLE RAIL
44
100
SECTION A-A
11
16 MATCH BOARDS
SECTION B-B
SKETCH OF BRACKET
14
57
CLIP FOR CONVERTING AN OPEN BRACKET INTO A CLOSED END TYPE.
38
TRACK
BRACKET
LINTOL
13
25
HANGER
HENDERSON DOOR GEAR DETAILS
44
13
GUIDE
19
MIN 10
ROLLER
CHANNEL
LUG
SHEET 47

Sheet 48.15 shows the small-scale key elevation of a partition with three full leaves and a half-leaf, which is always necessary in this type. All the full leaves are of equal width with the pivot half-leaf half the width of a full leaf, less the throw of the hinge shown at A in the plan, Sheet 48.16. The throw of the hinge is exactly half the finished thickness of the screen when the hinge is fitted with the centre pin in line with the face of the screen. A wicket door may be provided in any leaf not containing a bottom roller.

The formulae for the accurate width of the leaves for a screen 44 mm finished thickness, is as follows: add 22 mm to the width between the rebates of the end jambs for a screen with one pivot half leaf. Divide the result by twice the number of full leaves plus one for the half leaf. This gives half the width of a full leaf. Deduct 22 mm for the width of the pivot half leaf. Example:

$$2.733 \text{ m between rebates } \frac{2.733 \text{ m} + 22 \text{ mm}}{7} = 394 \text{ mm}$$

394 mm less 22 mm = 372 mm width of pivot half leaf
394 mm × 2 = 788 mm width of three full leaves.

The joiner's work in whichever type of design is used must be most accurately done so that the edges of all leaves are perfectly parallel with the ends square. The edges of the leaves may be rebated or tongued and grooved, Sheet 48.16. An enlarged detail of the latter is shown in Sheet 48.17 where it will be seen that to avoid cutting through the tongue and groove, the hinges as well as the tongues are offset. Broad hinges, fixed back-flap may be used as an alternative, Sheet 48.18, but this method does not give as neat a finish on the face side as hinges let in the edges of the leaves.

The fittings, apart from the hinges, are ball-bearing bottom rollers, Sheet 48.19, with guide rollers at the top, each of these is fixed centrally on flush plates let into the top and bottom edges of the leaves. A brass bottom rail is let in and screwed to a hardwood sill with a 5 mm clearance at the bottom. It is most important that the track is fixed level and perfectly straight. As can be seen, the top guide track is of a different section from any previously used and is fixed and cased in when the screens are in position.

Flush bolts with flush pulls over are fitted on alternate leaves and end leaves, with a rebated cylinder lock on the wicket to complete the partition.

Sliding End-folding Partition

Sheet 48.20 shows the small-scale key elevation and plan of an end-folding top-hung partition with four leaves hinged to a jamb. An even number of leaves, not exceeding six and hinged together as a unit, should be used whenever possible. All the leaves are of equal width except the leaf hinged to the jamb. This is narrower by half the thickness of the screen plus 38 mm. Notice that the screen in this case is 38 mm thick.

The formulae for the accurate width of the leaves for such a screen is as follows:

Add 56 mm (A + B), Sheet 48.22, to the width between the rebates, divide the total by the number of leaves to give the width of the equal leaves. The pivot or hinged leaf is 56 mm less in width.

The joinery work and preparation of the screen are similar to the previous example except that being top hung, special angle apron hangers, Sheet 48.21, and guide plates are fitted to the edges of the stile leaving clear flush faces to the screen.

Sheet 48.22 shows the plan of the doors folded back with the jointing and hinging of the leaves. A section showing the door gear with the fixing and finishings is shown in Sheet 48.23.

Bank Entrance Doors and Vestibule

Sheet 49.1 shows the plan and external elevation of the entrance doors and surrounding stone dressings to a bank. The joiner's work in this example is often termed traditional work, but even in new bank buildings this type of entrance is still to be found, apart from restoration work and work in extensions or additions to existing buildings. Certainly it represents high-class joiner's work in polished hardwood, and for this reason is being dealt with here.

The front doors consist of a pair of 56 mm thick folding leaves each 900 mm wide and 2.850 m high, which fold back into a vestibule when the premises are open. Above the doors can be seen a bronze grille which has a 32 mm metal frame behind glazed with 6 mm plate glass. The whole is contained within a 133 mm by 100 mm frame with a 175 mm by 100 mm transom. Sheet 49.2 shows a section through the vestibule with an entrance leaf folded back and a ceiling light above. Vestibule doors are usually swing doors and generally swing or open both ways on pivots or spring hinges. Examples of these can be seen in the chapter dealing with special ironmongery. The elevation of the vestibule doors with a glazed fanlight over is shown in Sheet 49.3.

Sheet 50, figs. 4, 5 and 6 show the details of the joiner's work for the bank entrance doors and vestibule. It will be seen that this is all moulded on the solid, with the exception of the entrance doors, which have bolection moulds on the face side.

The horizontal section in Sheet 50.4 shows one of the entrance doors folded back so that when open it is flush with the frame, the joint being masked by the roll moulding on the stile. In this case the door rotates on centres at the bottom and top. The side screens to the vestibule are framed and

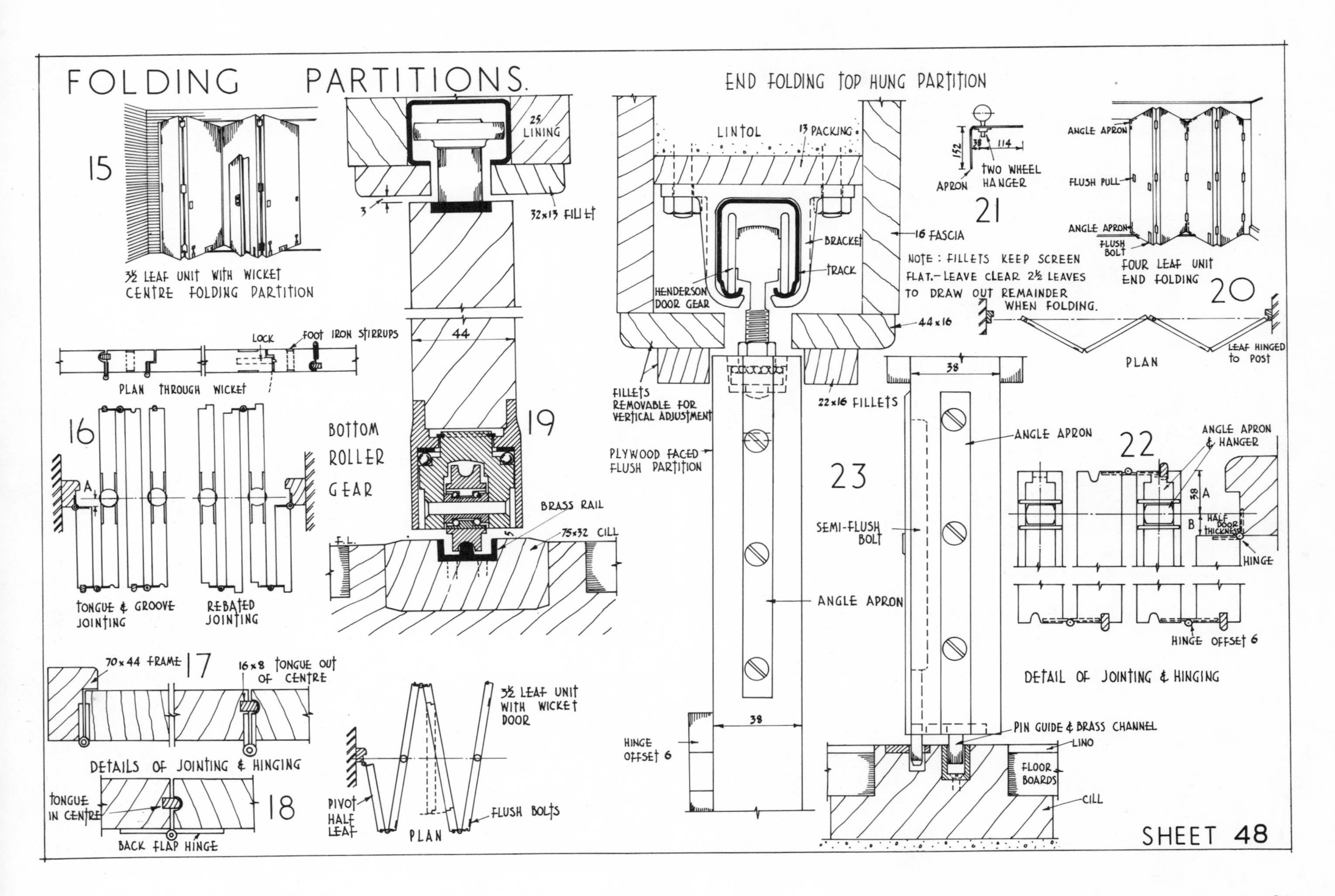
FOLDING PARTITIONS.
15
3½ LEAF UNIT WITH WICKET
CENTRE FOLDING PARTITION
25 LINING
3
32×13 FILLET
44
PLAN THROUGH WICKET
LOCK
FOOT IRON STIRRUPS
16
A
TONGUE & GROOVE JOINTING
REBATED JOINTING
BOTTOM ROLLER GEAR
19
BRASS RAIL
75×32 CILL
F.L.
70×44 FRAME
17
16×8 TONGUE OUT OF CENTRE
DETAILS OF JOINTING & HINGING
TONGUE IN CENTRE
18
BACK FLAP HINGE
3½ LEAF UNIT WITH WICKET DOOR
PIVOT HALF LEAF
PLAN
FLUSH BOLTS
END FOLDING TOP HUNG PARTITION
LINTOL
13 PACKING
BRACKET
TRACK
16 FASCIA
HENDERSON DOOR GEAR
44×16
FILLETS REMOVABLE FOR VERTICAL ADJUSTMENT
22×16 FILLETS
PLYWOOD FACED FLUSH PARTITION
23
ANGLE APRON
38
HINGE OFFSET 6
152
38
114
APRON
TWO WHEEL HANGER
21
NOTE: FILLETS KEEP SCREEN FLAT.– LEAVE CLEAR 2½ LEAVES TO DRAW OUT REMAINDER WHEN FOLDING.
ANGLE APRON
FLUSH PULL
ANGLE APRON
FLUSH BOLT
FOUR LEAF UNIT END FOLDING
20
PLAN
LEAF HINGED TO POST
38
ANGLE APRON
SEMI-FLUSH BOLT
PIN GUIDE & BRASS CHANNEL
LINO
FLOOR BOARDS
CILL
22
ANGLE APRON & HANGER
A
B
HALF DOOR THICKNESS
HINGE
HINGE OFFSET 6
DETAIL OF JOINTING & HINGING
SHEET 48

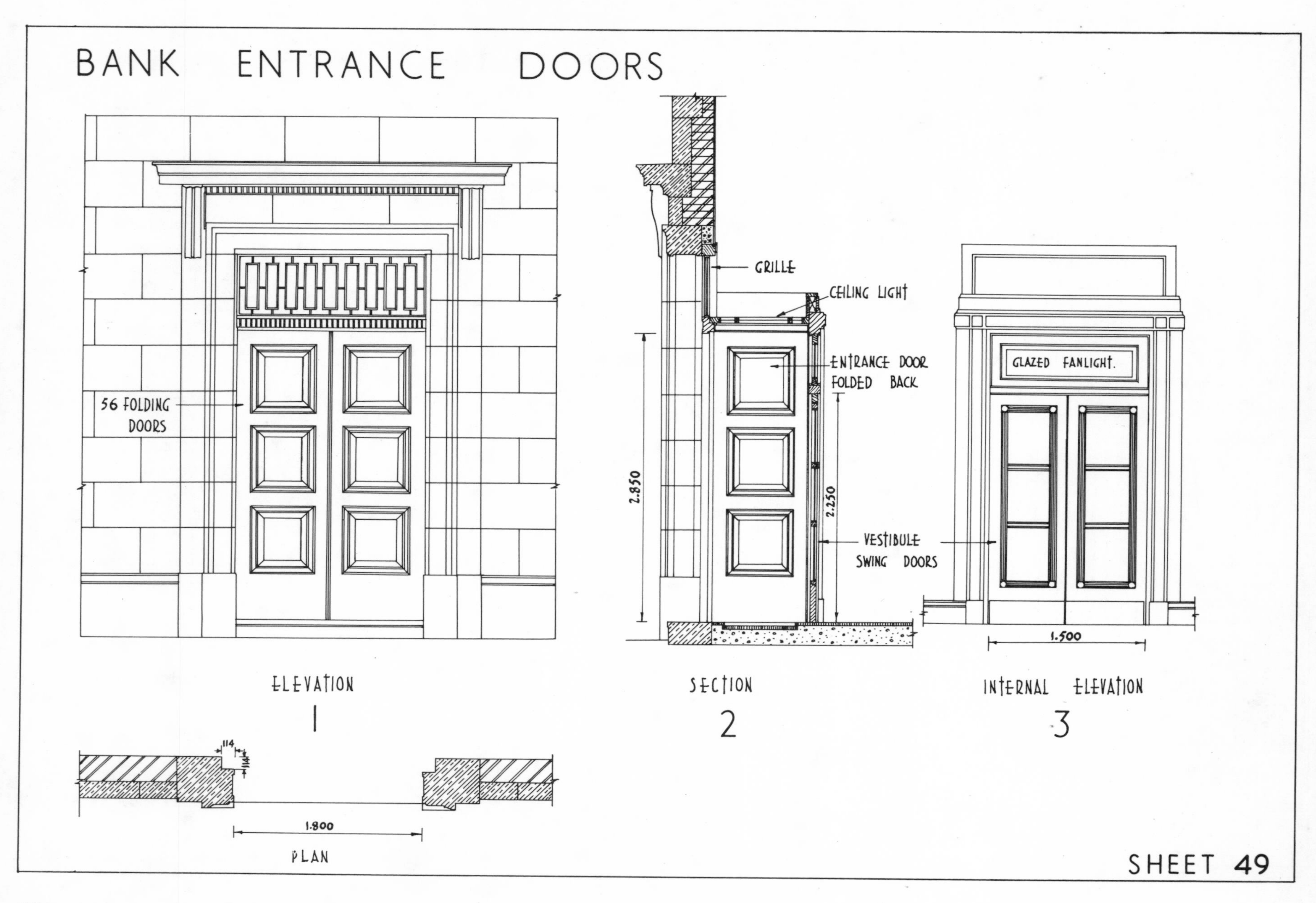
BANK ENTRANCE DOORS
56 FOLDING DOORS
GRILLE
CEILING LIGHT
ENTRANCE DOOR FOLDED BACK
2.850
2.250
VESTIBULE SWING DOORS
GLAZED FANLIGHT.
1.500
ELEVATION
1
SECTION
2
INTERNAL ELEVATION
3
114
114
1.800
PLAN
SHEET 49

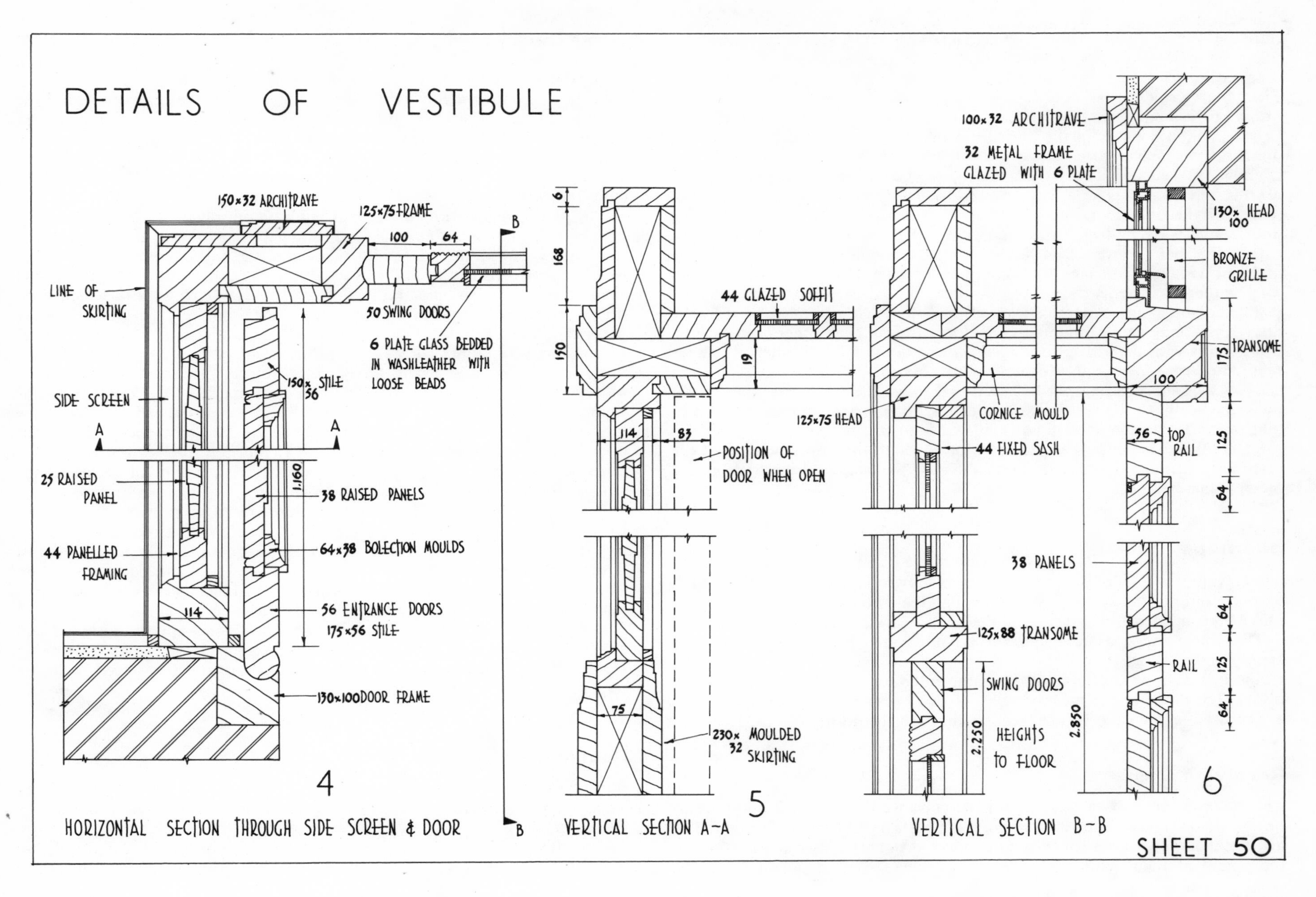
DETAILS OF VESTIBULE
150×32 ARCHITRAVE
125×75 FRAME
100
64
B
LINE OF SKIRTING
50 SWING DOORS
6 PLATE GLASS BEDDED IN WASHLEATHER WITH LOOSE BEADS
150×56 STILE
SIDE SCREEN
A
A
25 RAISED PANEL
1.160
38 RAISED PANELS
44 PANELLED FRAMING
64×38 BOLECTION MOULDS
114
56 ENTRANCE DOORS 175×56 STILE
130×100 DOOR FRAME
4
HORIZONTAL SECTION THROUGH SIDE SCREEN & DOOR
B
6
168
150
44 GLAZED SOFFIT
19
114
83
POSITION OF DOOR WHEN OPEN
75
230×32 MOULDED SKIRTING
5
VERTICAL SECTION A-A
100×32 ARCHITRAVE
32 METAL FRAME GLAZED WITH 6 PLATE
130×100 HEAD
BRONZE GRILLE
TRANSOME
175
100
125×75 HEAD
CORNICE MOULD
56
TOP RAIL
125
44 FIXED SASH
64
38 PANELS
125×88 TRANSOME
64
RAIL
125
SWING DOORS
64
2.250
HEIGHTS TO FLOOR
2.850
6
VERTICAL SECTION B-B
SHEET 50

filled with raised panels. These are fixed in rebated end and corner frames with fillets. Moulded skirtings and architraves finish the vestibule as shown. The inner swing doors, of which only a part is shown, are fitted into a solid frame. The hanging stiles of the doors are worked to a segment of a circle which fits into a corresponding hollow on the frame. The centre of rotation should be kept half the thickness of the door from the face of the frame, to allow the doors to open at right-angles to the frame. The meeting stiles also are rounded; this curve is struck from the centre of rotation. The doors swing on centres, the lower one a floor spring and the upper one made to rise and fall to enable the door to be mounted, or on helical spring hinges. The swing doors have wide built-up stiles consisting of a reeded sub-frame tongued within the main frame. This allows for any swelling or shrinking of the material without disturbing the joinery as a whole. Plate glass is bedded in washleather and fixed within the sub-frame with beads. This method is used to prevent breakage of large sheets of glass through any movement in the door which may cause binding or the doors to foul each other when in use. Flush bolts with a mortise lock secure these doors while flush bolts and a rebated mortise lock secure the entrance doors.

Sheet 50.5 shows a part vertical section through the side frame and part of the glazed soffit to the vestibule. Artificial lighting is fixed above the soffit so that the vestibule can be brilliantly lit without glare.

Sheet 50.6 shows a part vertical section through the upper portion of the vestibule, entrance doors and frame showing the bronze grille with glazed metal frame behind.

Showroom Doors

The elevation of a set of doors suitable for a showroom is shown in Sheet 51.1. Two are sliding and two are fixed. Although these are classified as doors they may be grouped with windows.

The track and gear arrangement is similar to the details in previous examples.

Sheet 51.2 shows the detail of the frame and door stile of one of the sliding doors. The detail at the meeting stiles of two doors is also shown.

Sheet 51.3 shows the vertical section through the doors.

Revolving Doors

This type of door is very suitable for the entrance halls to such buildings as hotels, banks, restaurants, etc., where it is necessary to exclude draught. The arrangement consists of four leaves connected in various ways according to the patent at their centre. They are constructed so that openings can be thrown open by folding the leaves to the centre or to one side, to leave an unobstructed entrance for admitting goods. To do this, various mechanical devices are used, the leaves being hinged to fold up.

Sheet 52.1 shows a four compartment plan of revolving doors within the entrance to a building. The four leaves collapse and fold to one side as shown.

Sheet 52.2 shows a three-compartment plan, with side lights between the circular enclosing screen and the vestibule walls. The three leaves collapse and fold against one side giving a clear opening. They are suitable for entrances where small doors only can be accommodated.

The plan and elevation, Sheet 52.3, show 38 mm thick doors 2.100 m high, contained within a flush circular enclosing screen, 1.800 m diameter and 35 mm thick. The doors, normally made in oak, walnut or mahogany and polished, are constructed in the normal way and fitted with plate glass above the middle rail, push bars and kicking plates. The hard wear to which these doors are subject requires mechanism of a substantial character. They are suspended from a trolley mounted inside two steel joists and this part of the mechanism is mounted out of sight, in the head soffit. A 13 mm spindle running on a large ball-bearing thrust race projects through the soffit and is mortised between the two stiles of the doors marked 'A', Sheet 52.4. The doors are held together by means of the hinge wings, the centre block being tapped for this purpose. There are three centre blocks in the height of the doors, which are drilled vertically for the 13 mm rod to pass through for the purpose of lifting from the floor socket, the bottom pivot releasing the trolley from the head track.

The hinged doors are held to the fixed doors by cross connections, having a turnbuckle and quick release knob shown in the plan. When folding the doors to the side, the connections are first released, doors 'B' are folded against doors 'A' when an automatic catch engages and locks the doors together. The centre rod is next lifted by the projecting handle and the whole assembly pushed to one side. Flush bolts lock the doors when not in use.

Between the doors and screen there should be a clearance of about 13 mm along the vertical edge for safety, but this space is closed by flexible rubber trailers fixed to the outer stiles and rubbing against the flush-faced enclosing screen.

Double-Margin Door

Sheet 53.1 shows the elevation of a double-margin door 2.100 m by 1.350 m. This type of door is made wide and in such a manner as to match with ordinary folding doors.

The door is first framed in two separate leaves which are secured together with a through top rail the full width of the door, brindle jointed at the centre

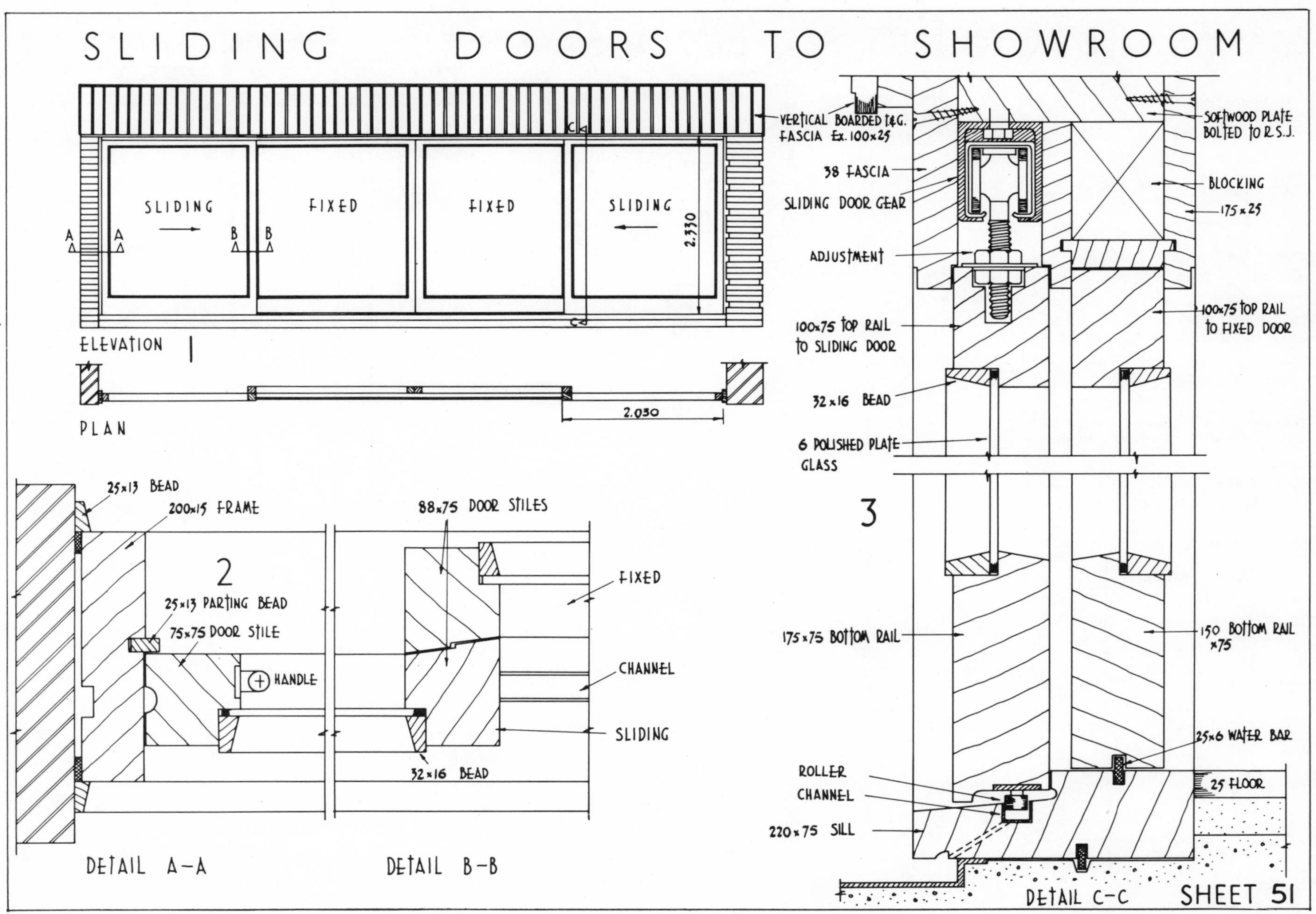
SLIDING DOORS TO SHOWROOM
SLIDING
FIXED
FIXED
SLIDING
2.330
ELEVATION 1
PLAN
2.030
VERTICAL BOARDED T&G. FASCIA Ex. 100×25
38 FASCIA
SLIDING DOOR GEAR
ADJUSTMENT
100×75 TOP RAIL TO SLIDING DOOR
32×16 BEAD
6 POLISHED PLATE GLASS
SOFTWOOD PLATE BOLTED TO R.S.J.
BLOCKING
175×25
100×75 TOP RAIL TO FIXED DOOR
3
175×75 BOTTOM RAIL
150 BOTTOM RAIL ×75
25×6 WATER BAR
ROLLER
CHANNEL
220×75 SILL
25 FLOOR
DETAIL C–C
SHEET 51
25×13 BEAD
200×15 FRAME
2
25×13 PARTING BEAD
75×75 DOOR STILE
HANDLE
88×75 DOOR STILES
FIXED
CHANNEL
SLIDING
32×16 BEAD
DETAIL A–A
DETAIL B–B

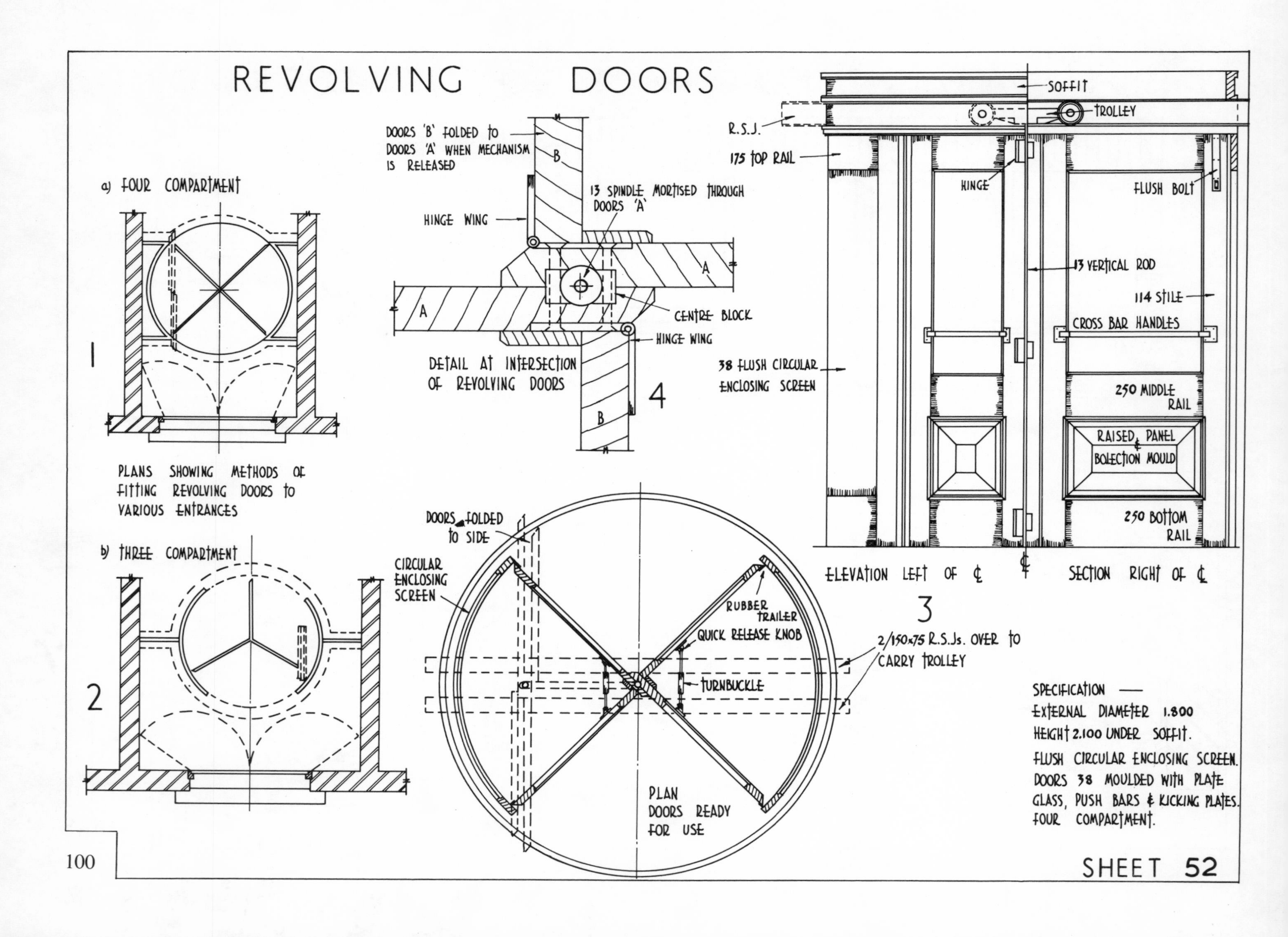
REVOLVING DOORS
a) FOUR COMPARTMENT
1
PLANS SHOWING METHODS OF FITTING REVOLVING DOORS TO VARIOUS ENTRANCES
b) THREE COMPARTMENT
2
DOORS 'B' FOLDED TO DOORS 'A' WHEN MECHANISM IS RELEASED
B
HINGE WING
13 SPINDLE MORTISED THROUGH DOORS 'A'
A
A
CENTRE BLOCK
HINGE WING
DETAIL AT INTERSECTION OF REVOLVING DOORS
4
B
SOFFIT
TROLLEY
R.S.J.
175 TOP RAIL
HINGE
FLUSH BOLT
13 VERTICAL ROD
114 STILE
CROSS BAR HANDLES
38 FLUSH CIRCULAR ENCLOSING SCREEN
250 MIDDLE RAIL
RAISED PANEL & BOLECTION MOULD
250 BOTTOM RAIL
ELEVATION LEFT OF ℄
SECTION RIGHT OF ℄
3
DOORS FOLDED TO SIDE
CIRCULAR ENCLOSING SCREEN
RUBBER TRAILER
QUICK RELEASE KNOB
TURNBUCKLE
2/150×75 R.S.Js. OVER TO CARRY TROLLEY
PLAN DOORS READY FOR USE
SPECIFICATION —
EXTERNAL DIAMETER 1.800
HEIGHT 2.100 UNDER SOFFIT.
FLUSH CIRCULAR ENCLOSING SCREEN.
DOORS 38 MOULDED WITH PLATE GLASS, PUSH BARS & KICKING PLATES.
FOUR COMPARTMENT.
SHEET 52

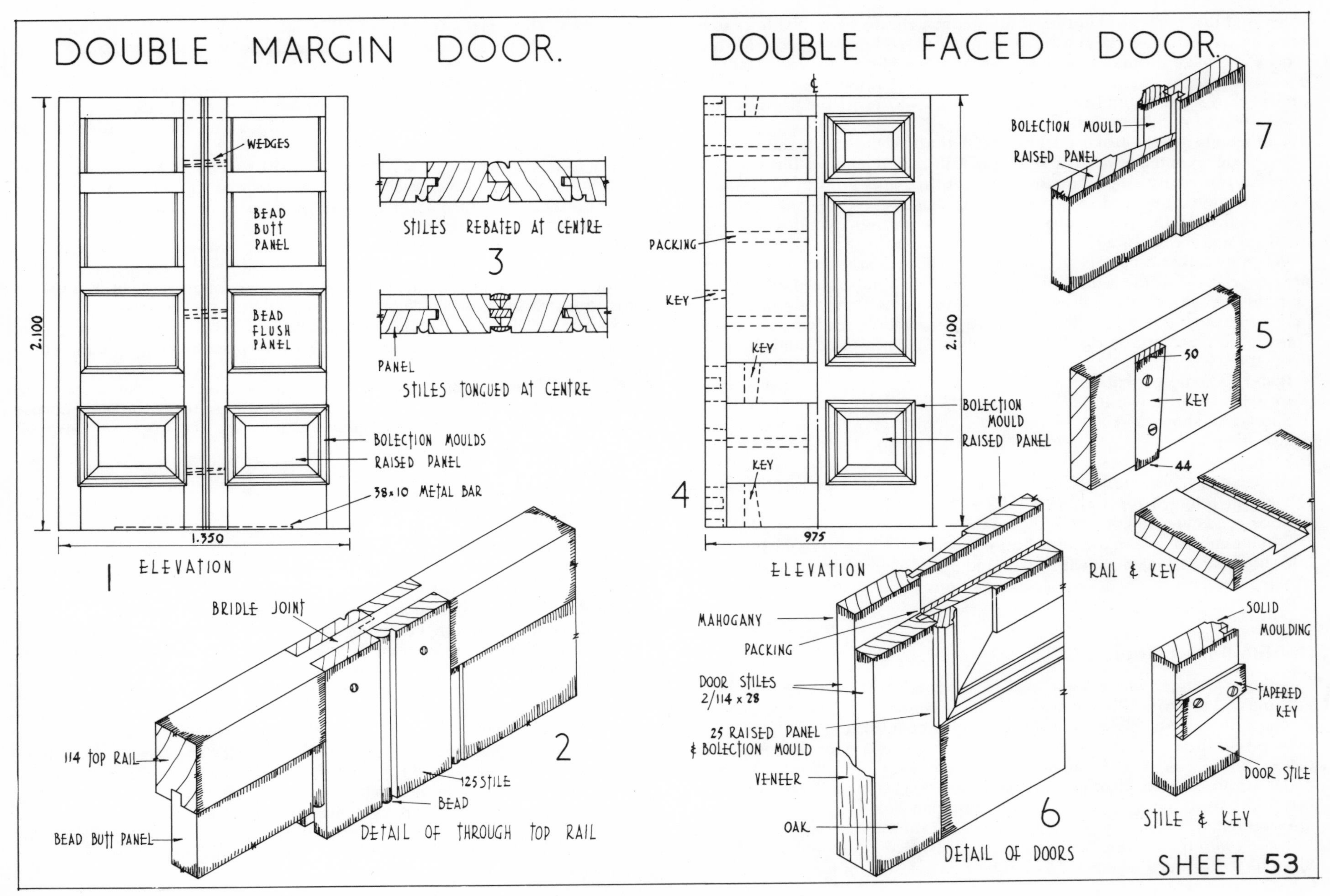
DOUBLE MARGIN DOOR.
DOUBLE FACED DOOR.
WEDGES
BEAD BUTT PANEL
BEAD FLUSH PANEL
2.100
BOLECTION MOULDS
RAISED PANEL
38×10 METAL BAR
1.350
1
ELEVATION
STILES REBATED AT CENTRE
3
PANEL
STILES TONGUED AT CENTRE
BRIDLE JOINT
114 TOP RAIL
BEAD BUTT PANEL
125 STILE
BEAD
2
DETAIL OF THROUGH TOP RAIL
PACKING
KEY
KEY
KEY
4
2.100
975
ELEVATION
BOLECTION MOULD
RAISED PANEL
BOLECTION MOULD
RAISED PANEL
7
5
50
KEY
44
RAIL & KEY
MAHOGANY
PACKING
DOOR STILES 2/114 x 28
25 RAISED PANEL & BOLECTION MOULD
VENEER
OAK
6
DETAIL OF DOORS
SOLID MOULDING
TAPERED KEY
DOOR STILE
STILE & KEY
SHEET 53

stiles, and three pairs of hardwood fox wedges, shown by dotted lines in the drawing. A metal bar 38 mm by 10 mm is housed and screwed with brass screws into the bottom rails after the door has been fitted, for added stiffness. The wedges are positioned at distances shown, so that there may be sufficient timber between them and the ordinary mortises to resist the pressure when these wedges are tightened up.

The tenons are the usual thickness, one-third of the stiles, and the hardwood wedges are the same. The stiles and rails are prepared in the usual way, with the two centre stiles mortised through the panel grooves for the extra keys. The doors are then glued, without the panels and top rail, care being taken that only the centre stile and rail ends that go into this are glued, cramped and wedged.

When both leaves have been dealt with in this way, the ends of the tenons are dressed off and the centre joint prepared. This is next glued and wedged, and any dressing off of wedges in the grooves done. Panels are then put in and the through top rail glued and dowelled in position, ready for the outside stiles to be glued and wedged. Panel moulds are then fitted, and the whole dressed off.

It should be noted that alternative panel finishings, suitable for this type of work, have been shown in the elevation.

Sheet 53.2 shows a sketch of the bridle jointing between the top rail and centre stiles. A bead stuck on the solid breaks the joint of the two leaves at the centre.

Alternative centre jointing is shown in Sheet 53.3 with centre stiles either rebated or tongued together with a loose tongue. When the latter is used, planted beads in a rebate break the joint.

Double-margin doors are very heavy because of their size and thickness, and should always be hung with a pair and a half of 100 mm butt hinges, half of the hinge sunk in the door edge and half into the door frame.

Double-Faced Door

In public buildings, where much high-class joiner's work is to be found, it is not unusual to have different wall finishings between adjoining rooms, and between rooms and corridors, where each may be finished in a different kind of wood. In such cases it is necessary that each side of a door shall match the other fixings, oak to oak, mahogany to mahogany, etc. If flush-veneered doors are to be used, then there is no difficulty, each side being finished to match the surrounding work. Where framed panelled doors with mouldings are required, then a door made of different kinds of wood on each face is used.

The elevation of a double-faced door, 2.100 m by 900 mm is shown in Sheet 53.4. The door, which is 56 mm thick, is made up of two parts, each 28 mm thick. These are secured together by keys in the positions shown in the drawing. The keys are tapered in length and have bevelled edges, being about 11 mm thick.

The keys are fitted into one member and then laid on the other, carefully marked and taken out, glued and screwed to the corresponding piece, Sheet 53.5. When this has been done the two members with the faces for the joint are glued and put together.

The panels, which are 19 mm thick, are kept apart by packing pieces, shown by the dotted lines in the elevation, and are inserted to keep the panels from warping. They are glued and pinned to the back of one panel, giving a solid bearing for the other panel.

When the door has been assembled, the edges should be veneered. The closing stile of the door should be finished with the same material as the fittings in the room, while the hanging stile corresponds to the fittings on the side to which the door leads off.

Sheet 53.6 shows a sketch of part of the finished door with raised panels and bolection moulds on the outside, and raised panels with a solid moulding on the inside.

Sheet 53.7 shows a sketch of the stile, mould and panel. It should be noted that the 10 mm groove is out of the centre of the stile, so that the 10 mm mortise and tenon joints between the muntin and rails and the stiles and rails will likewise be out of centre.

Fire-Check Doors

This type of door is being extensively used today in modern building. Hospitals, factories, boiler houses, warehouses and workshops are examples of where they may be used. They may be constructed in a number of ways. Framed with solid panels in hardwood or as an alternative a solid door built up of three layers of tongue and groove boards, the surfaces then covered with facing sheets of aluminium or galvanized steel. Certain manufacturers have developed alternative constructions. One of these has a core frame similar to normal flush door construction with an infilling of compressed straw slab. There are two types (a) half-hour resistance, (b) one-hour resistance, and they are highly efficient fire-barriers which fulfil fire-check requirements as laid down by B.S.S. 459, Part III. The facings are 6 mm. plywood to half-hour type 44 mm finished thickness, and 6 mm metal-faced plywood to one-hour type 56 mm finished thickness.

Regulations concerning fire-resistance in doors approve certain hardwoods —oak, teak, jarrah karri and others, provided the door is solid and the finished thickness is not less than 44 mm. All metal and metal-faced doors are approved.

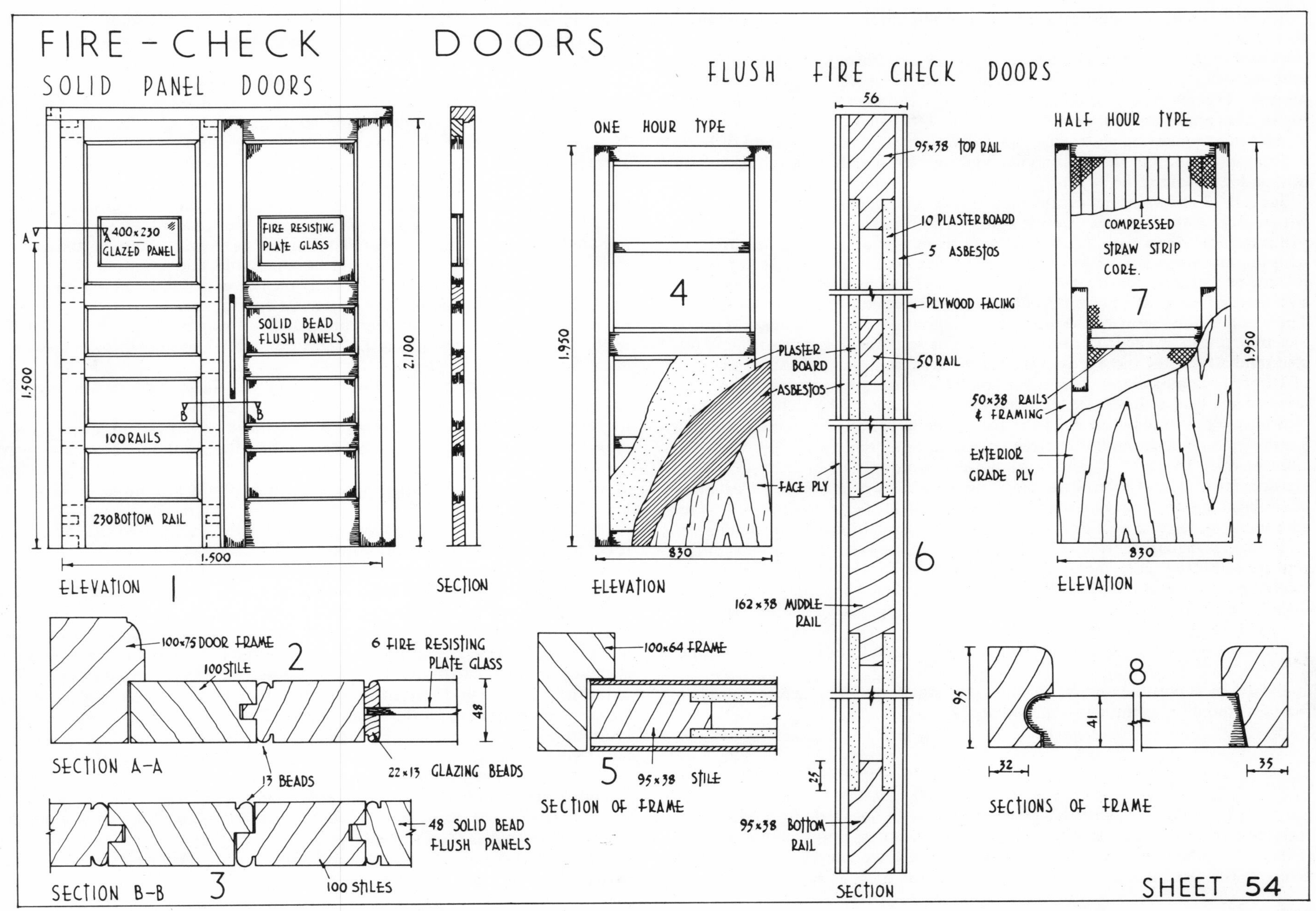
FIRE-CHECK DOORS
SOLID PANEL DOORS
FLUSH FIRE CHECK DOORS
400x230 GLAZED PANEL
FIRE RESISTING PLATE GLASS
SOLID BEAD FLUSH PANELS
100 RAILS
230 BOTTOM RAIL
1.500
2.100
ELEVATION 1
SECTION
ONE HOUR TYPE
4
PLASTER BOARD
ASBESTOS
FACE PLY
1.950
830
ELEVATION
56
95x38 TOP RAIL
10 PLASTERBOARD
5 ASBESTOS
PLYWOOD FACING
50 RAIL
6
162x38 MIDDLE RAIL
25
95x38 BOTTOM RAIL
SECTION
HALF HOUR TYPE
COMPRESSED STRAW STRIP CORE.
7
50x38 RAILS & FRAMING
EXTERIOR GRADE PLY
1.950
830
ELEVATION
100x75 DOOR FRAME
100 STILE
2
6 FIRE RESISTING PLATE GLASS
48
13 BEADS
22x13 GLAZING BEADS
SECTION A-A
48 SOLID BEAD FLUSH PANELS
100 STILES
SECTION B-B 3
100x64 FRAME
5
95x38 STILE
SECTION OF FRAME
95
41
8
32
35
SECTIONS OF FRAME
SHEET 54

Sheet 54.1 shows the elevation and vertical section of a solid panel door 48 mm finished thickness. The solid panels may have a 13 mm bead worked in the solid on the edges as bead flush panels, or be finished with loose beads and tongues. The panels may be built up in widths of four to five inches with a similar type of bead to each joint. This helps to reduce swelling and shrinking in the panels, but is more expensive. The top panel is prepared to receive a panel of fire-resisting plate glass as an observation panel, at a height of 1.500 m from the ground to the centre. Glazing beads fixed with cups and screws on both sides surround the panel.

Sheet 54.2 shows a section through the door frame and stile of the door with a section through the meeting stiles shown in Sheet 54.3.

Sheet 54.4 shows the elevation of a flush fire-check door one-hour type, with the facing cut away to show the construction. The core-frame is rebated to receive a plasterboard protective filling and covered with sheet asbestos and finally faced with plywood. The finished thickness is 56 mm as shown. The whole frame is glued and pressed together and no metal fastenings in the face plies are to be used. The timber should have a moisture content of 14 per cent, and all frames and exposed timber must be impregnated with a solution of 18 per cent mono-ammonium phosphate in water. A section through the door frame and stile of the door is shown in Sheet 54.5. The maximum clearance allowed between the door and frame in this type of construction is 3 mm. Sheet 54.6 shows the vertical section through the door.

Sheet 54.7 shows a patent flush fire-check door half-hour type manufactured by Linden Doors Ltd. The construction is of stiles and cross-rails of timber with an infilling core of closely packed strips of compressed straw with lock blocks on each side. The middle rail allows for the fitting of a letter-box. Facings are of exterior grade plywood. The fire-resisting qualities of the compressed straw core make it possible to offer a 41 mm thick plywood-faced flush door that gives a 'half-hour' fire resistance, when hung in a suitable frame. This construction gives a firm core that supports the facings and will not twist or set up internal stresses and results in a properly stressed skin design. The core is of sufficient weight to give good sound reduction and freedom from drumming.

Sheet 54.8 shows the section through the door frame designed to complete the unit, the door being left-hand hung.

Church Doors

Sheet 55.1 shows the elevation of a pair of oak church doors and frame, suitable either for internal or external use. The stiles, muntins and shaped rails are moulded on the solid. Typical sections are shown in Sheet 55.2 grooved for the panels. The bottom rail, Sheet 55.3, is machined with a chamfer on the top edge over which the mouldings of the stiles and muntins are scribed. On external doors the chamfered bottom rail assists in weathering. The shaped top or head rails are cut out of the solid, and care should be taken when marking out the plank to see the run of the grain is such that as little short grain as possible occurs in the finished rail. The joints between the shaped rails and stiles in the doors and in the frame occur at the springing. Here, the jointing is made using a single hammer-headed key, worked on the stiles with cross tongues at the shoulders to prevent twisting, as shown in Sheet 55.4. The joint between the shaped heads of the door frame is made using a dovetail key, being cut from hardwood and mortised into the shaped heads as shown in Sheet 55.5. It is glued and allowed to set before the remainder of the frame is assembled. Similarly the joints between the shaped rails and stiles of the doors should be glued and allowed to set before the assembly of the doors, as some difficulty may be experienced in pulling up the joint. The most suitable joint at this point is a loose tenon inserted and well glued into the slotted rail and pinned as shown in Sheet 55.6. Normal wedging can then be carried out. A section through the rebated door stiles is shown in Sheet 55.7.

Sheet 55.8 shows a single door with shaped head suitable for the side entrance to a church or as an internal door. The construction is very similar to that in the previous example. At the head a hardwood dovetail key mortised into the shaped rails is used. A handrail bolt and dowels may be used at this point as an alternative method of jointing. The jointing at the springing between the shaped head and door stile is a hammer-headed key with hardwood wedges.

Sheet 55.9 shows a part section with solid panels beaded on the inside and tongued into the stiles and muntins.

A sketch of the joint between the chamfered bottom rail and door stile is shown in Sheet 55.10. It will be seen that here, as with the muntins, the mould is scribed over the chamfer as in the previous example. The muntins are stub-tenoned into the bottom rail and also into the shaped top rails where the moulds are mitred. The width and length of the material required to cut the shaped rail is indicated and also the jointing, shown by dotted lines.

Framed, ledged and braced doors for external use may be used as an alternative construction to those illustrated. They are filled with tongued and grooved boards and are usually hinged on bands and gudgeons or plain cast butt hinges. The bands are usually a special decorative feature of this type of door. The construction of the doors follows that of the ordinary framed, ledged, braced and battened door, apart from the jointing of the shaped head and the stiles of the door. Here the outside stile is allowed to run beyond the springing, so that the top rail is tenoned and wedged at the springing without a further joint in the stile at this point. This method ensures a sound jointing between the members of the doors.

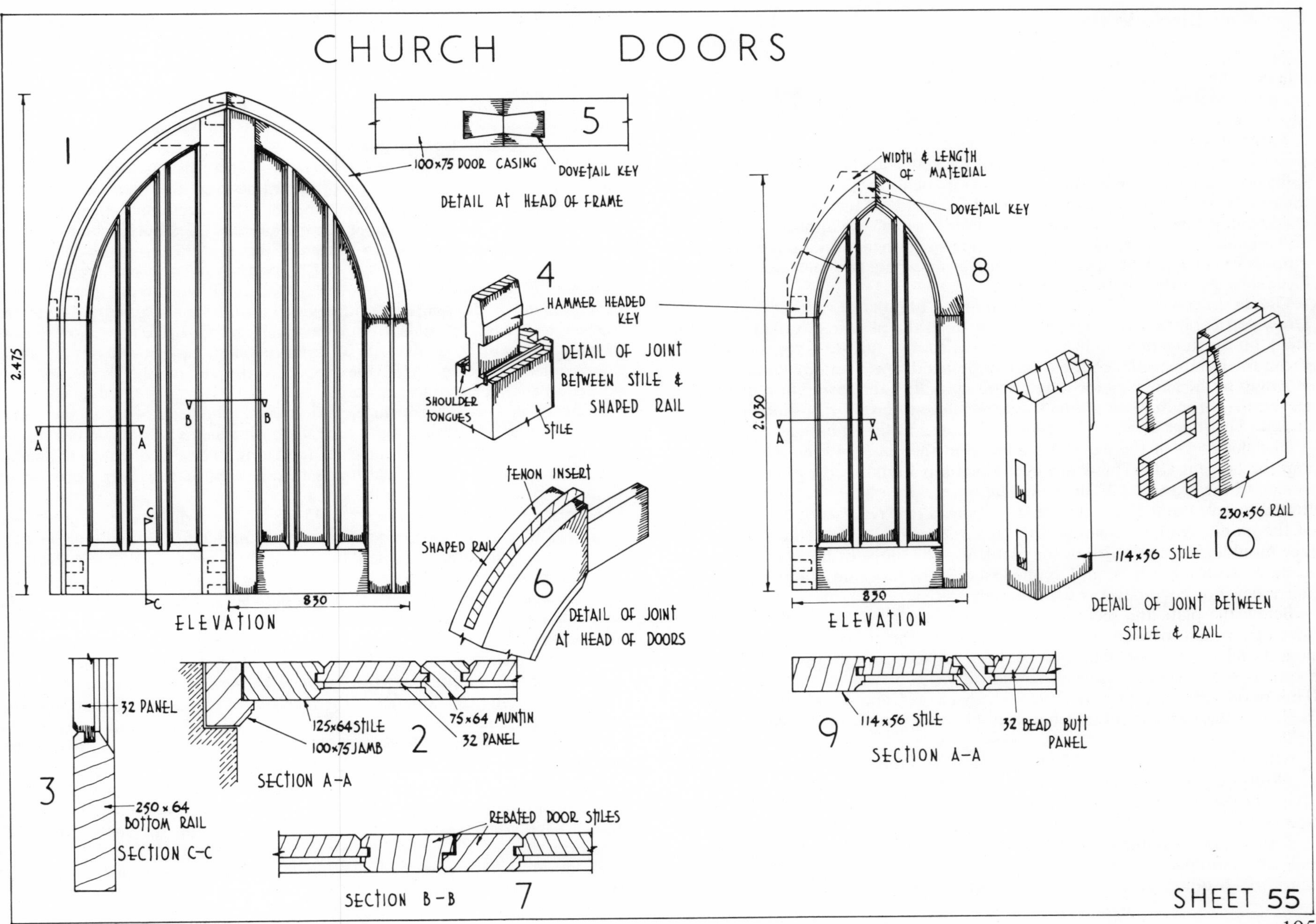
CHURCH DOORS
1
2.475
100×75 DOOR CASING
DOVETAIL KEY
5
DETAIL AT HEAD OF FRAME
4
HAMMER HEADED KEY
DETAIL OF JOINT BETWEEN STILE & SHAPED RAIL
SHOULDER TONGUES
STILE
TENON INSERT
SHAPED RAIL
6
DETAIL OF JOINT AT HEAD OF DOORS
830
ELEVATION
WIDTH & LENGTH OF MATERIAL
DOVETAIL KEY
8
2.030
830
ELEVATION
114×56 STILE
10
230×56 RAIL
DETAIL OF JOINT BETWEEN STILE & RAIL
32 PANEL
125×64 STILE
75×64 MUNTIN
100×75 JAMB
2
32 PANEL
SECTION A–A
3
250×64 BOTTOM RAIL
SECTION C–C
9
114×56 STILE
32 BEAD BUTT PANEL
SECTION A–A
REBATED DOOR STILES
SECTION B–B
7
SHEET 55

Circle-on-Circle Work—1

Work of double curvature is the description applied to construction which is curved both in plan and elevation. Examples of this type of work are to be found in door and window frames where the accurate geometrical setting out of certain bevels and moulds or templets is necessary.

Sheet 56.1 shows the plan and elevation of a semicircular headed frame which is segmental in plan. The jambs of the frame are parallel in plan and the soffit is level at the crown, the head being made in two sections with joints at the positions shown. As both sections of the head are identical, only one half need be developed. The necessary face moulds which are applied to the inside and outside faces of the plank are the same when the jambs are parallel. Both moulds have been developed in the illustration where two methods are shown of dealing with the plank as follows:

Divide the outside curve in elevation into a number of parts 1, 2, 3, 4, 5, 6. Draw vertical lines from these points through the plan to cut the face of the plank on the tangent C–D in 1, 2, 3, 4, 5. The tangent C–D is parallel to A–B. From the points on the chord AB erect perpendiculars and make them equal in length to the height of the corresponding points above the springing line in elevation; through these points trace the curve to give the outline of the face mould. The inner edge of the mould is treated in the same way.

The thickness of the plank T is shown in the plan between the chord AB and the tangent CD. The width of the plank is found by drawing ab at right angles to the tangent CD in plan. The inner edge of the face mould is continued to give point E. A line drawn joining E to a represents the inside edge of the plank. A line drawn parallel to Ea, touching the outside edge of the face mould, gives the width of the plank W. The shape of the plank being cut as at W, with the face mould applied, shows the foot and plumb bevels. The mitre bevel or edge bevel is seen in plan.

Below the plan, the second method is shown with the width of material taken from first method. The bevels applied to the plank are shown, but the edge bevel in this case has to be developed because the top surface of the plank is not parallel to the horizontal plane as it was in the first method. This bevel is similar to that shown in Sheet 56.4 and is dealt with later. A sketch of the face mould and bevels applied to the plank is shown in Sheet 56.2.

After shaping the plank to the face moulds, falling moulds for the inside and outside surfaces are required to mark the surfaces and give the correct shape in plan. The method of finding the moulds is the same for the inside and outside.

The points dividing the outside curve in elevation are again used and projected into plan. From point B draw a line at right-angles to the jamb and mark off distances 1′, 2′, 3′, 4′, 5′, 6′ equal to the elevation curve distances 1, 2, 3, 4, 5, 6. Drop perpendiculars from these points and intersect them with lines drawn from the corresponding numbered points on the plan of the frame. Free-hand curves drawn through the points of intersection will give the shape of the falling mould.

The development of the inside falling mould is shown in Sheet 56.3 and found in the same way; half the mould only is shown projected on the inside of the jamb due to lack of space.

Sheet 56.4 shows the plan and elevation of a frame semicircular in elevation and segmental in plan. The jambs of the frame radiate in plan and the soffit is level at the crown, the head being made in two sections with joints at the positions shown. In this example, the geometrical solid upon which the frame is based is termed a cuneoid. A sketch of this is shown in Sheet 56.5. As the surface of the cuneoid has a varying amount of twist in its length, it is not possible to accurately develop the surface. An approximate geometrical development of the templet to lay over the outer surface of the frame after shaping to the face moulds, to mark the surfaces and give the required shape in plan will later be explained. In practice this templet need not be used if dimensions transferred from the plan to ordinates on the frame and from points on the face mould are used.

In Sheet 56.4 one half of the elevation shows the head of the frame and the other the cuneoid. With radiating jambs both inside and outside face moulds are required to apply to the faces of the plank. These are clearly shown and along with bevels, and the width and thickness of material are obtained in a similar way to the last example.

To obtain the mitre or edge bevel, the edges of the two face moulds are projected to cut the edge of the plank in M and N. Point M is projected at right angles across the thickness of the plank to point O. Angle MNO gives the edge bevel.

When constructing the elevation, the plan should be drawn first. It will be seen that the face edge of the cuneoid is also shown and is in line with the outer edges of the frame. The left elevation of the cuneoid is a semicircle and is represented on the right by the dotted line. The outer arris of the head is set out by dropping points from the curve on to the face of the cuneoid in plan. These points are then radiated to the plan centre and where they cut the outer edge of the frame in plan give points to project upwards to intersect with corresponding points on the cuneoid face, giving the outer arris. These projections have been omitted so as not to confuse the drawing.

The cuneoid, Sheet 56.5, shows the ordinates on the face which are projected with level lines to the rear of the solid giving the heights a, b, c, d, e, f, g on the back edge. These same points can be seen on the centre line of the elevation.

To develop approximately the top curved surface or templet a line is drawn at 90 degrees in plan to the generator Oa. The vertical heights b, c, d,

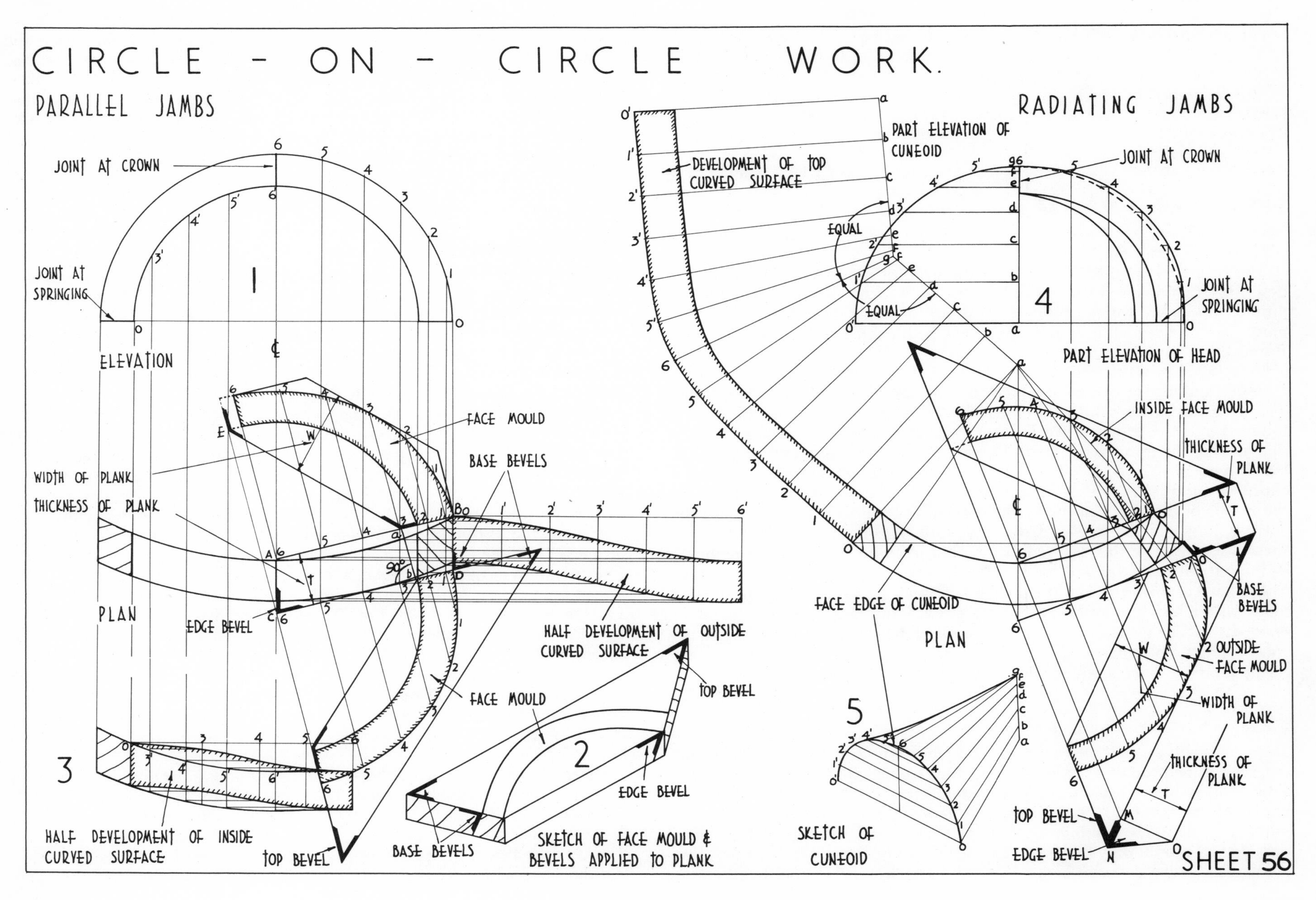
CIRCLE - ON - CIRCLE WORK.
PARALLEL JAMBS
RADIATING JAMBS
JOINT AT CROWN
JOINT AT SPRINGING
ELEVATION
PLAN
FACE MOULD
BASE BEVELS
WIDTH OF PLANK
THICKNESS OF PLANK
EDGE BEVEL
HALF DEVELOPMENT OF OUTSIDE CURVED SURFACE
TOP BEVEL
HALF DEVELOPMENT OF INSIDE CURVED SURFACE
SKETCH OF FACE MOULD & BEVELS APPLIED TO PLANK
PART ELEVATION OF CUNEOID
DEVELOPMENT OF TOP CURVED SURFACE
EQUAL
PART ELEVATION OF HEAD
INSIDE FACE MOULD
OUTSIDE FACE MOULD
FACE EDGE OF CUNEOID
SKETCH OF CUNEOID
SHEET 56

e, f are marked off on this line. From point O describe an arc with radius equal to the length from O′–1′ in the elevation. From point b describe an arc with radius equal to a–0 in plan. The intersection of these two arcs will give point 1 in the development. Each pair of points on the level lines is treated in the same way. Curves through the points obtained will give the approximate development of the templet.

Circle-on-Circle Work—2

Sheet 57.6 shows the plan and elevation of a circle-on-circle door frame with transom and radiating bars.

The frame has parallel jambs with the head of the frame similar to that described in the previous example. The method of determining the face mould and bevels for the head is therefore the same. The face moulds for marking out the radial bars at A and D will be the same, as will those for bars B and C.

Sheet 57.7 shows a vertical section taken at the centre through the head, transom and centre boss into which the bottom ends of the bars are fitted. The radiating bars are flush with the inside of the frame and the centre boss is tongued into the transom. To find the face mould to be applied across the thickness of the radial bars A and D, Sheet 57.6, divide the centre line of the bar into equal parts 0, 1, 2, 3, 4, and project these points into plan cutting a chord line in 0″, 1″, 2″, 3″, 4″. Erect perpendiculars upon the centre line from the points of division, and make them equal in length to the distance of the corresponding points in plan from the chord line. Draw the curve through these points to give the required face mould. Additional length for tenoning of the bars at each end will have to be allowed.

The head of the frame is formed in two sections, the joint at the crown being made with a handrail bolt and dowels, and at the springing with a hammer-headed key. The transom, which is in line with the springing, is jointed to the stiles using double tenons, with the projecting front edge housed into the face of the jamb to make a watertight joint as shown in Sheet 57.8.

Circle-on-Circle Frame with Fanlight

Sheet 57.9 shows to the left the outside elevation, and to the right, a line diagram of the head of a circle-on-circle door frame with radiating jambs and soffit level at the crown. A fanlight is fitted above the transom. A half plan above the transom and a half plan diagram are shown. The construction and jointing of the frame is similar to that described in the previous example. The outside face mould only has been developed with the bevels omitted, as these are similar to the detail in Sheet 56.4 and have been fully described.

The development of the moulds for the shaped head of the fanlight is the same as that for the frame and will be set out with it. The bottom rail and the transom, being curved in one direction only on plan, need no development.

Where a fanlight is employed over a door, it is usual to arrange the springing of the head of the frame and the top edge of the sash bottom rail to take the same line, so that the mortises for the transom fall below the springing. This is shown clearly in Sheet 57.10, where a hammer-headed key joint is also shown between the shaped head and stile.

Splayed Linings

Sheet 58.1 shows the part plan and elevation of a window with splayed soffit and jamb linings. The linings are tongued and grooved together. To obtain the bevel required for the shoulder of the jamb and the groove in the soffit, the lining should be developed. AB represents the face of the jamb in the plan splayed at 60 degrees to the face of the frame. The edges are projected into elevation and the head lining drawn in to intersect on the mitre line CD. With point A as centre and radius AB, describe the arc BB′ bringing the edge B into the same plane as A. Draw a perpendicular from B′ in plan into the elevation cutting the top edge of the soffit lining produced in C′. Join C′D and the contained angle is the bevel for the top of the jamb. As the soffit is splayed at the same angle as the jambs, the same bevel will answer for both. If the angle is different as in the next example, then the soffit also must be turned into the vertical plane to give the bevel for grooving the soffit.

The method of obtaining the angle which the edge cut makes with the face of the jamb lining is shown in Sheet 58.2. As a geometrical problem it is the determination of the angle between the two planes. CD is the elevation of the line of intersection of these two planes and CE its true length. From point 5 a line is drawn at right angles to CD cutting the edges of the linings in points 2 and 3. With 5 as centre, an arc is drawn tangential to the true length CE cutting the intersection in point 4. From points 2 and 3 draw lines to point 4. Angle 2–4–3 is the dihedral angle. If the corners of the linings are mitred then angles 2–4–D and 3–4–D are the bevels required. If they are to be tongued and grooved then angle 3–4–F is the bevel to be applied to the vertical lining. In this example, only the side bevel for the vertical lining and the mitre bevel have been marked.

Sheet 58.3 shows the true shape of the jamb lining for a window or door opening having a segmental head. The jamb lining is splayed at 45 degrees. The method of obtaining the true shape is similar to the previous example and is clearly shown.

Sheet 58.4 shows the part elevation and plan of a window having a semi-circular head with a framed splayed lining. The stiles to the head lining are

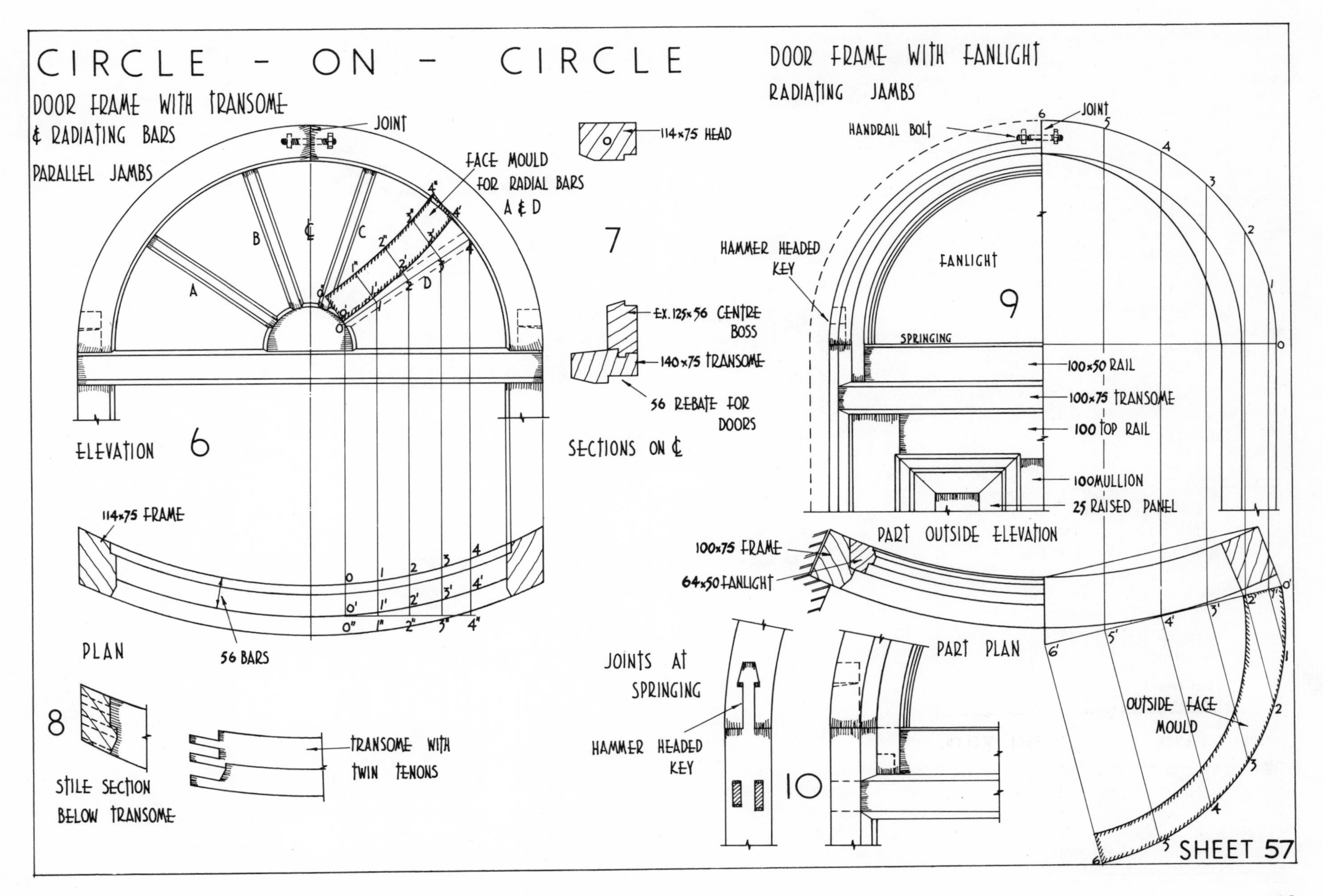
CIRCLE - ON - CIRCLE
DOOR FRAME WITH TRANSOME
& RADIATING BARS
PARALLEL JAMBS
JOINT
FACE MOULD FOR RADIAL BARS A & D
114×75 HEAD
7
ELEVATION 6
EX. 125×56 CENTRE BOSS
140×75 TRANSOME
56 REBATE FOR DOORS
SECTIONS ON ℄
114×75 FRAME
PLAN
56 BARS
8
STILE SECTION BELOW TRANSOME
TRANSOME WITH TWIN TENONS
DOOR FRAME WITH FANLIGHT
RADIATING JAMBS
HANDRAIL BOLT
HAMMER HEADED KEY
FANLIGHT
9
SPRINGING
100×50 RAIL
100×75 TRANSOME
100 TOP RAIL
100 MULLION
25 RAISED PANEL
PART OUTSIDE ELEVATION
100×75 FRAME
64×50 FANLIGHT
JOINTS AT SPRINGING
HAMMER HEADED KEY
10
PART PLAN
OUTSIDE FACE MOULD
SHEET 57

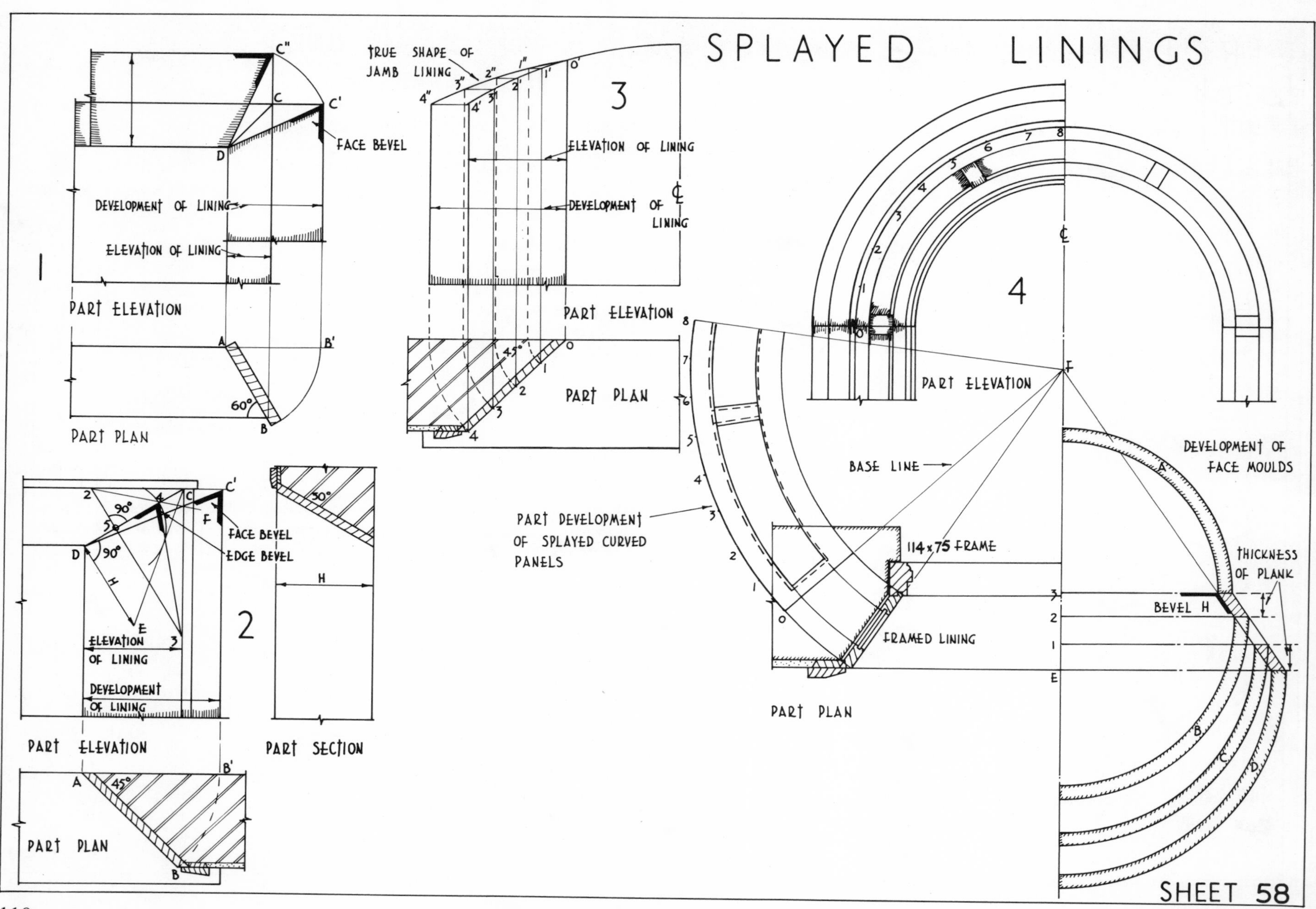
SPLAYED LININGS
1
PART ELEVATION
PART PLAN
DEVELOPMENT OF LINING
ELEVATION OF LINING
FACE BEVEL
60°
3
TRUE SHAPE OF JAMB LINING
ELEVATION OF LINING
DEVELOPMENT OF LINING
PART ELEVATION
PART PLAN
45°
2
PART ELEVATION
PART SECTION
PART PLAN
FACE BEVEL
EDGE BEVEL
ELEVATION OF LINING
DEVELOPMENT OF LINING
90°
30°
45°
4
PART ELEVATION
PART PLAN
PART DEVELOPMENT OF SPLAYED CURVED PANELS
BASE LINE
114 x 75 FRAME
FRAMED LINING
DEVELOPMENT OF FACE MOULDS
THICKNESS OF PLANK
BEVEL H
SHEET 58

worked in the solid in two pieces, joined at the crown and springings. The method of obtaining the face moulds for the head lining is shown to the right in plan. The edges of the stiles are drawn across to the centre line to give the centres E, 1, 2 and 3 for describing the moulds A, B, C and D.

The thickness of the plank is equal to the distances E–1, and 2–3 as shown on which the face moulds, along with the bevel, are applied. The width of the plank is determined by laying the two face moulds for each stile side by side, and measuring the width required. Mould A is applied to the face of the plank and the ends cut to the mould square across the face. The bevel H in plan is applied on the square ends working from the face side, giving points on the inside to which mould B is placed and marked. The piece is cut and worked to these mould lines, the inside edge being squared from the face ready for jointing at the ends. The other stile is dealt with similarly. The joints at the springing and crown are prepared for handrail bolts.

After setting out and mortising for the rails, the stiles are grooved for the panels. The development of a curved panel is shown in plan Sheet 58.4. A curved edge in elevation is divided into eight equal parts and numbered. Point F in plan is the intersection of the two linings produced and represents the apex of the semi-cone. With the apex F as centre, the edges of the stiles are struck, and from a base line the points 0–8 are marked off to give half the total length of the panels.

Thin veneered ply, which will easily bend to the curve, may be used for the panels.

Ironmongery

Windows designed so that they can be shut to exclude draughts must be fitted with the types of hardware which are adequate not only to close the window, but to keep it properly closed when so desired. A wide range of hardware is available including concealed espagnolettes, friction pivots, friction stays, casement stays and espagnolette-handles of modern design.

Where windows are manufactured by specialist firms, normally the sashes are hung in their frames with all built-in hardware, such as pivots, espagnolettes, hinges, sash couplers, fitted in position. This particular range of hardware is rust-proof and therefore should not be painted as its efficiency may be impaired. Surface hardware such as handles, friction stays and safety chains should be fixed after final completion of decoration.

Various types of hardware are illustrated on page 112. Sheet 59.1 shows a caulking lock with handle. These locks are fitted with a dead-locking lever and are particularly suitable for small windows, basement and stair windows, shutters, cupboard doors, etc. The lock is mortised into the stile or rail of the window, the receiver fixed to the frame with the handle screwed on the inside of the window. This handle is also used on horizontal pivot hung windows. Sheet 59.2 shows a friction stay. The friction bearing is screwed to the sash with the pin mounting screwed to the frame. The stay makes use of friction to hold the sash in any desired position, the holding power of the stay is the same at all angles. The amount of friction can be adjusted by a screw so designed to prevent it being actuated by the movement of the arm.

The stay can be used on bottom hung, top hung and pivot hung sashes, as well as side hung.

Sheet 59.3 shows two types of espagnolette bolts. The type on the left is for windows and that on the right for doors. All bolts can be made to any desired length with the handle placed in any position. Normally it is placed in the centre of normal length units. The espagnolette is housed into the edge of the stile of doors or windows so that when closed only the handle is visible. The bar stiffens the stile and at a turn of a single handle its bolts lock simultaneously at several points, forcing the windows or doors against the frame, thus ensuring that they are safeguarded against warping and are completely draught-proof.

A key-operated lock can also be fitted to any espagnolette where the timber sections allow for a lock housing.

Sheet 59.4 shows an espagnolette with three side levers and a key-lock.

Sheet 59.5 shows a 'round closing' espagnolette with locking points at each corner for horizontal pivot-hung windows.

Sheet 59.6 shows a three-section brass coupling screw for fastening sashes in double glazed windows. An alternative sash coupler is shown in Sheet 59.7. The coupler ensures that all double sashes are secured in strict alignment, with the requisite amount of air space between them. The male half of the coupler is surface-mounted on one sash, whilst the female half is countersunk into the other sash. The female half incorporates a spring-loaded catch which greatly facilitates the separation or coupling of sashes.

Sheet 59.8 shows a friction sash pivot for side fixing on flush lights. A face-fixing pivot on flush lights is shown in Sheet 59.9. A semi-flush pattern for fixing on recessed lights is shown in Sheet 59.10. The pivot will hold the sash open, safely and securely, in any desired position. It is adjustable; the amount of friction can be easily increased or decreased by the adjustment nut and screw. Once the correct frictional strength is obtained, the pivot remains locked to this adjustment. It enables the window to be cleaned on both sides from inside the room.

A further type of pivoting window bearing for vertical pivoted sashes is shown in Sheet 59.11.

Sheet 59.12 shows another type of friction pivot for horizontal pivoted sashes rotating through 180 degrees.

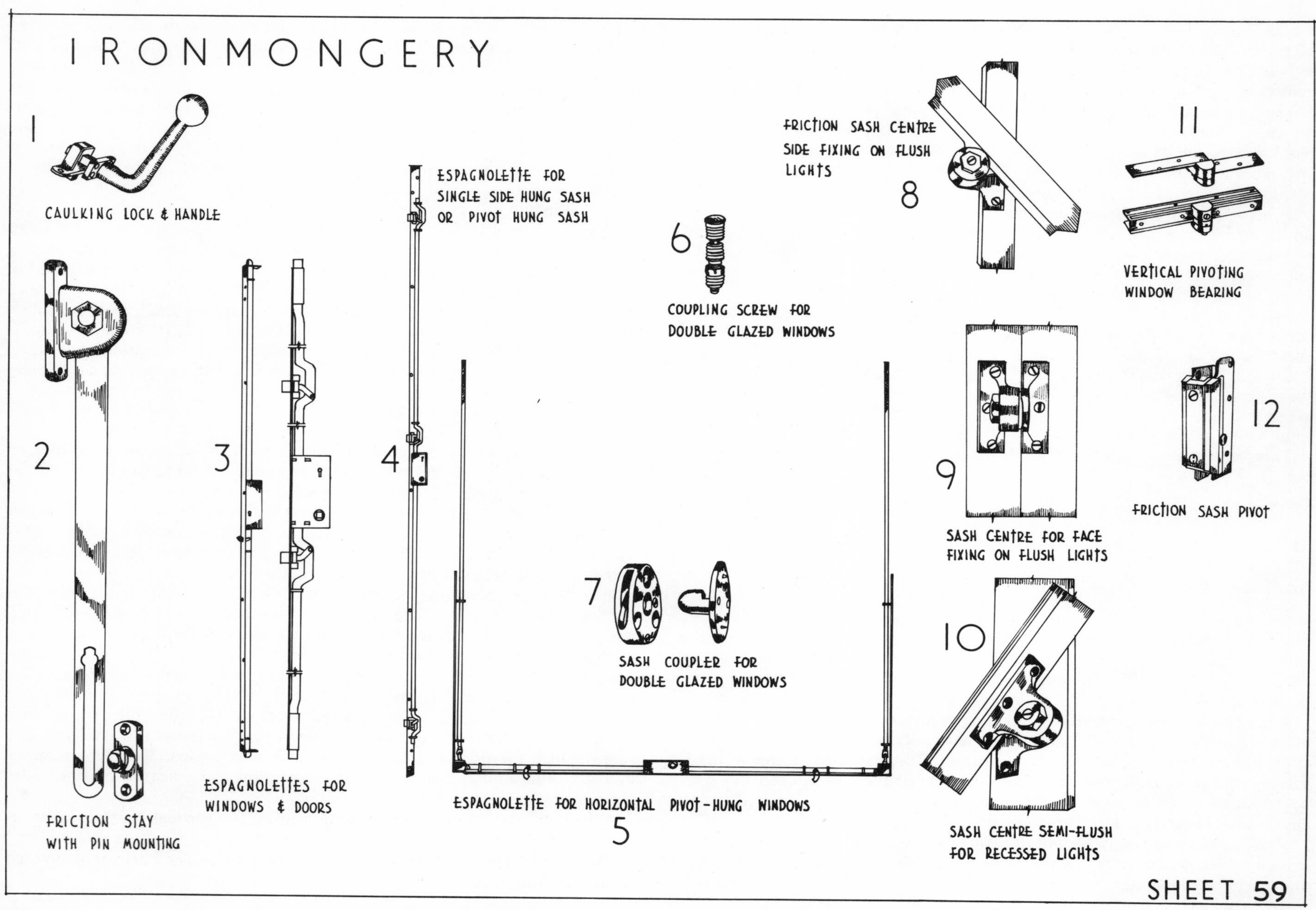
IRONMONGERY
1
CAULKING LOCK & HANDLE
2
FRICTION STAY
WITH PIN MOUNTING
3
ESPAGNOLETTES FOR
WINDOWS & DOORS
4
ESPAGNOLETTE FOR
SINGLE SIDE HUNG SASH
OR PIVOT HUNG SASH
5
ESPAGNOLETTE FOR HORIZONTAL PIVOT-HUNG WINDOWS
6
COUPLING SCREW FOR
DOUBLE GLAZED WINDOWS
7
SASH COUPLER FOR
DOUBLE GLAZED WINDOWS
8
FRICTION SASH CENTRE
SIDE FIXING ON FLUSH
LIGHTS
9
SASH CENTRE FOR FACE
FIXING ON FLUSH LIGHTS
10
SASH CENTRE SEMI-FLUSH
FOR RECESSED LIGHTS
11
VERTICAL PIVOTING
WINDOW BEARING
12
FRICTION SASH PIVOT
SHEET 59

IRONMONGERY

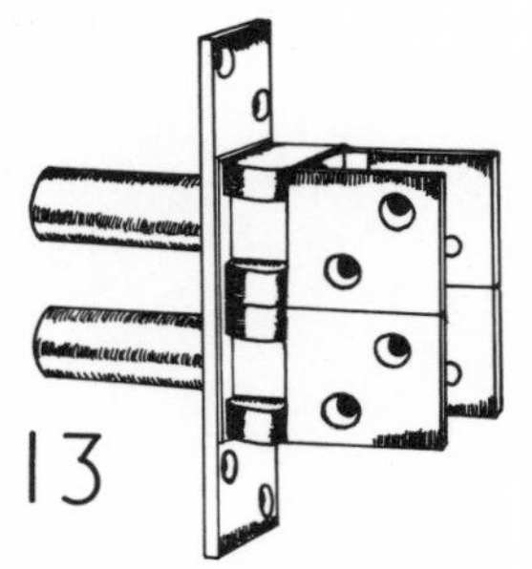

13

DOUBLE SPRING HINGE

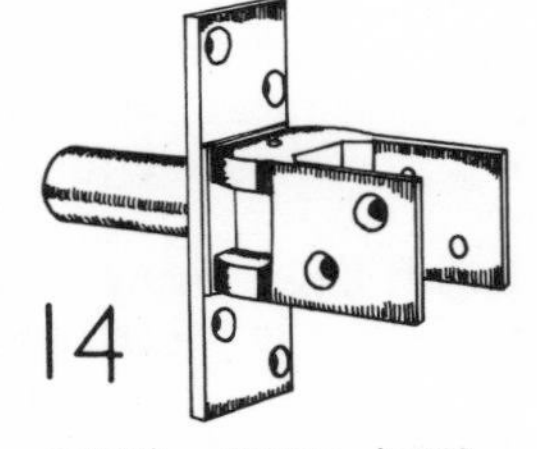

14

SINGLE SPRING HINGE

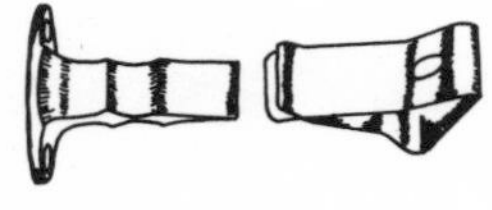

SPRING CLIP

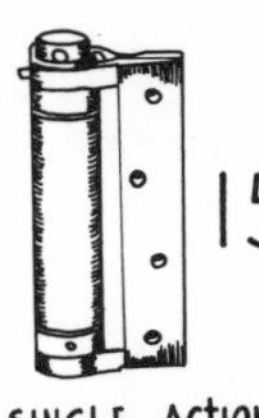

15

SINGLE ACTION SPRING HINGE

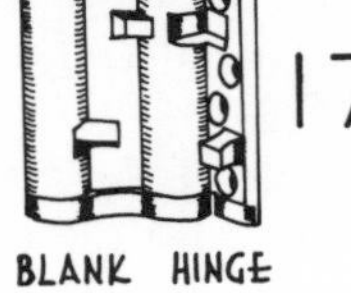

17

BLANK HINGE

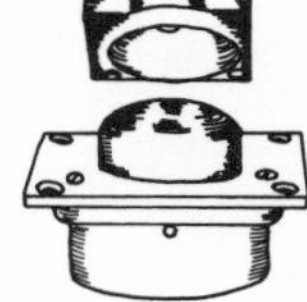

25

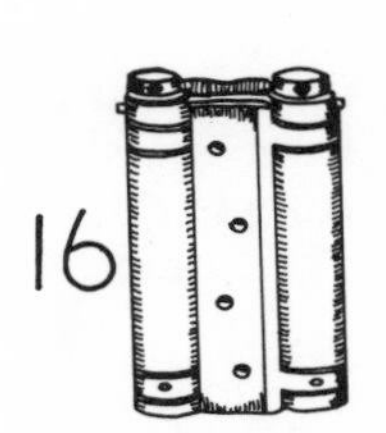

16

DOUBLE ACTION SPRING HINGE

19

DOOR CLOSER

DOOR HOLDERS

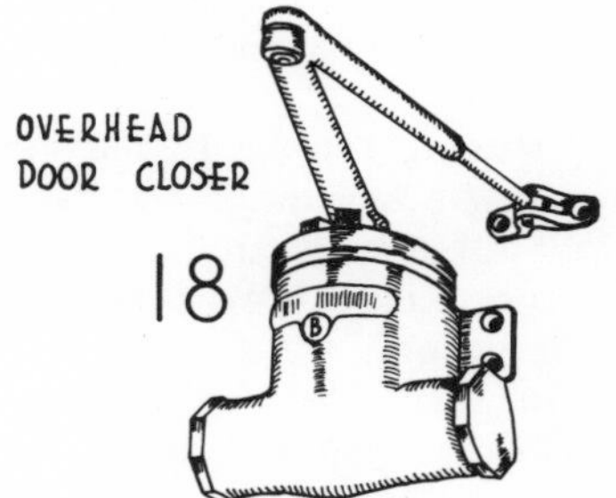

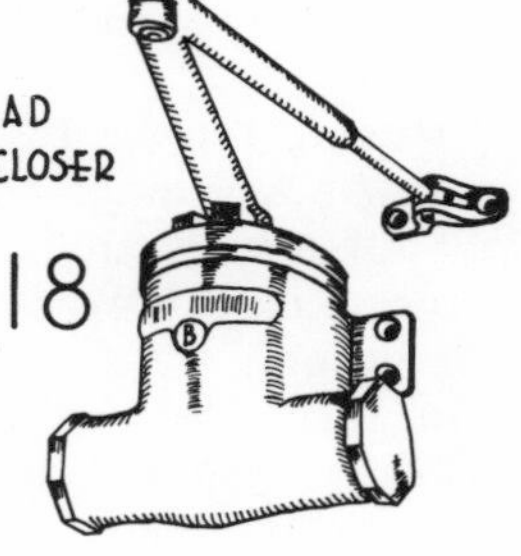

OVERHEAD DOOR CLOSER

18

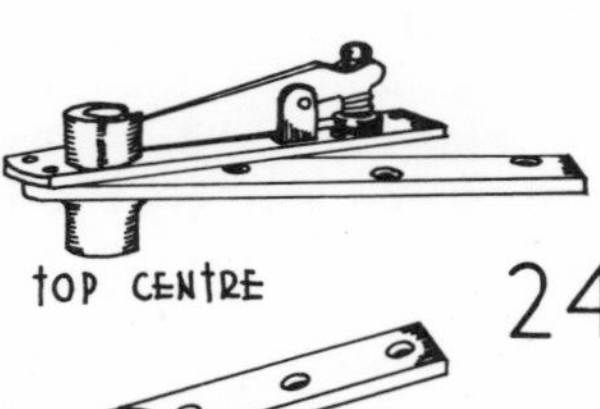

TOP CENTRE

24

TOP CENTRE FOR SINGLE ACTION FLOOR SPRINGS

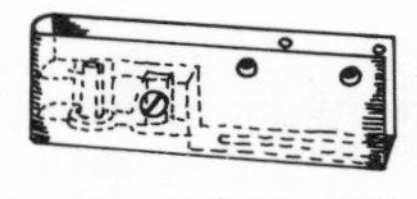

22

DOUBLE-ACTION SHOE

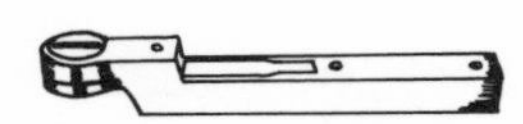

DOUBLE-ACTION STRAP

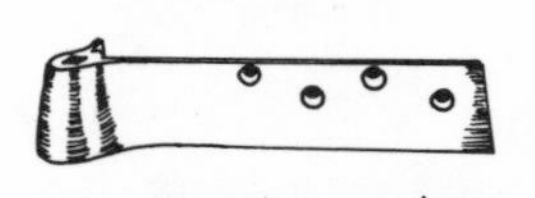

23

SHOE FOR SINGLE-ACTION FLOOR SPRINGS

20

CONCEALED DOOR CLOSER

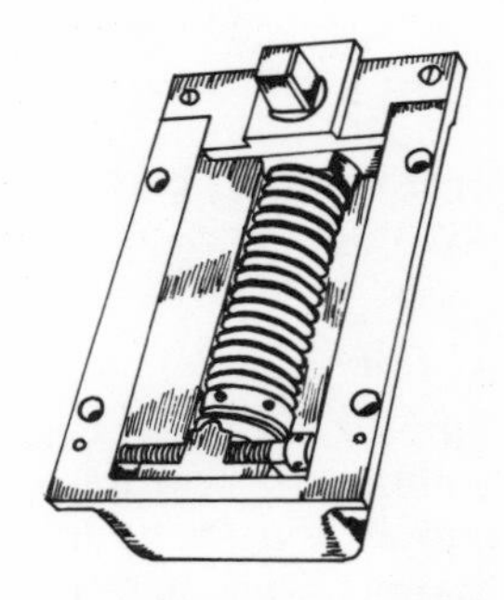

DOUBLE ACTION FLOOR SPRING

21

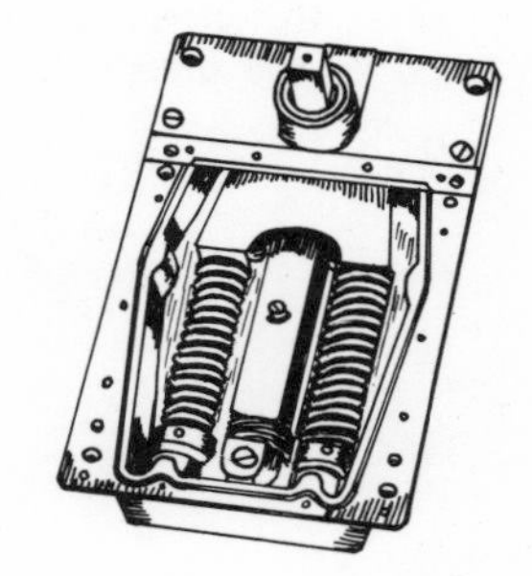

DOUBLE ACTION FLOOR SPRING WITH HYDRAULIC CHECK

SHEET 60

Spring Hinges

For doors which are required to close automatically, some kind of spring hinge is used. They are required to stand heavy wear, so their quality is most important. Various types of spring hinges are available for single-action doors and also for double-action or swing doors.

Sheet 60.13 shows a twin type of spring hinge for swing doors with a single type of the same hinge shown in Sheet 60.14. Two single type hinges are used for doors weighing up to 22.70 kg in weight. For doors between 22.70 and 45.4 kg in weight, one single type and one twin type are used. Two twin type hinges are used for doors weighing over 45.40 kg in weight.

A single-action helical spring hinge is shown in Sheet 60.15 suitable for doors up to 64 mm thick. A double-action hinge is shown in Sheet 60.16; this is also made in sizes to fit doors up to 64 mm thick and is suitable for swing doors. The barrels contain springs which are regulated by inserting a steel pin in the collar, when the hinge is fixed, one flap to the door and the other flap to the frame. In hanging a door, only one spring hinge is used at the top, the other hinge being a 'blank', Sheet 60.17, which has no springs. The spring and blank hinges are sold together as a pair.

Overhead Door Closers

Two examples of this type of fitting are shown in Sheet 60, figs. 18 and 19. They combine the features of a door check and spring for closing doors without noise, and are hydraulic. They may be obtained with either single or double action; single-action closers are made either to push or pull the door closed. Because it is sometimes necessary for the door to remain open for ventilation purposes, or while things are carried through the doorway, a closer which will hold the door open at 90 degrees may be fitted, being released with a slight pull.

The body of the closer is fixed to the top rail of the door and the arm bracket fixed to the face of the frame. With the door hung on normal butt hinges, the standard arms will allow the door to open through 120 degrees. If the door is hung on projecting hinges, the angle of opening will be reduced, subject to the amount of hinge projection. In some instances the door is required to open 180 degrees and longer arms become necessary.

A concealed door closer is shown in Sheet 60.20. These units are door closers specially designed for use where the first consideration is to maintain the aesthetic qualities of the door and surround. The model is primarily intended to control interior doors not exceeding 54.48 kg in weight. Apart from the slim arms of the fitting on the hinge side of the door, there is complete concealment, the body being recessed into the top edge of the door. The fitting is suitable for a door required to open to 90 or 180 degrees.

Floor Springs

Floor springs can be either single or double action for doors opening in one direction or both. The single-action type have either a hydraulic or a pneumatic check action which avoids slamming. The springs are let into the floor and take the weight on the pivot, Sheet 60.21. The springs for double-action or swing doors are provided with shoes and top centres. The shoe, Sheet 60.22, is fitted and screwed to the heel of the door. A square hole in the bottom of the shoe fits over the pivot which is actuated by a spring in the box. The adjustable shoe is fitted with an adjustable bush which can be regulated to align doors which may not be straight. A shoe for a single-action floor spring is shown in Sheet 60.23.

Sheet 60.24 shows two types of top centre used with floor springs, which are housed and fixed one part into the edge of the door, the other into the frame. Adjustable top centres are available which are fitted with regulators to allow adjustment to counteract any drop in the door.

Various types of door holders are shown in Sheet 60.25. The steel spring clip type is fixed to the door frame with the plunger screwed to the door. The other patterns are housed and fixed in the floor.

Chapter Six: Joinery

Double-Glazed Windows

Restrictions imposed on the use of timber during the war years, and for some time afterwards, delayed the development in this country of double-glazed timber windows.

In addition to the basic requirements of admitting the optimum amount of light and conforming to certain aesthetic standards, double-glazed windows pay due regard to the ever-increasing importance of heat loss, sound transmission, the inclusion of pleated blinds of the most modern appeal, and methods of opening, the latter being influenced by ventilation requirements and the ease of indoor window cleaning.

Heat Loss

There has been, until quite recently, a general disregard of thermal efficiency in some aspects of building design and construction. Deterioration in the supply of fuel, and its increased cost, have already compelled a far more serious attitude towards the problem of heat loss through roofs and walls. Logically, the same serious consideration should be given to heat loss through windows. Such losses can be controlled by paying due regard to the following factors: (1) fit between sash and frame; (2) closing mechanisms; (3) double glazing.

Fit Between Sash and Frame

While considerable heat loss takes place through a single pane of glass, as compared with double-glazed windows, an even greater loss occurs through gaps between the sashes and frames. Due to the natural movement of timber in changing moisture conditions, and to the need of space for paint, a fair clearance must be left between the edges of the sashes and frames. Air and moisture can best be prevented from entering by providing a form of packing at the joint; wind-pressure and capillary-action can be minimized by ensuring that grooves and rebates are of correct size and shape. If the packing is to serve as a draught excluder, then it is necessary for it to be sufficiently flexible to allow it to be compressed.

Closing Mechanisms

The types of hardware fitted must be adequate not only to close the window but to keep it properly closed when so desired. Examples of the range of hardware available are detailed on Sheet 59, to which reference should be made.

Double Glazing

Sheet 61.1 shows the elevation of a 'Tomo' double-glazed window. The second pane of glass is carried in a separate sash which is coupled to the other sash in such a way that adequate air filtration can take place to prevent condensation between the panes. This is obtained by using sash-couplers or coupling-screws which allow the windows to be cleaned between panes, and permits access to pleated blinds which may be housed between the double glazing. To separate the two sashes they have to be rotated through 180 degrees. The coupled sashes act as one and are just as easy to handle for ventilation purposes as any good single-glazed window. A round closing espagnolette with four locking points, one at each corner, which are actuated simultaneously by the one handle, is fitted to each window.

Details of the various sections are shown in Sheet 61, figs. 2, 3, 4 and 5.

Alternative methods of providing double glazing may be employed. These include sealed edge double glazing units and glazing frame assemblies in flexible plastic or rubber sections arranged to hold the two panes of glass with a 19 mm cavity between. They may be secured to wood or metal windows using various methods and the sections may be readily removed if required.

Combined Pivot-Hung and Side-Hung Casement

This 'Austin' manufactured window is shown in the elevation Sheet 61.6, and is one of a wide range of combined windows for use in houses, flats, schools, hospitals, offices, etc. The pivot sash is controlled by a friction stay whose position is to be determined so that the window opens normally through approximately 30 degrees. A single pivot-hung sash is shown in the sketch, Sheet 61.11. The pivot beads are of hardwood glued and screwed to the sash or frame as shown. Back flap hinges are fitted to the pivoted light with a pair of friction stays, and closure is obtained using a caulking lock and handle.

The details of the various sections are shown in Sheet 61, figs. 7, 8, 9 and 10.

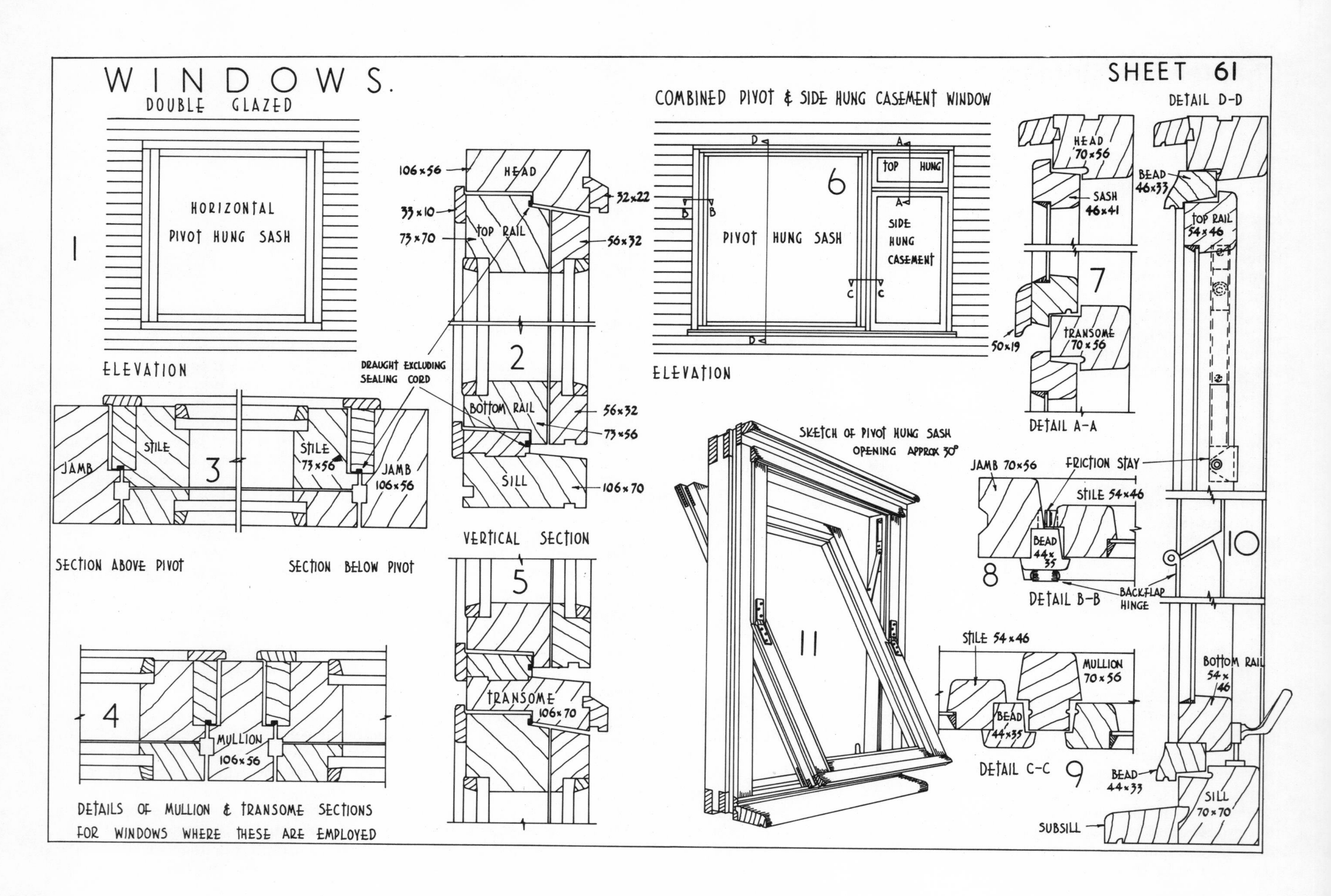
WINDOWS.
DOUBLE GLAZED
SHEET 61
HORIZONTAL PIVOT HUNG SASH
1
ELEVATION
106×56
HEAD
33×10
32×22
73×70
TOP RAIL
56×32
2
DRAUGHT EXCLUDING SEALING CORD
BOTTOM RAIL
56×32
73×56
SILL
106×70
VERTICAL SECTION
JAMB
STILE
3
STILE 73×56
JAMB 106×56
SECTION ABOVE PIVOT
SECTION BELOW PIVOT
5
TRANSOME 106×70
4
MULLION 106×56
DETAILS OF MULLION & TRANSOME SECTIONS FOR WINDOWS WHERE THESE ARE EMPLOYED
COMBINED PIVOT & SIDE HUNG CASEMENT WINDOW
TOP HUNG
6
PIVOT HUNG SASH
SIDE HUNG CASEMENT
ELEVATION
SKETCH OF PIVOT HUNG SASH OPENING APPROX 30°
11
HEAD 70×56
SASH 46×41
7
50×19
TRANSOME 70×56
DETAIL A-A
JAMB 70×56
FRICTION STAY
STILE 54×46
BEAD 44×35
8
DETAIL B-B
BACKFLAP HINGE
STILE 54×46
MULLION 70×56
BEAD 44×35
DETAIL C-C
9
DETAIL D-D
BEAD 46×33
TOP RAIL 54×46
10
BOTTOM RAIL 54×46
BEAD 44×33
SILL 70×70
SUBSILL

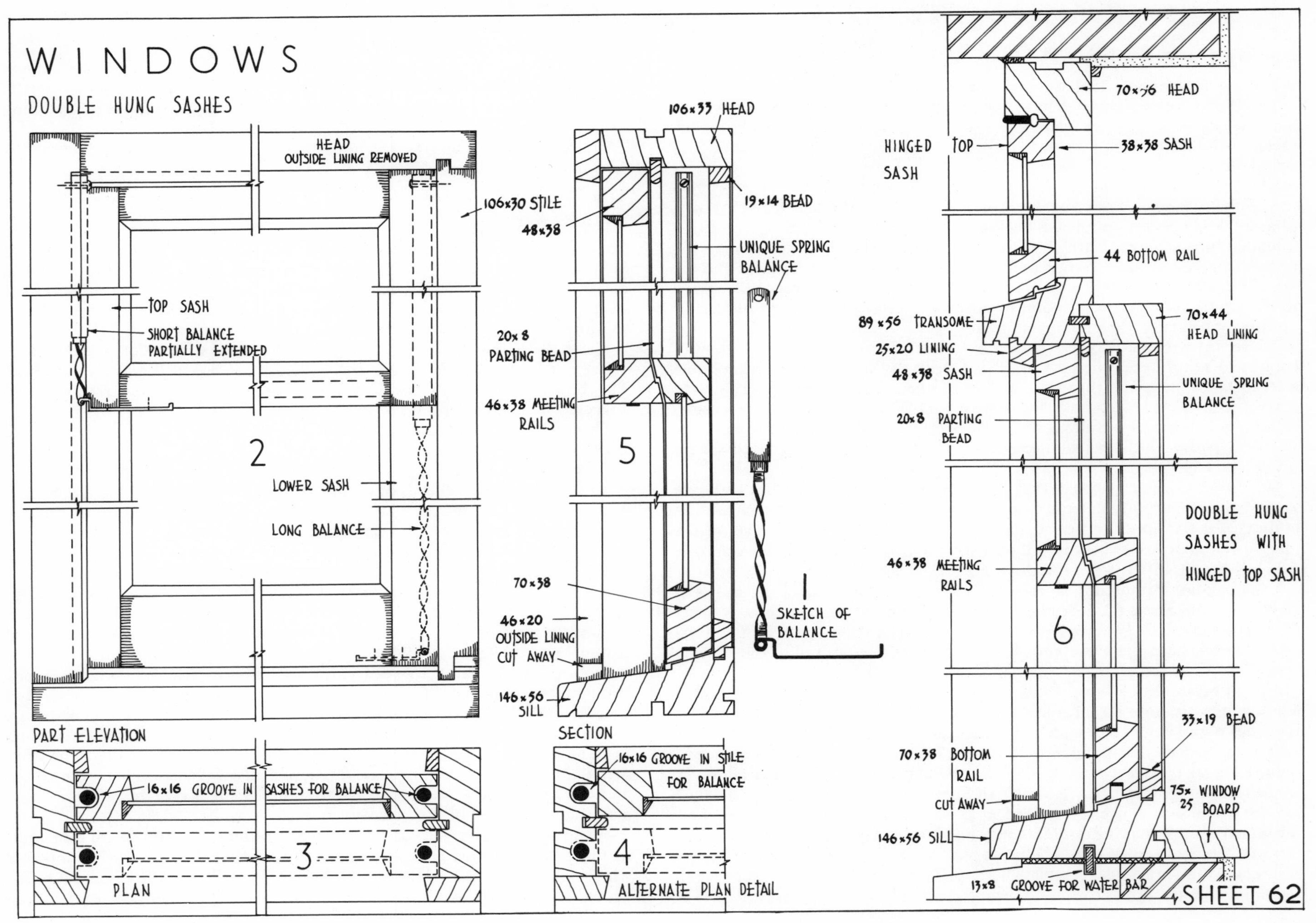
WINDOWS
DOUBLE HUNG SASHES
HEAD
OUTSIDE LINING REMOVED
106x30 STILE
TOP SASH
SHORT BALANCE PARTIALLY EXTENDED
2
LOWER SASH
LONG BALANCE
PART ELEVATION
16x16 GROOVE IN SASHES FOR BALANCE
3
PLAN
106x33 HEAD
19x14 BEAD
48x38
UNIQUE SPRING BALANCE
20x8 PARTING BEAD
46x38 MEETING RAILS
5
70x38
46x20 OUTSIDE LINING CUT AWAY
146x56 SILL
SECTION
1
SKETCH OF BALANCE
16x16 GROOVE IN STILE FOR BALANCE
4
ALTERNATE PLAN DETAIL
HINGED TOP SASH
70x76 HEAD
38x38 SASH
44 BOTTOM RAIL
89x56 TRANSOME
70x44 HEAD LINING
25x20 LINING
48x38 SASH
UNIQUE SPRING BALANCE
20x8 PARTING BEAD
DOUBLE HUNG SASHES WITH HINGED TOP SASH
46x38 MEETING RAILS
6
33x19 BEAD
70x38 BOTTOM RAIL
CUT AWAY
75x25 WINDOW BOARD
146x56 SILL
13x8 GROOVE FOR WATER BAR
SHEET 62

Double-Hung Sashes

Work on traditional double-hung sashes in cased or boxed frames with counterbalancing weights will have been dealt with in earlier years. A development which is now more generally used is the spiral sash balance which replaces the lead or iron weights formerly used. For this reason, details of double-hung sashes using spiral sash balances have been included here as modern practice, where this type of window is used.

Sheet 62.1 shows a sketch of the 'Unique' spiral balance which consists of a torsion spring and a spiral rod enclosed by a metal tube. The spiral rod is threaded through a bush attached to the spring and thereby causes the spring to be wound or unwound when the sash is raised or lowered. The pitch of the spiral rod is varied in such a manner that it equalizes the tension of the spring to the same amount at every point of operation, thus the weight of the sash is uniformly balanced throughout its travel.

There are two main types of balances. Each is available in a variety of lengths to suit different heights of sashes. Both can be adjusted within limits after installation to suit the actual weight of sash, but it is important that the correct size of balance, both for weight and length, is used. One type is suitable for domestic windows with sashes up to 13.60 kg glazed weight, the second type for all other installations where the weight of the glazed sash is up to 34.02 kg.

Sheet 62.2 shows a part external elevation of double-hung sashes with one short balance partially extended to the top sash. A similar balance will be required for the opposite side of the sash, but this is not shown. One of the long lower balances is shown on the right, and again the one for the opposite side has been omitted from the drawing.

Each balance is housed in a groove. This may be in the stile of the sashes as shown in plan, Sheet 62.3, or alternatively in the outer frame jambs as Sheet 62.4. The top of the balance is fixed to the stile by a drive screw nail and the sash attachment at the lower end is screwed to the bottom rail of the moving sash.

As can be seen, this type of balance has many advantages. The simplified window construction requires less timber and less joinery labour in making. Balances permanently eliminate the dangers of cord breakages and more glass area is provided, due to narrower frames being used. No maintenance is required, and if fixed correctly they can be expected to last indefinitely. A section through the window is shown in Sheet 62.5, where the spring balance for the lower sash can be seen fixed in position. Sheet 62.6 shows a section through a window having double-hung sliding sashes combined with a hinged top sash above to open out. The arrangement for the vertical sliding sashes is similar to the previous example where spring balances are used. In order to accommodate the upper top-hung sash, the outer frame jambs are reduced in width and extended to the full height and tenoned to the head. A transom tongued to the head lining is weathered and throated on the top side to receive the bottom rail of the hinged sash. The transom is grooved on the underside to receive a tongued outside lining and guide for the vertical sliding sashes.

The outer frame linings are housed into the head lining at their intersections, and the transom tenoned into the outer frame.

Dormer Windows

These are used to give light and ventilation to rooms formed in the roofs of buildings. An architectural feature is often made of this type of roof light in modern buildings. The sides and front of the dormer may be glazed into fixed sashes, but the front is usually arranged with either pivot-hung or side- or top-hung sashes. They may have various forms of roof finish.

Sheet 63.1 shows a sketch of a dormer arranged with a flat roof for sheet copper or lead covering. The sides or cheeks are boarded and also covered. The boarded roof falls towards the front edge where the rain-water is collected in a gutter and discharged to the main roof.

Sheet 63.2 shows a sketch of a small dormer having a pitched roof covered with either slates or tiles.

Sheet 63.3 shows a sketch of a similar dormer with a hipped roof.

Sheet 63.4 shows a sketch of a dormer with a shaped roof. When a dormer with a shaped roof intersects a main pitched roof, it is necessary to find the true shape of the opening to be formed in the main roof. This geometrical construction is shown on Sheet 38.

Sheet 63.5 shows a section through a dormer with a flat roof having a casement with a pivot-hung opening light with the cheeks prepared for boarding and finished in sheet copper.

Sheet 63.6 shows a part elevation of the roof light to the right and the roof trimming to the left.

The opening in the main roof is formed between two 100 mm by 75 mm trimmers, with the feet of the roof spars birds-mouthed over the top one and housed into the lower one and nailed. Two 100 mm by 75 mm trimming rafters are substituted for 100 mm by 50 mm common rafters at the sides, and the trimmer set vertically between them at the head of the opening. The trimmer at the foot of the opening may be placed as shown or a deeper trimmer used and placed vertically. This is allowed to stand 75 mm above the rafters to form a seating for the sill of the casement frame.

When the trimmer is fixed square to the rafters, 100 mm by 50 mm studs or uprights supported from the floor below carry a 100 mm by 75 mm head which supports the sill of the frame to the dormer front. The casement frame provides for a pivot-hung opening light in the centre, the others are fixed.

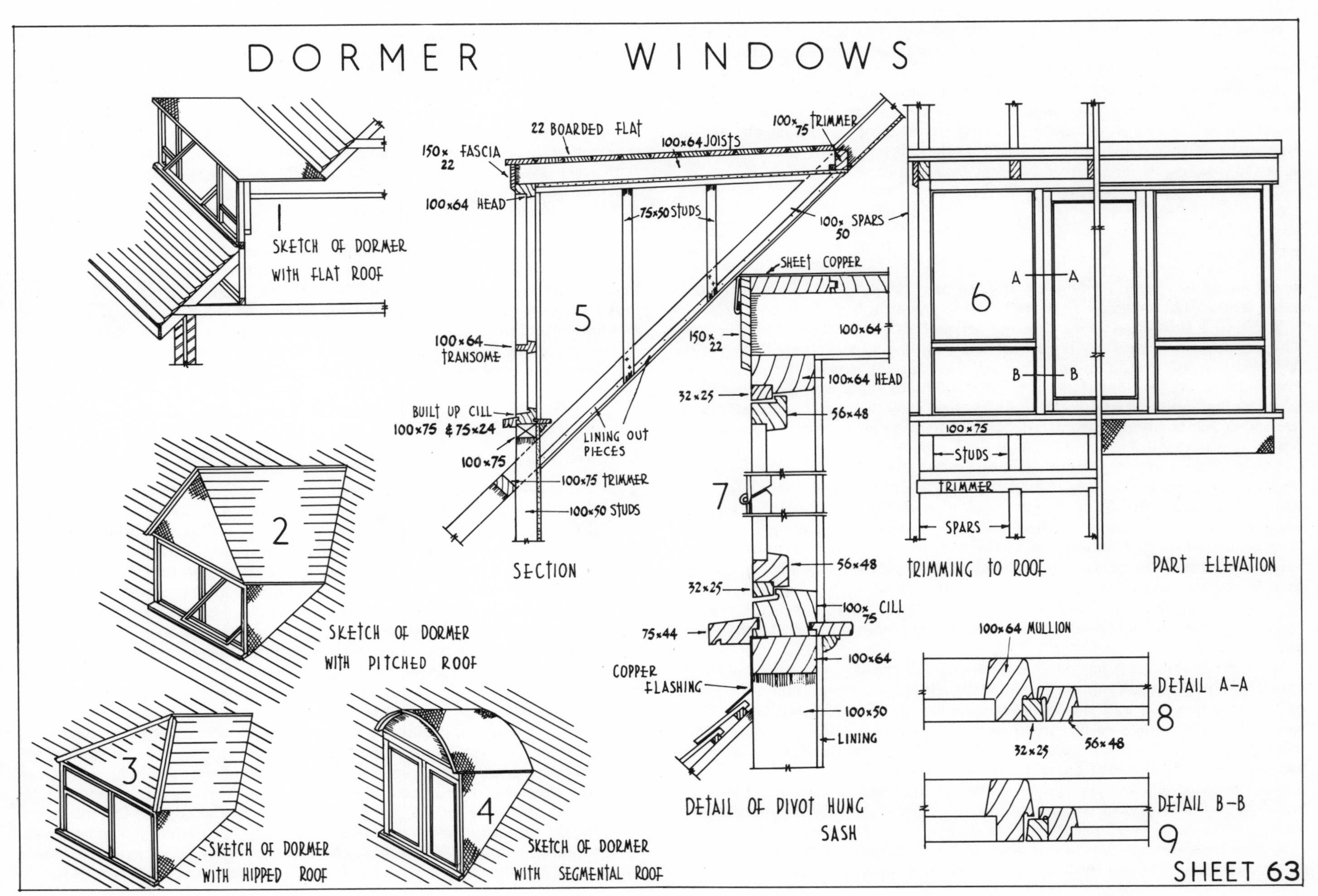
DORMER WINDOWS
1
SKETCH OF DORMER
WITH FLAT ROOF
22 BOARDED FLAT
150× 22 FASCIA
100×64 JOISTS
100× 75 TRIMMER
100×64 HEAD
75×50 STUDS
100× 50 SPARS
5
100×64 TRANSOME
BUILT UP CILL
100×75 & 75×24
100×75
LINING OUT PIECES
100×75 TRIMMER
100×50 STUDS
SECTION
SHEET COPPER
150× 22
100×64
100×64 HEAD
32×25
56×48
7
56×48
32×25
100× 75 CILL
75×44
100×64
COPPER FLASHING
100×50
LINING
DETAIL OF PIVOT HUNG SASH
6
A A
B B
100×75
STUDS
TRIMMER
SPARS
TRIMMING TO ROOF
PART ELEVATION
2
SKETCH OF DORMER
WITH PITCHED ROOF
3
SKETCH OF DORMER
WITH HIPPED ROOF
4
SKETCH OF DORMER
WITH SEGMENTAL ROOF
100×64 MULLION
DETAIL A–A
8
32×25
56×48
DETAIL B–B
9
SHEET 63

Sheet 63.7 shows an enlarged section through the front of the dormer showing details of the pivot-hung sash. Typical sections taken above and below the pivot of the sash are shown in Sheet 63, figs. 8 and 9.

The roof joists are fixed from the top trimmer to the head of the casement to carry a lined ceiling of insulation board, and a boarded flat which falls towards the front edge. A fascia is mitred round the face and cheeks of the dormer to finish the flat with a gutter to collect the rain-water and discharge it to the main roof.

Skylights

These are sashes fixed on inclined roofs to provide additional lighting to the space below. They may be fixed or made to open for ventilation purposes.

Sheet 64.1 shows a section through a framed skylight hinged at the top. The opening in the rafters is formed by trimming in the usual way to the required size. A curb or lining is framed, tongued and grooved at the angles, and fixed within the opening. The curb is 50 mm thick and 275 mm deep, standing high enough above the main roof to provide an adequate gutter at the top of the skylight. The curb is covered with a flashing on the outside to render it watertight.

The skylight rests on the curb and is hinged at the top. It is at this point where maintaining a watertight joint presents a problem. Alternative finishes will be dealt with later. A 50 mm projection on all four sides over the curb is necessary, with a throating on the underside, on all edges. The light consists of two stiles, top and bottom rails. The bottom rail of the skylight is kept below the level of the glass so that the glass runs over it. Bars are sometimes used to reduce glass sizes. A wide bottom rail is used, giving a wider projection over the curb on the inside to accommodate the opening gear. A sketch of this is shown in Sheet 64.2 which is of the quadrant type operated by cord gear and worm.

Sheet 64.3 shows a horizontal section through the skylight where glazing bars are shown. An alternative finish on the left is also shown. Here, the curb or lining is tongued and the sash correspondingly grooved. A fillet is also tongued on the underside of the projecting sash to assist in keeping out the driving rain. These are mitred at the corners. Haunched mortise and tenons are used between the top rail and stiles. A barefaced tenon is used for the bottom rail with glazing bars stub-tenoned at the top and bottom. It should be noted that the top rail is grooved to receive the glass with all other members rebated.

As the insulation properties of glass are low, condensation which takes place on the underside of the glass in cold weather has to be allowed to escape. To allow this to happen the face of the bottom rail is kept below the rebate line of the stiles by 10 mm as shown, leaving a gap between the glass and bottom rail. Fillets fixed to the bottom rail will help to check any draughts and assist in channelling the condensation clear as shown in Sheet 64.4. When skylights are hinged, non-corrosive butts should be used, fixed first to the curb and then to the sash.

Sheet 64.5 shows a section through a skylight in an asbestos-covered roof with steel trusses, using patent glazing bars. Although this work is not the work of the carpenter and joiner, Sheet 64.6 shows an example where he has to make provision for the fixing by others of this type of glazing in a timber roof.

The trimming is done as for a timber skylight. At the bottom, a filler piece having a bevelled top edge is fixed to the trimmer to the required height and a fascia or lining, for fixing the tops of the bars, fixed on the inside.

A section of a glazing bar is shown in Sheet 64.7. This consists of a steel tee bar galvanized and clothed with a lead sheath sealed at both ends. The two lead wings make a watertight joint with the glass seating on an imperishable, oiled asbestos cord cushion.

Lantern Lights

This form of light is used on flat roofs to give top lighting to staircases and rooms. They also provide extra ventilation from vertical side lights. The outline of the plan may be square, rectangular, circular or polygonal, and the roof covering may be formed with inclined lights or a lead flat.

Sheet 65.1 shows the half plan and half horizontal section of a lantern light having glazed hipped roof lights. A half sectional elevation and half elevation are also shown. The roof lights are constructed of four frames which are mitred and tongued and grooved at their intersections. The lantern is raised above the main roof on a curb, resting on the trimming joists of the lead flat. The side lights are 600 mm deep and may be top hung to open out or pivoted or fixed. The details of the vertical lights are similar to those for a square bay window.

The corner posts and mullions are rebated in the solid. The head is bevelled to the slope of the roof on the top edge, which may have a tongue worked on it with a corresponding groove in the bottom rail of the roof light. The sill is mitred at the corners and fixed by handrail bolts. The stiles are tenoned into the sill and head, the tenons being set in line with the edge of the rebates.

When the lantern is fixed, a lining, which may be framed and panelled or of hardboard, is fixed to mask the curb and trimming. To prevent condensation falling into the room below, condensation gutters are planted and flashed,

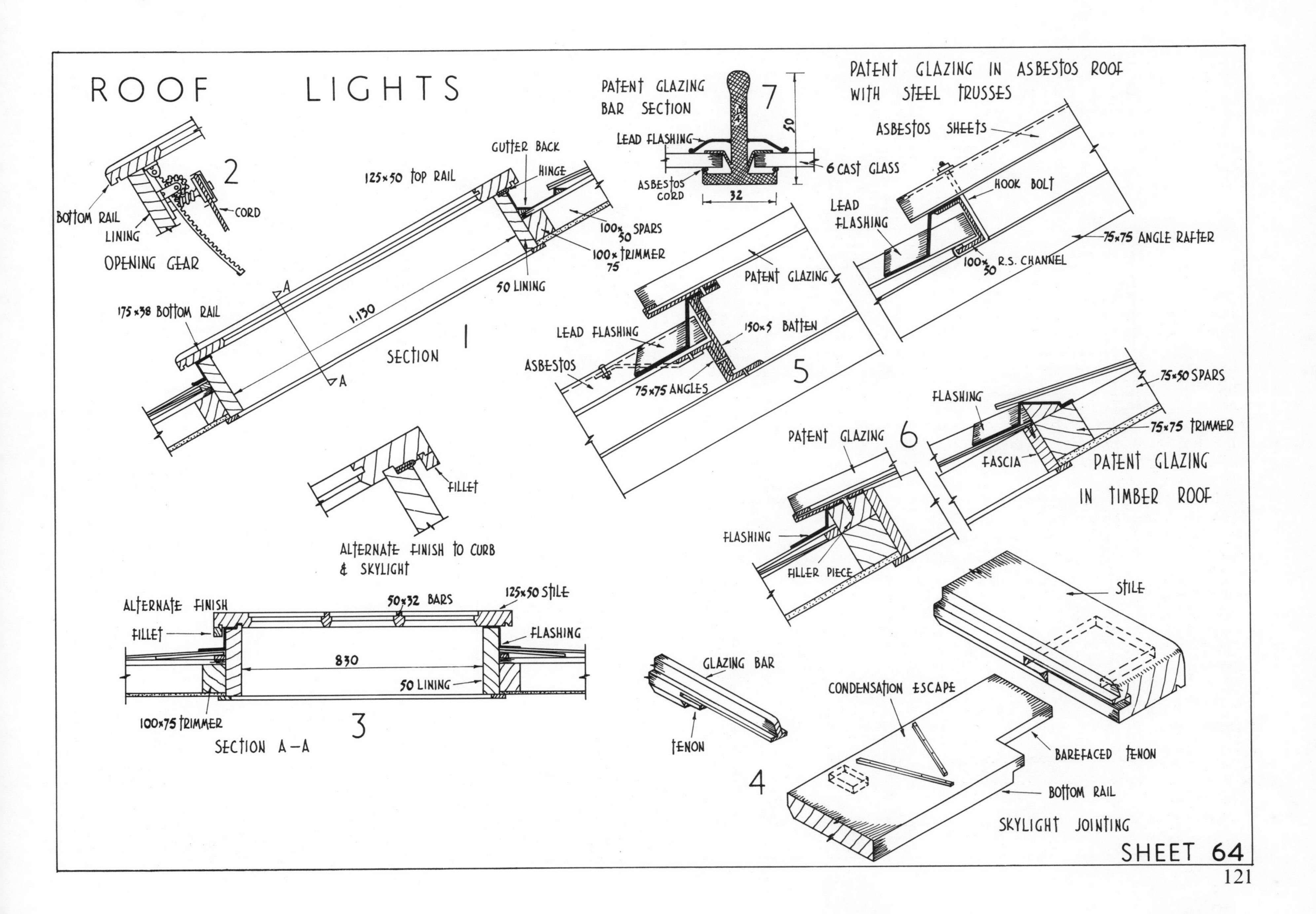
ROOF LIGHTS
2
BOTTOM RAIL
LINING
CORD
OPENING GEAR
125×50 TOP RAIL
GUTTER BACK
HINGE
100×50 SPARS
100×75 TRIMMER
50 LINING
175×38 BOTTOM RAIL
1.130
A
A
SECTION 1
FILLET
ALTERNATE FINISH TO CURB & SKYLIGHT
PATENT GLAZING BAR SECTION 7
LEAD FLASHING
ASBESTOS CORD
6 CAST GLASS
50
32
PATENT GLAZING
LEAD FLASHING
150×5 BATTEN
ASBESTOS
75×75 ANGLES
5
PATENT GLAZING IN ASBESTOS ROOF WITH STEEL TRUSSES
ASBESTOS SHEETS
HOOK BOLT
LEAD FLASHING
75×75 ANGLE RAFTER
100×50 R.S. CHANNEL
75×50 SPARS
FLASHING
75×75 TRIMMER
PATENT GLAZING 6
FASCIA
PATENT GLAZING IN TIMBER ROOF
FLASHING
FILLER PIECE
ALTERNATE FINISH
50×32 BARS
125×50 STILE
FILLET
FLASHING
830
50 LINING
100×75 TRIMMER
SECTION A–A 3
GLAZING BAR
TENON
CONDENSATION ESCAPE
STILE
BAREFACED TENON
BOTTOM RAIL
4
SKYLIGHT JOINTING
SHEET 64

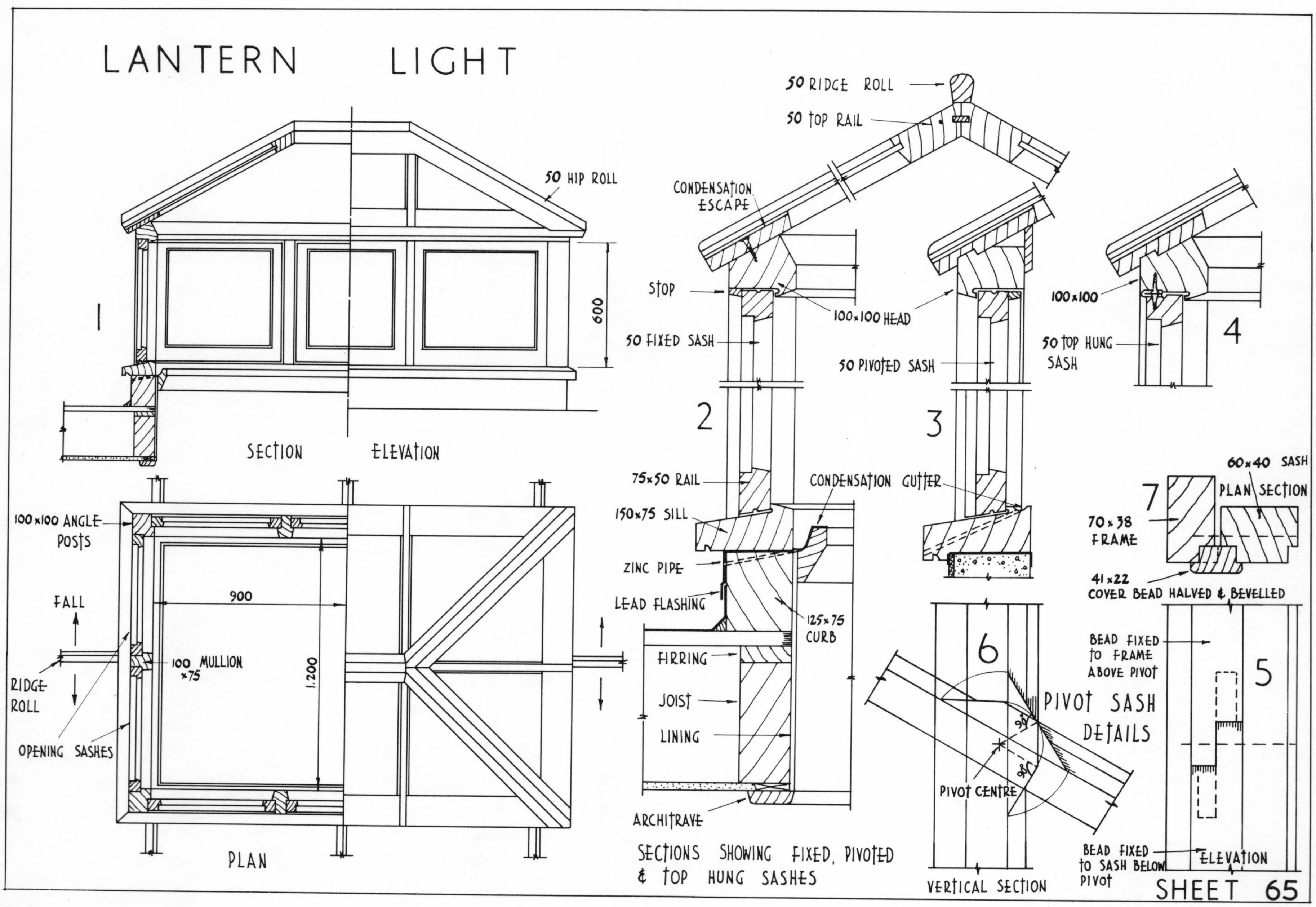

LANTERN LIGHT
50 RIDGE ROLL
50 TOP RAIL
50 HIP ROLL
CONDENSATION ESCAPE
600
1
STOP
100×100 HEAD
50 FIXED SASH
50 PIVOTED SASH
100×100
50 TOP HUNG SASH
4
2
3
SECTION
ELEVATION
75×50 RAIL
CONDENSATION GUTTER
60×40 SASH
PLAN SECTION
7
70×38 FRAME
150×75 SILL
100×100 ANGLE POSTS
ZINC PIPE
41×22 COVER BEAD HALVED & BEVELLED
FALL
900
LEAD FLASHING
125×75 CURB
RIDGE ROLL
100 MULLION ×75
1.200
FIRRING
JOIST
LINING
6
BEAD FIXED TO FRAME ABOVE PIVOT
5
PIVOT SASH DETAILS
OPENING SASHES
PIVOT CENTRE
ARCHITRAVE
PLAN
SECTIONS SHOWING FIXED, PIVOTED & TOP HUNG SASHES
BEAD FIXED TO SASH BELOW PIVOT
ELEVATION
VERTICAL SECTION
SHEET 65

with outlets bored through the curb as shown. Hip and ridge rolls are mitred together at their intersections and fixed ready for flashing.

Sheet 65.2 shows an enlarged detail through the lantern showing a fixed sash in position.

Sheet 65.3 shows a section through a concrete flat roof with the curb formed also in concrete and the whole covered with asphalt. It will be noted that the sill shown has a condensation groove worked in the solid. The head is tongued with the bottom rail grooved to receive this, and a lining fixed on the inside.

Sheet 65.4 shows a detail of the head prepared for side lights, top hung to open out.

When the lantern is roofed with four sashes or lights as shown in the illustration, the mitre bevel required is half the dihedral angle. The method of finding this is dealt with under the heading of 'roof bevels' on Sheet 33.

An alternative method of roofing a lantern is to use moulded bars. The hip, ridge and other bars will be of different sections, owing to their varying inclinations. Reference should be made to the section dealing with the work on inclined mouldings to lantern, on Sheet 92.

A further method of covering a lantern is by using a flat roof. In this case, joists, having a fall in two directions, are boarded and covered with lead and finished with a fascia and gutter all round.

Pivot Sashes

Pivot-hung windows are now widely used in all types of modern buildings including offices, flats, hospitals, schools and stores. In curtain wall construction pivot windows designed for horizontal and vertical swinging often form a special feature of the architectural design. The kinds of pivots available are many and varied, the more recent of these friction type pivots shown on Sheet 59 have been in use on the Continent some long time.

The construction of the window is governed by a number of factors: (1) the type of pivot, (2) the method of opening—horizontally or vertically, (3) the amount of swing, (4) the position of the sash in relation to the frame.

As there are various types of pivot, so there are varying timber sections used in the construction of the window. Manufacturers have their own sections and architects specify many others in which the joiner's work in making the window and in the hanging of the sashes requires special attention.

The section of such a window is shown in Sheet 65.7. In this detail the sash is flush with the frame on the outside. Both members are of stock sections and rebated on all edges to receive a rebated cover bead, screwed to the frame above the pivot and to the sash below the pivot. The bevels for cutting the beads are shown in Sheet 65.6, which as well as being bevelled are halved as seen in the elevation, Sheet 65.5. The sash in this example opens to an angle of 30 degrees to the horizontal.

Chapter Seven: Joinery

Prefabricated Timber Buildings

The post-war shortage of certain types of buildings, particularly schools, has largely been responsible for the high quality prefabricated timber buildings now being widely used.

Prefabrication means to manufacture standardized sections of a structure for rapid assembly on the site. This pre-assembly of work in the shop may be either partial or complete. Complete structures may be built and assembled, depending on their size, in the workshop or only sections of the walls, partitions, floors and roof sections, ready for erection on the site. There are three groups of systems for prefabricating buildings: timber, metal and concrete. The components of the different systems differ greatly in the structural materials. In timber construction, each of the firms specializing in prefabricated buildings has its own system. Some use square timber columns as structural members with panels bolted or nailed on, other makers use built-up hollow posts of machined locking members, yet the square columns are themselves stiffened by the panel designs interlocked for rigidity.

To facilitate the prefabrication of the sections in the shop and to ensure that the varying forms of these units will fit together on assembly on the site, the units are based on a common dimension or module. This is taken from the centre lines of the units horizontally. The A75 prefabricated building system illustrated is that developed by A. H. Anderson Ltd.

This system has been devised for the speedy erection of complete permanent single- or two-storey buildings. It is based on a horizontal centre line module of 1.875 m and vertical increments of 600 mm. Structural components are available for single-storey buildings of up to 4.800 m ceiling height, two-storey buildings, and any other desired combination. Structures may have pitched roofs, using roof trusses with roof panels and felt finish, or flat roofs, using plywood box beams with roof panels and similar finish.

With this system there is unusual freedom of action. The system can be employed, if desired, in conjunction with other methods of construction. Considerable variations in elevational treatment and in finish are possible. Sheet 66 shows a sketch of a layout of single- and two-storey structures which can be altered or dismantled, and re-erected with little waste of materials.

External wall panels consist of a basic frame jointed in the usual way, with a variety of infilling sub-assemblies. The principal types in the 2.400 m and 3.000 m range are shown in Sheet 66, figs. 2 and 3. Detail sections of the panel framing members are shown in Sheet 66.4.

Typical details of wall panel junctions are shown in Sheet 66.5, 13 mm bolts are used to connect the wall panels. The end stile is reinforced at the beam entry position and a longer bolt is used. Hardwood tongues are used vertically between panels, and horizontally between panel sill and slate sill and between coupled panels. Hardwood cover strips are fixed on the outer face of panel junctions to provide additional weathering protection.

Sheet 66.6 shows a detail of the roof finish, 50 mm thick solid boarded panels spanning between beams are covered with three layers of felt roofing and 13 mm chippings, and an aluminium perimeter trim.

A part elevation of a two-storey structure is shown in Sheet 67.7. Sheet 67.8 shows a part section of the same structure.

A combination of 3.000 m and 4.200 m structures is shown in section in Sheet 67.9. The details of foundations are shown in Sheet 67.10.

Sheet 68.11 shows an isometric view of a typical single-storey assembly. 2.400 m partition frame range is shown in Sheet 68.12. These, together with timber connecting pieces, have been designed for use with 64 mm Paramount dry partitioning.

Details at the head and floor of a partition are shown in Sheet 68, figs. 13 and 15. A section through a door and head lining are shown in Sheet 68.14.

Timber Frame Construction

This is a system of building which has been in use in Canada and the U.S.A. for over a century. Eighty per cent of house building follows timber frame technique. Frame construction has certain advantages over traditional building. Work can be completed in less time, which means labour costs are lower, is cheaper and production is consequently higher.

Frame houses are better insulated than traditional houses, timber framed walls can provide upwards of twice the insulation of brick cavity walls and are therefore warmer. The structure can be erected and roofed-in very rapidly so that all finished work takes place under cover.

Walls and partitions can be of dry construction so that interiors are decorated without delay. Drying out, condensation and cracking are avoided. Dwellings are habitable the day the builder leaves.

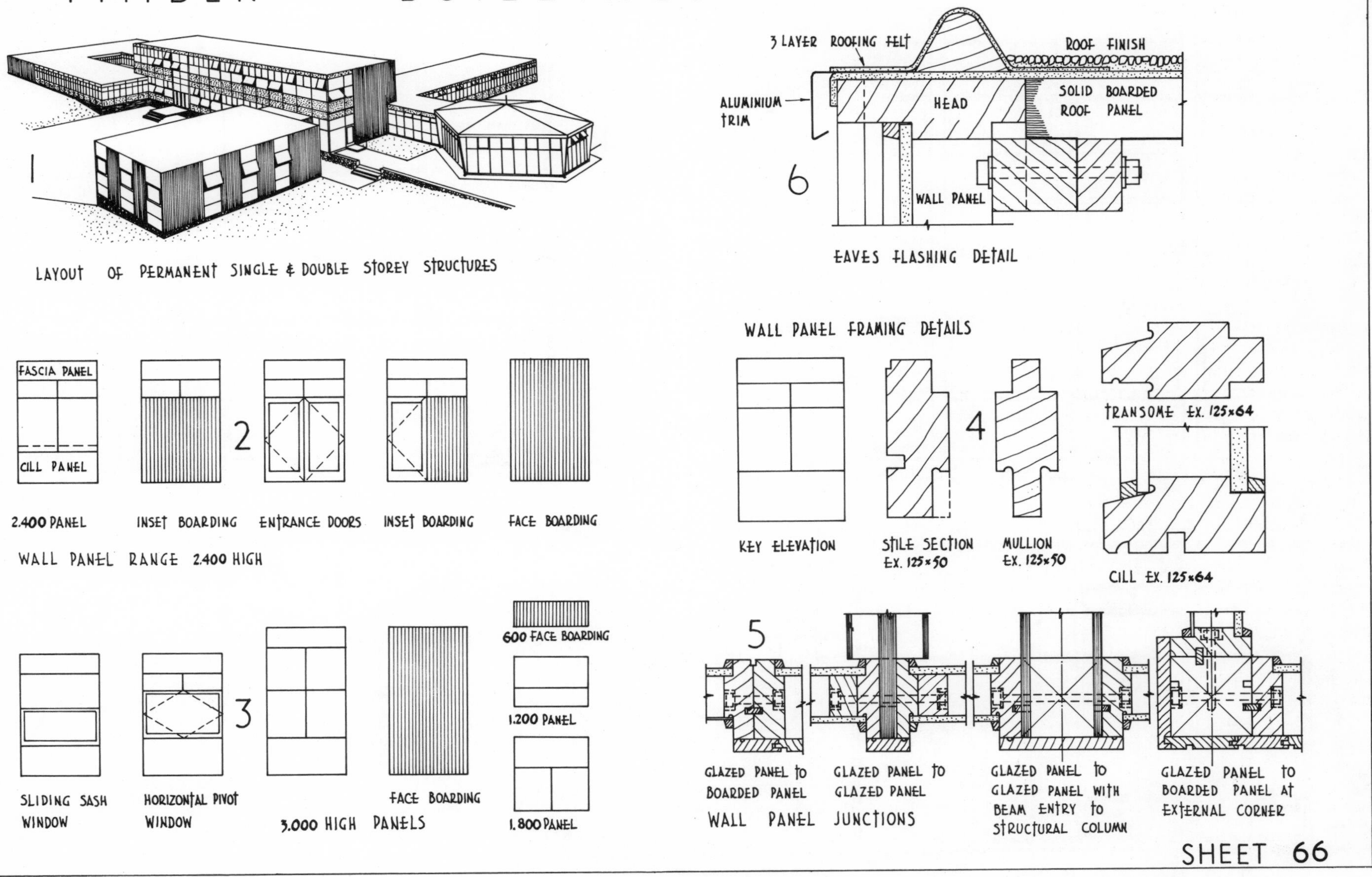
TIMBER BUILDINGS
1
LAYOUT OF PERMANENT SINGLE & DOUBLE STOREY STRUCTURES
3 LAYER ROOFING FELT
ROOF FINISH
ALUMINIUM TRIM
HEAD
SOLID BOARDED ROOF PANEL
6
WALL PANEL
EAVES FLASHING DETAIL
FASCIA PANEL
CILL PANEL
2
2.400 PANEL
INSET BOARDING
ENTRANCE DOORS
INSET BOARDING
FACE BOARDING
WALL PANEL RANGE 2.400 HIGH
WALL PANEL FRAMING DETAILS
4
TRANSOME EX. 125×64
KEY ELEVATION
STILE SECTION EX. 125×50
MULLION EX. 125×50
CILL EX. 125×64
600 FACE BOARDING
3
1.200 PANEL
SLIDING SASH WINDOW
HORIZONTAL PIVOT WINDOW
3.000 HIGH PANELS
FACE BOARDING
1.800 PANEL
5
GLAZED PANEL TO BOARDED PANEL
GLAZED PANEL TO GLAZED PANEL
GLAZED PANEL TO GLAZED PANEL WITH BEAM ENTRY TO STRUCTURAL COLUMN
GLAZED PANEL TO BOARDED PANEL AT EXTERNAL CORNER
WALL PANEL JUNCTIONS
SHEET 66

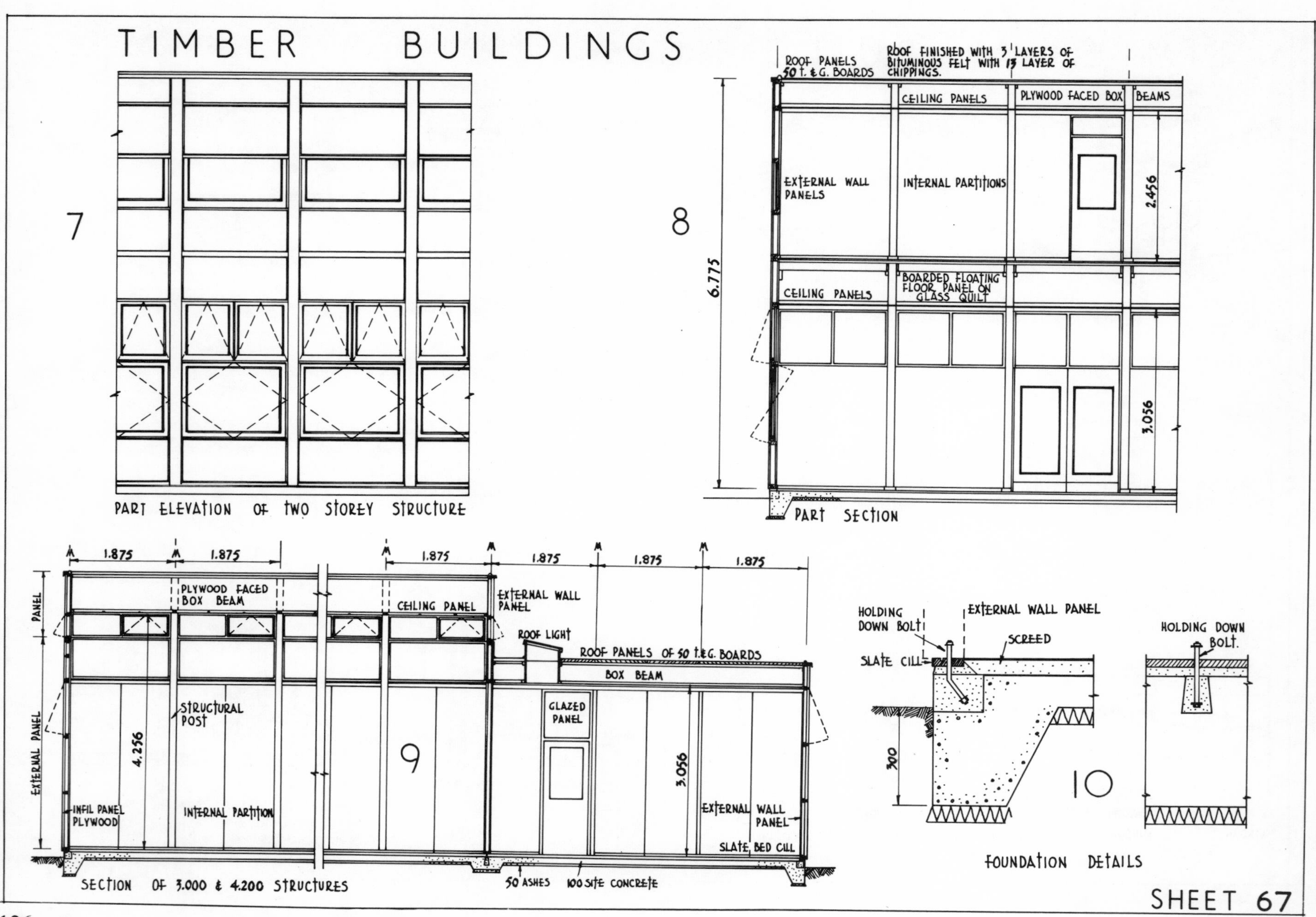
TIMBER BUILDINGS
7
PART ELEVATION OF TWO STOREY STRUCTURE
8
ROOF PANELS 50 T. & G. BOARDS
ROOF FINISHED WITH 3 LAYERS OF BITUMINOUS FELT WITH 13 LAYER OF CHIPPINGS.
CEILING PANELS
PLYWOOD FACED BOX BEAMS
EXTERNAL WALL PANELS
INTERNAL PARTITIONS
2.456
6.775
CEILING PANELS
BOARDED FLOATING FLOOR PANEL ON GLASS QUILT
3.056
PART SECTION
1.875
1.875
1.875
1.875
1.875
1.875
PANEL
PLYWOOD FACED BOX BEAM
CEILING PANEL
EXTERNAL WALL PANEL
ROOF LIGHT
ROOF PANELS OF 50 T.&G. BOARDS
BOX BEAM
EXTERNAL PANEL
STRUCTURAL POST
GLAZED PANEL
4.256
9
3.056
INFIL PANEL PLYWOOD
INTERNAL PARTITION
EXTERNAL WALL PANEL
SLATE BED CILL
SECTION OF 3.000 & 4.200 STRUCTURES
50 ASHES
100 SITE CONCRETE
HOLDING DOWN BOLT
EXTERNAL WALL PANEL
SCREED
SLATE CILL
300
10
HOLDING DOWN BOLT.
FOUNDATION DETAILS
SHEET 67

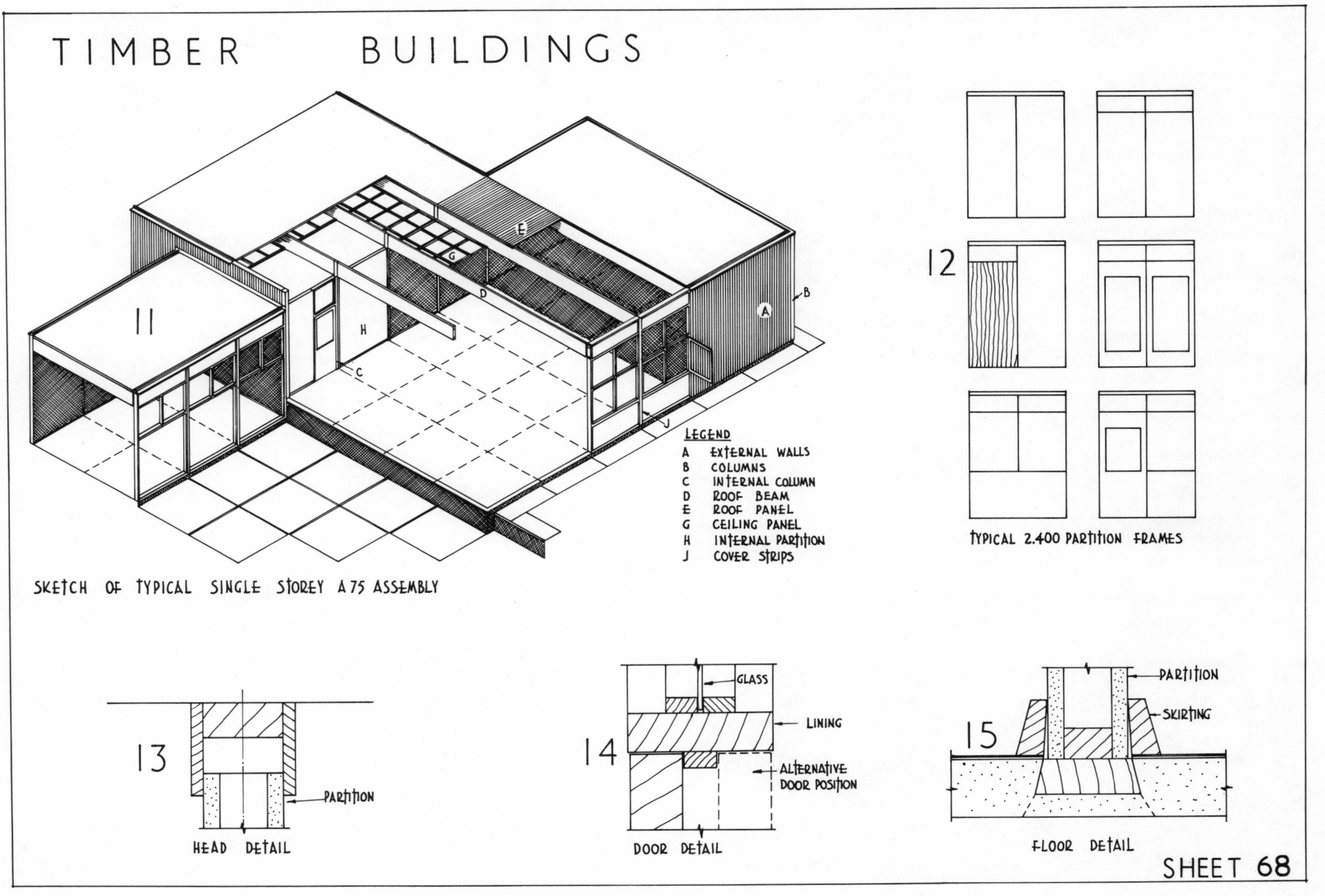
TIMBER BUILDINGS
11
A
B
C
D
E
G
H
J
LEGEND
A EXTERNAL WALLS
B COLUMNS
C INTERNAL COLUMN
D ROOF BEAM
E ROOF PANEL
G CEILING PANEL
H INTERNAL PARTITION
J COVER STRIPS
SKETCH OF TYPICAL SINGLE STOREY A75 ASSEMBLY
12
TYPICAL 2.400 PARTITION FRAMES
13
PARTITION
HEAD DETAIL
14
GLASS
LINING
ALTERNATIVE DOOR POSITION
DOOR DETAIL
15
PARTITION
SKIRTING
FLOOR DETAIL
SHEET 68

While house frames are usually fabricated on the site, where available with the aid of power tools, wall frame sections can be factory built with great economy.

With sound design and workmanship, timber frame houses will outlast many others. In frame construction all timber is isolated from brickwork and masonry and damp-proofed, so that the usual starting points of dry rot are eliminated.

In a timber frame house the basic framework of studding is a free-standing structure, acting as a braced and unified assembly which supports all loads. Since the exterior facing, brick or stone veneer, timber weather-boarding or cement rendering, performs no structural function other than that of providing additional bracing, it need not be started until the building has been closed in.

Two main forms of frame construction are in current use, platform and balloon.

Sheet 69.1 shows a section using platform construction. The ground floor extends to the outside edges of the building and provides a platform upon which exterior walls and interior partitions are erected. This system may be used for single- and double-storey buildings.

In balloon construction the studs are continuous from sill to eaves in a two-storey building.

A 150 mm by 50 mm treated timber sill laid on a bituminous damp-proof course and mortar bed is anchored by means of 13 mm bolts embedded in the foundation walls, as shown in Sheet 69.2. The ground floor joists are placed at 400 mm centres and braced at intervals with bridging pieces. Double joists are used under partition walls. Sub-flooring may be laid at right-angles to the joists or diagonally using square-edged or tongued and grooved boards, Sheet 69.3. Plywood sub-flooring may be used in 2.400 m by 1.200 m sheets, with blocking pieces fixed to afford edge nailing. The finished floor is not laid over the sub-floor until the building has been closed in.

When a timber frame house is to be built on a concrete slab, sub-flooring may be omitted and the finished flooring laid directly on pressure-treated timber sleepers, embedded in or anchored to the slab.

Normal wall frames in platform frame construction consist of storey-high 100 mm by 50 mm studs placed at 400 mm centres, to which top and bottom plates are nailed. Corners are formed by multiple studs. At door and window openings the wall studs are doubled.

In North America, timber or plywood sheathing is applied to the outside of the stud frames to stiffen the structure, to improve insulation and to provide an all-over nailing area for timber cladding or rendering. Building paper is applied to the outer face of the sheathing before the addition of exterior cladding.

Sheathing also figures in frame housing in this country, but many houses have been built without it. Where sheathing is not incorporated the building paper moisture barrier is applied directly to the studs.

In a two-storey dwelling built on platform frame principles, the joists of the upper floor are spiked to the top plates of the lower storey wall frames and boxed by headers as at ground floor level. Sub-flooring is then laid, after which preparations and erection of stud frames proceed as before.

Roofs for frame houses broadly conform to traditional practice and detailing differs little from that used in brick and stone houses; a detail at the eaves is shown in Sheet 69.4. Shallow mono-pitch and duo-pitch roofs are often formed by a system of widely spaced deep-section rafters or beams and structural roof planking of 50 mm. Western red cedar, a form of construction which ensures excellent thermal insulation. Roof planking and beams are left exposed for interior decorative effect and are overlaid above with additional rigid insulation and the outer weather skin. The standard pitched roof system designed by the Timber Research and Development Association can also be used in conjunction with frame house construction.

Most people who build timber frame houses like to have at least part of the structure clothed externally with wood as well. Brick, stone, rendering, tile and shingle can all be used alone or in combination with vertical or horizontal timber weather-boarding.

Post and Beam Construction

This is another method for framing floors and roofs, having been used in timber building for many years. In post and beam framing, plank subfloors or roofs, usually 50 mm thick, are supported on beams spaced 2.400 m apart. The ends of the beams are supported on posts or piers. Windows and doors should be located between posts in exterior walls to eliminate the need for headers over the openings. The wide spacing between posts permits the use of large glass areas.

A combination of platform frame and post and beam construction is often used. Sheet 69.5 shows an example of this form of combination. Brick facing has been used for the flank walls, with Western red cedar weather-boarding as a cladding material, with cement rendering, on the front and back elevations.

Cross Wall Construction

This construction is now widely used, not only for blocks of flats, but for terraced housing and semi-detached homes as well. The cross walls are, as a rule, the party walls between dwellings and the end walls. These load-bearing walls are normally of brick or concrete with the usual form of foundations,

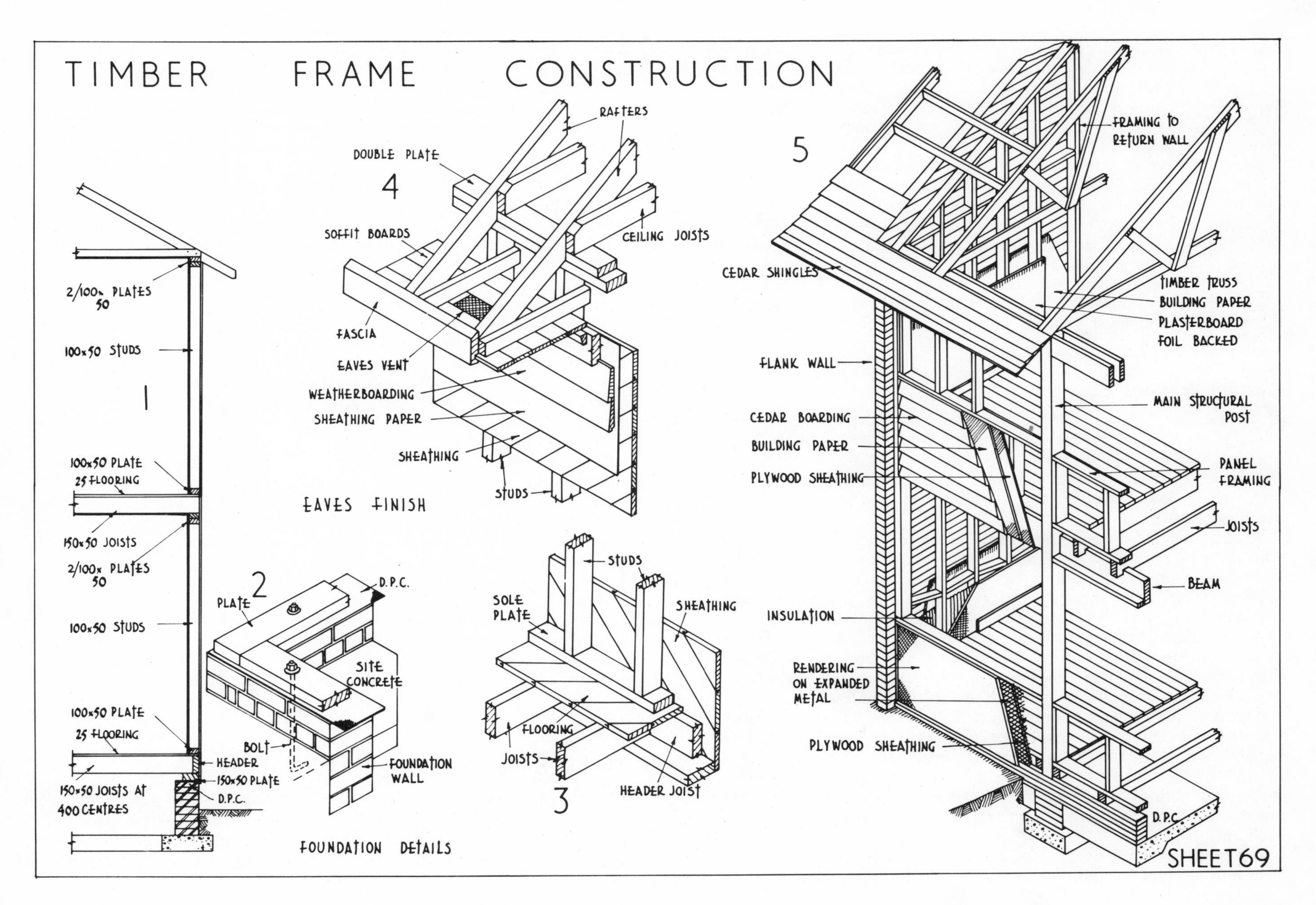

TIMBER FRAME CONSTRUCTION
1
2/100x50 PLATES
100x50 STUDS
100x50 PLATE
25 FLOORING
150x50 JOISTS
2/100x50 PLATES
100x50 STUDS
100x50 PLATE
25 FLOORING
HEADER
150x50 PLATE
150x50 JOISTS AT 400 CENTRES
D.P.C.
4
DOUBLE PLATE
RAFTERS
SOFFIT BOARDS
CEILING JOISTS
FASCIA
EAVES VENT
WEATHERBOARDING
SHEATHING PAPER
SHEATHING
STUDS
EAVES FINISH
2
PLATE
D.P.C.
SITE CONCRETE
BOLT
FOUNDATION WALL
FOUNDATION DETAILS
3
STUDS
SOLE PLATE
SHEATHING
FLOORING
JOISTS
HEADER JOIST
5
FRAMING TO RETURN WALL
CEDAR SHINGLES
TIMBER TRUSS
BUILDING PAPER
PLASTERBOARD FOIL BACKED
FLANK WALL
MAIN STRUCTURAL POST
CEDAR BOARDING
BUILDING PAPER
PLYWOOD SHEATHING
PANEL FRAMING
JOISTS
BEAM
INSULATION
RENDERING ON EXPANDED METAL
PLYWOOD SHEATHING
D.P.C.
SHEET69

and provide the structural supports for floors and roofs. The front and rear walls are non-load-bearing and may be treated as curtain walling problems. Prefabricated timber panels or timber frame construction may be used. Whichever method is employed, the site concrete slab serves as the foundation support to these walls. Sheet 70.6 shows a sketch of typical cross wall construction in terraced housing.

Sheet 70.7 shows various forms of horizontal weather-boarding with nailing positions indicated by dotted lines. Corrosion-resistant nails are used at 600 mm intervals when fixing to a groundwork of timber sheathing. Where sheathing other than timber boards is concerned, nails should be driven through into the studs at each bearing, and blocking between studs will be needed if cladding is applied vertically.

Sheet 70.8 shows various forms of vertical cladding. Where red cedar is used in exterior cladding, it may be left untreated. Those wishing to preserve the natural colour must resort to clear finishes or light pigmented stain.

Curtain Walling

A curtain wall is a wall of windows and spandrel panels in a framework which is not built up within the main structure of the building. These walls are not load-bearing, therefore any form of walling which is not load-bearing is a curtain wall. The illustrations in frame and cross wall construction on Sheets 69 and 70 show this type of construction where the enclosing walls are curtain walls.

Recent years have seen considerable changes in the design of buildings with the replacement of brick and masonry walls by thin sheet materials in large panels, or by light timber or metal frameworks with various forms of infilling. Usually the framework is suspended right across the face of the building, being held back to it at widely spaced points.

As the curtain wall supports only its own weight and its primary function is to keep out the weather, it can be light in weight and thin in construction. This relieves the building structure of a considerable dead load and makes it possible to economize in the foundations to the building.

Curtain walls are usually prefabricated. This means that with standardization of component parts assembly of comparatively large units in the shop is possible and, erected in this way, shows considerable saving in erection time. In spite of its thinness, curtain walling can provide better insulation than traditional masonry walls and examples of its use in schools, hospitals, offices and factories have been found to be completely satisfactory and economical.

The main materials used in the framework of curtain walling are timber, steel and aluminium and for the panel infilling, timber, multi-pane glass, aluminium sheet, enamelled glass, asbestos and thin stone are some of the materials employed. The design of the walling must provide for expansion and contraction both in itself and in movement in the building, and panels should be easily removed for repair or replacement.

Curtain walls in timber usually consist of a light framing in timber, of mullions spanning from floor to floor with transom members to build up the frame and panel infilling according to the type of building. The mullions run between sill and head members, where they are mortised and tenoned and wedged in the usual way. The fixing at the sill and head and intermediate floors is by mild steel clips with slotted bolt holes to allow for movement. Horizontal members in timber should be steeply sloped and deeply throated and should project sufficiently to throw the water clear of the wall face. Where large units are prefabricated in the shop they are left unglazed, hoisted complete into position on the site and fixed.

Timber panel infilling is carried out either vertically or horizontally in a wide variety of timbers and with many variations of joint and moulding. Exterior grade resin-bonded plywood is also used.

Suitable hardwoods for curtain walling are afrormosia, agba, idigbo, iroko, karri, meranti, European oak, African mahogany and teak. Softwoods include Western red cedar, Douglas fir, hemlock, European redwood, European whitewood and pitch pine.

Sheet 71.1 shows the elevation of a two-storey building clad with timber curtain walling.

The mullions shown in the details Sheet 71, figs. 2, 3 and 4 may be solid or glued laminated members rebated to receive the glazed panel frames.

Vertical cedar boarding is used between the two floors and also at the roof level, and finished with linseed oil, or a clear lacquer or varnish.

A vertical section through the lower storey of the curtain walling is shown in Sheet 71.5.

It will be seen that stock single and double rebated window sections are used which require fillets to be planted in the rebates of the vertical members where the cedar weather-boarding is fixed. The entrance doors are not in line with the glass rebates, requiring the sub-frames to be used at these points.

Timber Bridges

The use of timber as a structural engineering material has only become accepted in this country since the Second World War. Progress made both in North America and the Continent has been much more advanced in this field where timber is used in bridge construction as well as in methods of roofing.

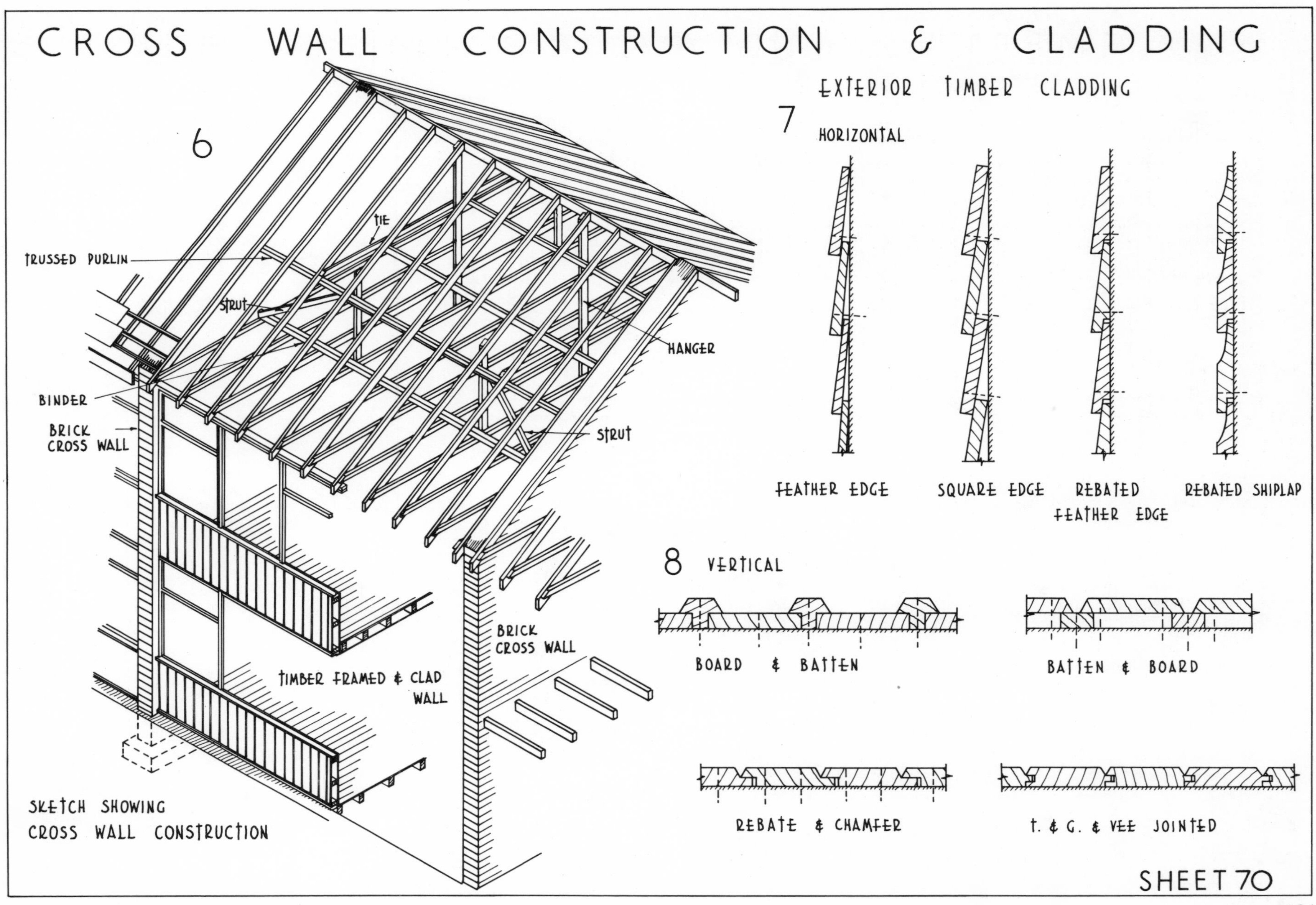
CROSS WALL CONSTRUCTION & CLADDING
6
TRUSSED PURLIN
TIE
STRUT
HANGER
BINDER
BRICK CROSS WALL
STRUT
BRICK CROSS WALL
TIMBER FRAMED & CLAD WALL
SKETCH SHOWING CROSS WALL CONSTRUCTION
EXTERIOR TIMBER CLADDING
7 HORIZONTAL
FEATHER EDGE
SQUARE EDGE
REBATED FEATHER EDGE
REBATED SHIPLAP
8 VERTICAL
BOARD & BATTEN
BATTEN & BOARD
REBATE & CHAMFER
T. & G. & VEE JOINTED
SHEET 70

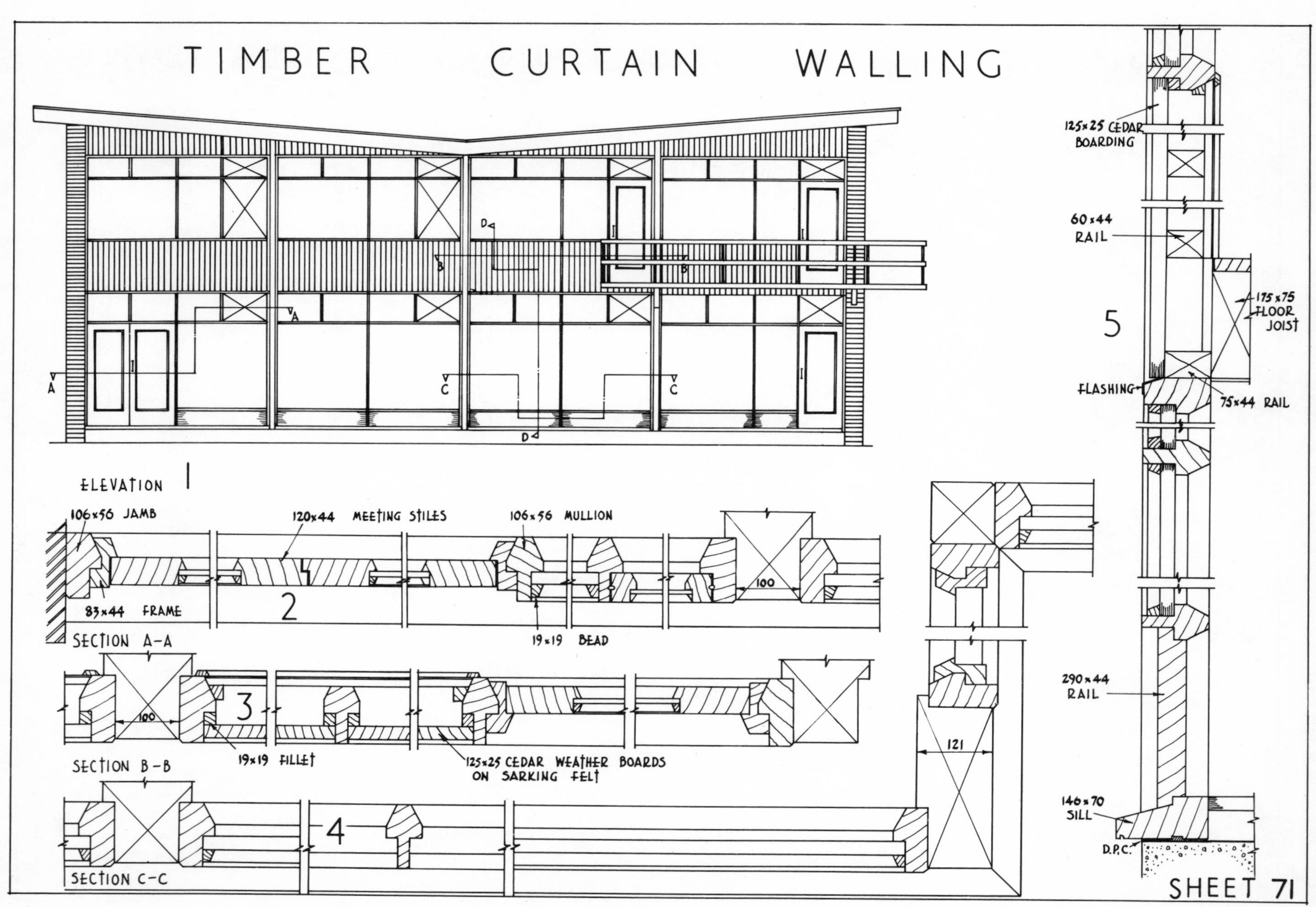

TIMBER CURTAIN WALLING
ELEVATION 1
106×56 JAMB
120×44 MEETING STILES
106×56 MULLION
83×44 FRAME
2
100
19×19 BEAD
SECTION A–A
3
100
19×19 FILLET
125×25 CEDAR WEATHER BOARDS
ON SARKING FELT
SECTION B–B
121
4
SECTION C–C
125×25 CEDAR
BOARDING
60×44
RAIL
175×75
FLOOR
JOIST
5
FLASHING
75×44 RAIL
290×44
RAIL
146×70
SILL
D.P.C.
SHEET 71

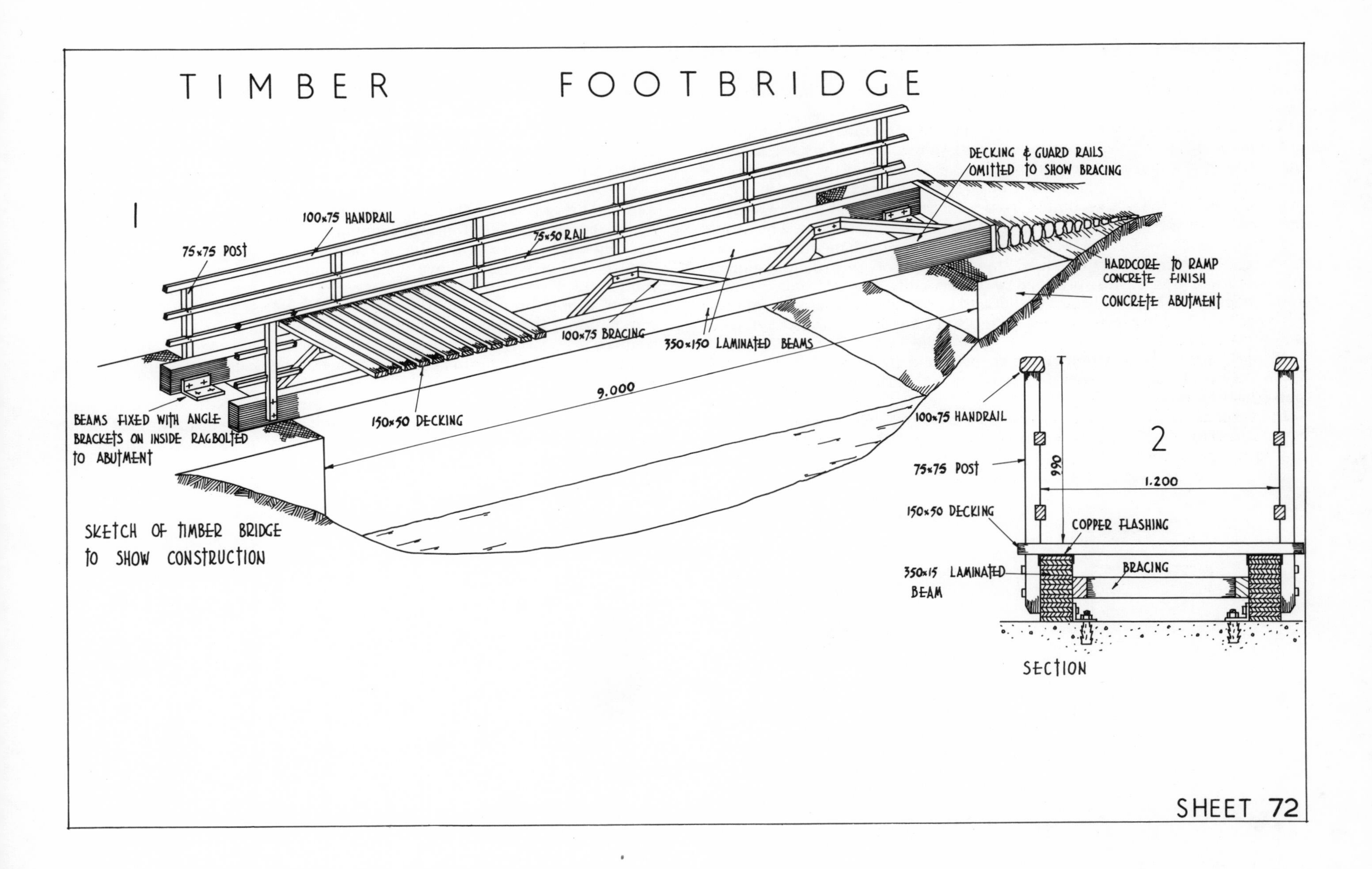
TIMBER FOOTBRIDGE
1
100×75 HANDRAIL
75×75 POST
75×50 RAIL
DECKING & GUARD RAILS OMITTED TO SHOW BRACING
HARDCORE TO RAMP CONCRETE FINISH
CONCRETE ABUTMENT
100×75 BRACING
350×150 LAMINATED BEAMS
9.000
BEAMS FIXED WITH ANGLE BRACKETS ON INSIDE RAGBOLTED TO ABUTMENT
150×50 DECKING
SKETCH OF TIMBER BRIDGE TO SHOW CONSTRUCTION
100×75 HANDRAIL
2
75×75 POST
990
1.200
150×50 DECKING
COPPER FLASHING
350×15 LAMINATED BEAM
BRACING
SECTION
SHEET 72

Timber bridges were in use by the railways in this country until the 1930s, where the components were limited in size, shape and quality available. Reinforced concrete and steel constructions replaced timber in this type of construction due to these limitations.

The techniques which have been developed to overcome these limitations are in the jointing together of members using timber connectors and laminating, using new waterproof adhesives. This has resulted in the production of members which have no limitations of size, shape or conditions of use. It may be said that the size is limited from a fabrication point of view only by the capacity of the laminating equipment used at the workshop. This would indicate a much wider use being made in the immediate future of prefabricated units similar to those used at present in roof construction, namely, laminated and box beams, plywood and diagonally boarded nailed girders and trusses, to be used in bridge design and construction. Examples of these have been covered in previous work on roofs in Chapter 3, to which reference should be made.

Bridges constructed in timber are to be found in North America up to 75.000 m span where two types are used. These are classed as (a) grid and (b) slab constructions.

In grid construction the load is carried by cross members spanning between longitudinal members. These are trusses in most cases, having glued laminated bowstrings and chords.

Slab construction is a combination of timber and concrete which forms the finished road surface.

A sketch of a timber footbridge classed as grid construction is shown in Sheet 72.1, having a span of 9.000 m and a footwalk 1.200 m wide. Two 350 mm by 150 mm laminated beams are used as the longitudinal members spaced 900 mm apart and fixed to the concrete abutments by angle plates, coach screwed to the beam and ragbolted to the concrete. The beams have copper flashings in the form of sleeves at each end for complete protection, and are braced with 100 mm by 75 mm bracing fixed on the insides of the beams as shown. The footwalk consists of 150 mm by 50 mm hardwood decking bolted through into the two beams.

A filler piece is used against the beam ends to receive the hardcore for the ramps which are finished in concrete.

The section through the bridge is shown in Sheet 72.2. The 100 mm by 75 mm handrail is carried on 75 mm by 75 mm posts bolted to the laminated beams with intermediate rails housed as shown. The decking is cut round the posts where these occur to allow firm fixing of the posts to the outside faces of the beams.

All the timber used in bridge construction should be treated with a preservative of some kind. The treatment of timber has been dealt with in Chapter 3. Likewise all ironwork, bolts, angles, etc. should be treated with a rust-preventing coat.

Chapter Eight: Joinery

Stairs

Building Regulations

The new Building Regulations made by the Minister of Public Building and Works have now replaced the Model Bye-Laws and apply to all building applications lodged with the relevant Authority, except Scotland which has its own regulations, and Inner London, where the London Building Acts still operate.

The existing bye-laws did not include requirements for the construction of stairs, staircases and handrails.

Interpretation

The Building Regulations differentiate between stairs for common use and stairs for private use.

Common stairway—a stairway of steps, all steps having straight nosings on plan, which is intended for common use within any building for occupation by more than one family.

Private stairway—a stairway of steps, all steps having straight nosings on plan, within a building and intended solely for use in connection with a dwelling occupied by one family only.

Private Stairways

Parallel step—a step of which the nosing is parallel to the nosing of the step or landing immediately above it as shown in Sheet 73.1.

Tapered step—a step of which the nosing is not parallel to the nosing of the step or landing immediately above it as shown in Sheet 73.2.

Pitch line—a notional or imaginary line drawn to connect all the nosings of the treads in a flight of stairs, shown in Sheet 73.3.

Going. The going of a step is measured on plan between the nosing of its tread and the nosing of the next tread or landing above. The methods of measurement of the rise and going of a step are shown in Sheet 73.4.

Width of stairway. This is measured from centre to centre of the handrails as shown in Sheet 73.5. If there is no handrail, it is to be measured to the surface of the wall, screen or balustrade facing the stairway or railing.

Notional width. Sheet 73.6 shows a stairway having several consecutive tapered steps of differing widths. In a stairway of this type, all such tapered steps are deemed to have a notional width equal to the width of the narrowest part of the tapered steps measured from the side of the stairway where the treads are narrowest.

General Requirements for Private Stairways

The rise must be the same for each step in a flight between consecutive floors as shown in Sheet 73.7.

The going must be the same for each parallel step in a flight between consecutive floors.

The headroom must not be less than 2.000 m measured vertically above the pitch line and the clearance must not be less than 1.500 m measured at right-angles to the pitch line as shown in Sheet 73.7.

In an open stair with no risers, the nosing of the tread of any step or landing must overlap the back edge of the tread of the step below it by at least 15 mm as shown in Sheet 73.8.

For any parallel step in a private stair the sum of its going plus twice its rise must not be less than 550 mm and not more than 700 mm.

The rise of any step in a private stair must not be more than 220 mm and the going not less than 220 mm with the maximum angle of pitch 42° to the horizontal as shown in Sheet 73.8.

Any tapered steps must comply to the regulations (see tapered steps).

Sheet 73.9 shows the graph from which the maximum and minimum permitted going and rise can be obtained within the shaded area for private stairways.

Common Stairways

The rise of any step in a common stairway must not be more than 190 mm and the going not less than 230 mm with the maximum angle of pitch 38° to the horizontal as shown in Sheet 74.10.

There must not be more than 16 risers in any flight.

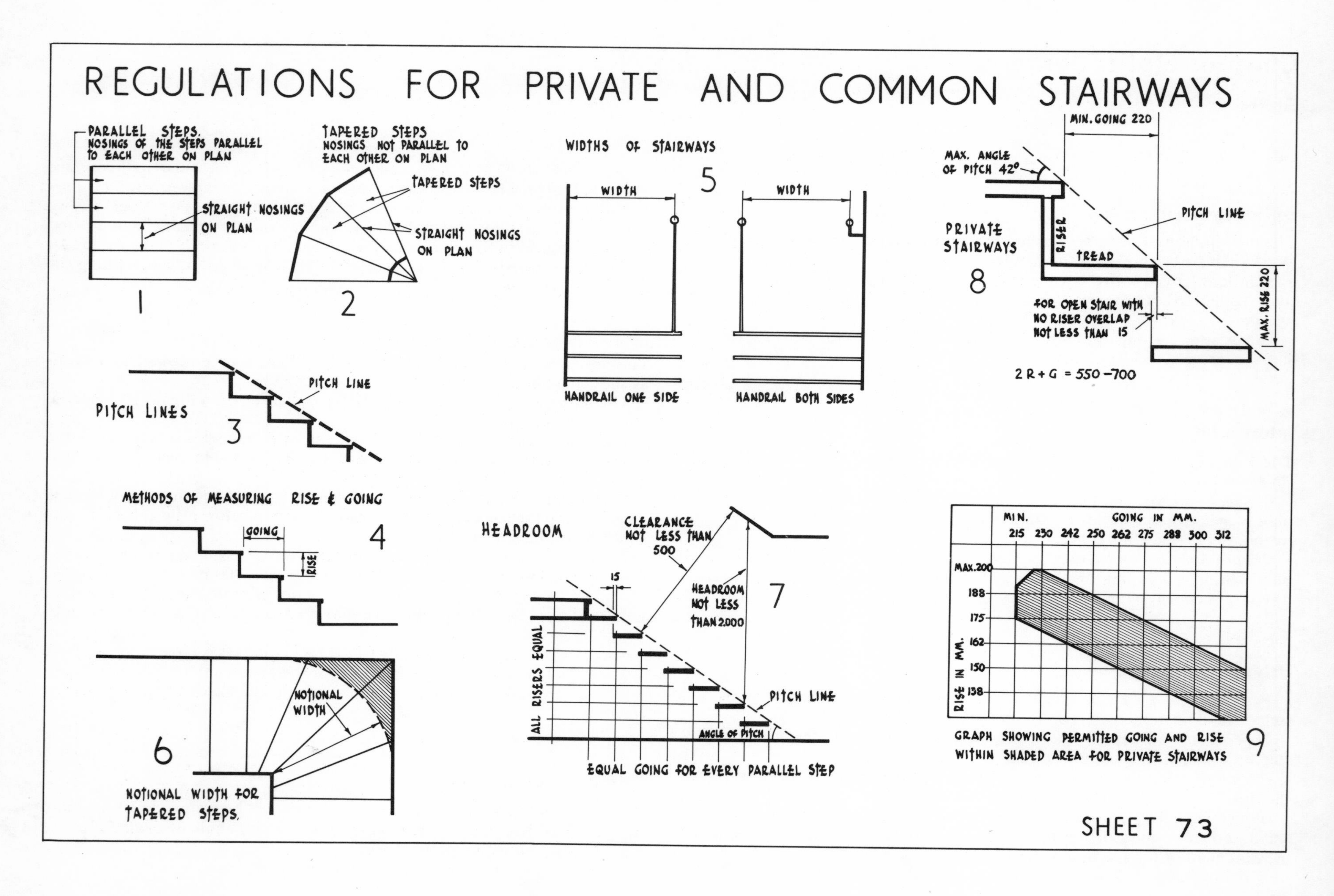
REGULATIONS FOR PRIVATE AND COMMON STAIRWAYS
PARALLEL STEPS.
NOSINGS OF THE STEPS PARALLEL TO EACH OTHER ON PLAN
STRAIGHT NOSINGS ON PLAN
1
TAPERED STEPS
NOSINGS NOT PARALLEL TO EACH OTHER ON PLAN
TAPERED STEPS
STRAIGHT NOSINGS ON PLAN
2
WIDTHS OF STAIRWAYS
5
WIDTH
WIDTH
HANDRAIL ONE SIDE
HANDRAIL BOTH SIDES
MIN. GOING 220
MAX. ANGLE OF PITCH 42°
PITCH LINE
PRIVATE STAIRWAYS
RISER
TREAD
8
MAX. RISE 220
FOR OPEN STAIR WITH NO RISER OVERLAP NOT LESS THAN 15
2 R + G = 550 –700
PITCH LINES
3
PITCH LINE
METHODS OF MEASURING RISE & GOING
GOING
RISE
4
HEADROOM
CLEARANCE NOT LESS THAN 500
15
HEADROOM NOT LESS THAN 2.000
7
ALL RISERS EQUAL
PITCH LINE
ANGLE OF PITCH
EQUAL GOING FOR EVERY PARALLEL STEP
NOTIONAL WIDTH
6
NOTIONAL WIDTH FOR TAPERED STEPS.
MIN.
GOING IN MM.
215 230 242 250 262 275 288 300 312
MAX.200
188
175
162
150
138
RISE IN MM.
GRAPH SHOWING PERMITTED GOING AND RISE WITHIN SHADED AREA FOR PRIVATE STAIRWAYS
9
SHEET 73

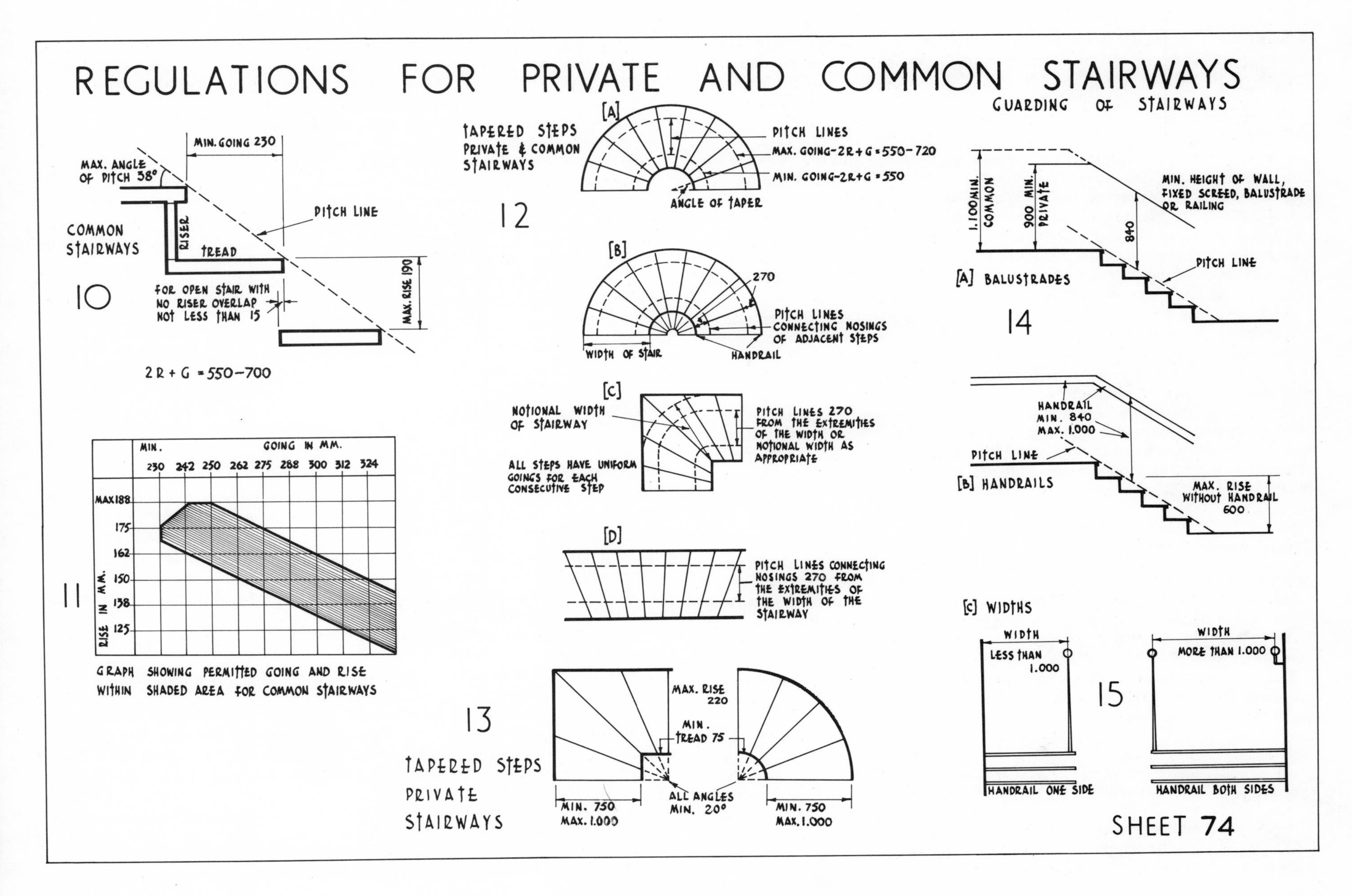
REGULATIONS FOR PRIVATE AND COMMON STAIRWAYS
MIN. GOING 230
MAX. ANGLE OF PITCH 38°
PITCH LINE
COMMON STAIRWAYS
RISER
TREAD
10
FOR OPEN STAIR WITH NO RISER OVERLAP NOT LESS THAN 15
MAX. RISE 190
2R + G = 550–700
TAPERED STEPS PRIVATE & COMMON STAIRWAYS
[A]
PITCH LINES
MAX. GOING–2R+G = 550–720
MIN. GOING–2R+G = 550
ANGLE OF TAPER
12
[B]
270
PITCH LINES CONNECTING NOSINGS OF ADJACENT STEPS
WIDTH OF STAIR
HANDRAIL
[C]
NOTIONAL WIDTH OF STAIRWAY
PITCH LINES 270 FROM THE EXTREMITIES OF THE WIDTH OR NOTIONAL WIDTH AS APPROPRIATE
ALL STEPS HAVE UNIFORM GOINGS FOR EACH CONSECUTIVE STEP
[D]
PITCH LINES CONNECTING NOSINGS 270 FROM THE EXTREMITIES OF THE WIDTH OF THE STAIRWAY
MIN. GOING IN MM.
230 242 250 262 275 288 300 312 324
MAX 188
175
162
150
138
125
RISE IN MM.
11
GRAPH SHOWING PERMITTED GOING AND RISE WITHIN SHADED AREA FOR COMMON STAIRWAYS
13
TAPERED STEPS PRIVATE STAIRWAYS
MAX. RISE 220
MIN. TREAD 75
ALL ANGLES MIN. 20°
MIN. 750 MAX. 1.000
MIN. 750 MAX. 1.000
GUARDING OF STAIRWAYS
1.100 MIN. COMMON
900 MIN. PRIVATE
MIN. HEIGHT OF WALL, FIXED SCREED, BALUSTRADE OR RAILING
840
PITCH LINE
[A] BALUSTRADES
14
HANDRAIL MIN. 840 MAX. 1.000
PITCH LINE
[B] HANDRAILS
MAX. RISE WITHOUT HANDRAIL 600
[C] WIDTHS
WIDTH
LESS THAN 1.000
WIDTH
MORE THAN 1.000
15
HANDRAIL ONE SIDE
HANDRAIL BOTH SIDES
SHEET 74

Sheet 74.11 shows the graph from which the maximum and minimum permitted going and rise can be obtained within the shaded area for common stairways.

Tapered Steps

The going of a tapered step varies throughout its width and the rules for the going of a parallel step, previously dealt with, cannot therefore apply. The regulations give rules which are to be applied to the design of tapered steps generally in both private and common stairways.

Sheet 74.12 shows how the proportions of tapered steps are regulated with two notional goings measured 270 mm from the extremities of the width or notional width of the stairway. These goings are to be measured in the vertical planes of the pitch lines connecting consecutive nosings.

The maximum going for tapered steps is twice the rise plus the going equals 720 mm except where the angle of taper is 10° or less, when the maximum going would be twice the rise plus the going equals 635 mm.

The minimum going is the same for any degree of taper and is twice the rise plus the going equals 550 mm.

Sheet 74.13 shows the application of the regulations to tapered steps in private stairways of width 750 mm to 1.000 m, the special limiting factors are, that the nosings must make a minimum angle of 20° on plan, and that the going must not be less than 75 mm at the narrowest point of the tapered step.

Guarding of Stairways

Any private or common stairway must have a wall, screen, balustrade or railing at least 840 mm in vertical height above the pitch lines of the stairway on each side as shown in Sheet 74.14. The guarding of a landing should be taken to a height of 1.000 m for a private stairway and 1.050 m if it is a common stairway.

The stairways shown in Sheet 74.15 require a handrail on one side only for stairways less than 1.000 m wide and a handrail on both sides for stairways more than 1.000 m wide.

Stairways

The types of staircase are known as Straight Flight, Dogleg, Open Newel and Geometrical. The first type it is assumed will have been covered in earlier work and only those remaining will be considered here. The type of staircase to be installed in any building will be decided at a very early stage, particular attention being paid in the planning of the stair to step proportions, headroom above the staircase and also beneath the landings.

The examples which have been detailed demonstrate the approach to given situations, but it should be appreciated that many other arrangements would be equally satisfactory.

In the dogleg and open newel stairs, newel posts are placed at the head and foot of each flight. The newels are therefore a conspicuous feature.

In the geometrical stair both the strings and the handrails are continuous, there are no newels, however many turns the stair may take. A newel may, for reasons of design, be introduced at the bottom and top of such a stair, but it is not an essential part of the construction.

Stairs may be further described by the number of turns made by the stairs, by the type of strings, by the landings or whether tapered steps are used in the construction.

Quarter-turn Stair

This type of stair changes its direction either by a quarter space landing or by tapered steps.

Half-turn Stair

The direction is reversed in this stair either by a half-space landing, or two quarter-space landings, or a quarter-space landing and a quarter space of tapered steps, or completely by tapered steps. Newel stairs are confined to rectangular plans but geometrical stairs may be arranged upon rectangles, circles, ellipses or polygons.

Sheet 75, figs. 1, 2 and 3 are line diagrams of the three types of stairs which will be dealt with here.

The three usual methods of constructing the steps are shown in Sheet 75, figs. 4, 5 and 6.

In Sheet 75.4 the bottom edges of the risers are tongued into the treads, the top edges are square edged and pocket screwed to the bottom side of the tread with the scotia moulding housed as shown. The steps are housed and wedged into the stringboards with glue blocks rubbed between them and the strings. The nosing line and margin are also indicated.

Sheet 75.5 is similar to the above except that the tread is tongued into the riser.

In Sheet 75.6 the riser is tongued into the two treads both at the top and bottom. No scotia is used in this construction.

An alternative nosing detail is shown in Sheet 75.7.

Sheet 75.8 shows the section through the lower steps in a stair with the

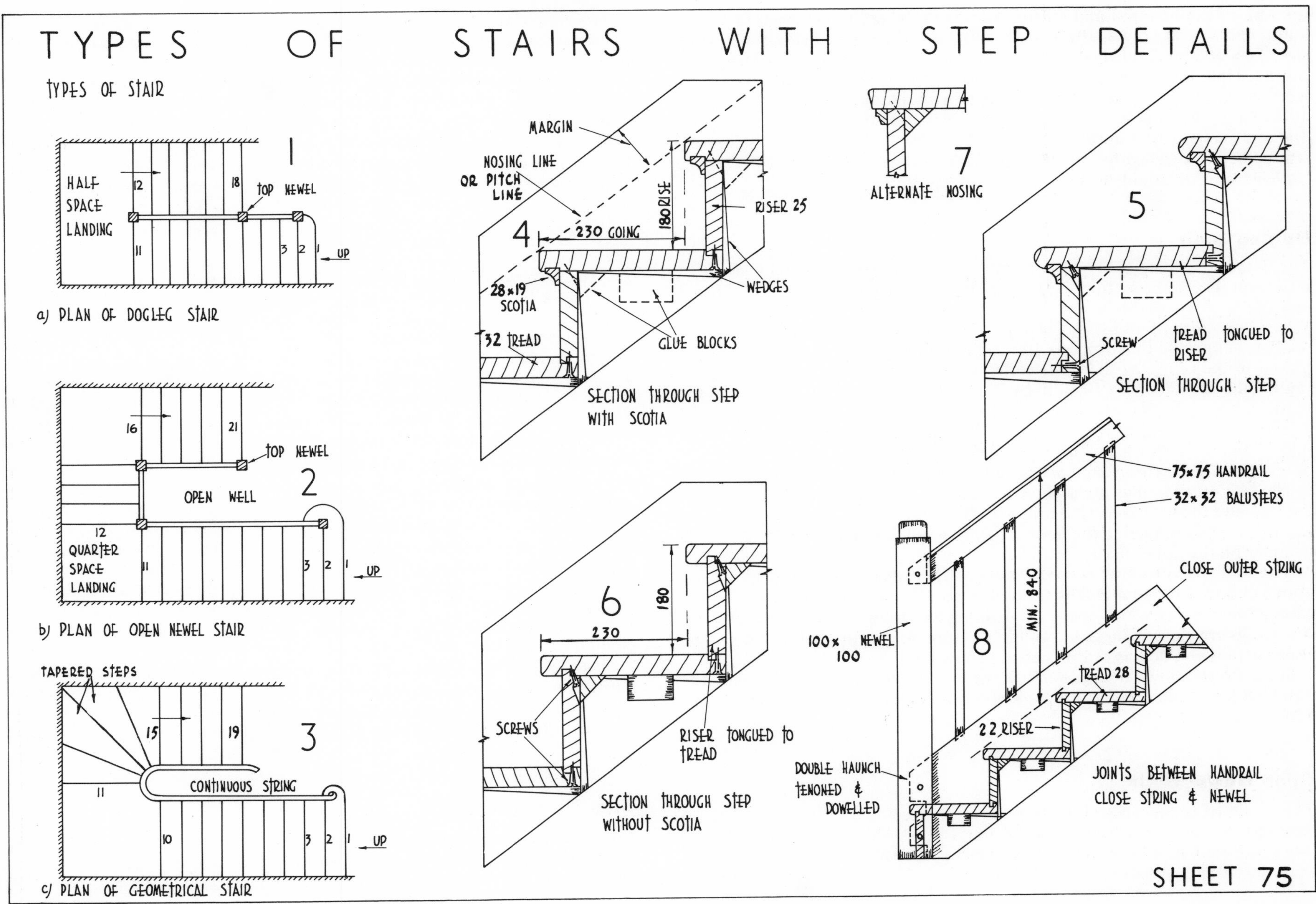

TYPES OF STAIRS WITH STEP DETAILS
TYPES OF STAIR
1
HALF SPACE LANDING
12
18
TOP NEWEL
11
3 2 1
UP
a) PLAN OF DOGLEG STAIR
16
21
TOP NEWEL
2
OPEN WELL
12
QUARTER SPACE LANDING
11
3 2 1
UP
b) PLAN OF OPEN NEWEL STAIR
TAPERED STEPS
15
19
3
CONTINUOUS STRING
11
10
3 2 1
UP
c) PLAN OF GEOMETRICAL STAIR
MARGIN
NOSING LINE OR PITCH LINE
180 RISE
4
230 GOING
RISER 25
28 x 19 SCOTIA
WEDGES
GLUE BLOCKS
32 TREAD
SECTION THROUGH STEP WITH SCOTIA
7
ALTERNATE NOSING
5
SCREW
TREAD TONGUED TO RISER
SECTION THROUGH STEP
6
180
230
SCREWS
RISER TONGUED TO TREAD
SECTION THROUGH STEP WITHOUT SCOTIA
75 x 75 HANDRAIL
32 x 32 BALUSTERS
CLOSE OUTER STRING
MIN. 840
100 x 100 NEWEL
8
TREAD 28
22 RISER
DOUBLE HAUNCH TENONED & DOWELLED
JOINTS BETWEEN HANDRAIL CLOSE STRING & NEWEL
SHEET 75

jointing of the handrail and stringboard to the newel post shown in dotted lines. The newels are usually square, 100 mm by 100 mm being the normal size. Two oblique haunched tenons are formed in the centre of the string and mortised into the newel, the joint is then secured by draw-boring, using a hardwood dowel at each tenon. Barefaced tenons should not be used at this point as the strength of the stair depends a good deal upon the rigidity of the newels and the method of jointing the strings to them. The handrail is housed, tenoned and dowelled by draw-boring to the newel, which is finished at the top with a solid moulded cap.

Dogleg Stair

The details of this type of stair are shown in Sheet 76.9, so called because of its appearance in the sectional elevation. It is used when the going is restricted and the space available is equal to the combined widths of the two flights. Where the width of a flight is 900 mm or more it is desirable to use an intermediate support in the form of a 100 mm by 75 mm bearer or carriage. This is placed in the centre of the stair, nailed at the foot and its upper end birds-mouthed and nailed to a 100 mm by 75 mm trimmer which is tenoned to the newel at one end and carried by the wall at the other. Rough brackets, shaped as shown, are nailed to the carriage with their upper edges cut square to fit tightly under the treads for further support; these brackets are fixed on alternate sides of the carriage and are usually short ends of floorboard.

The two outer strings are secured to 100 mm by 100 mm newels placed at the foot and the head of each flight, with the centre newel fixed to the landing trimmer and continued to the floor. The top newel is notched and fixed to the trimmer on the upper floor.

The half-space landing is constructed of 100 mm by 50 mm joists housed at one end to a 100 mm by 75 mm trimmer and supported by the wall at the other. The trimmer which spans the opening is carried by the outer walls. All the strings are close strings and the flight commences with a splayed step. The construction and details of this step are shown on Sheet 79.

Sheet 76.10 shows a detail of the step and carriage supports with the moulded handrail and finish to the newel cap shown in Sheet 76, figs. 11 and 12.

Open Newel Stair

The details of an open newel stair are shown in Sheet 77.13. This type of newel stair differs from the dogleg in that it has a well or open space between the two outer strings. This requires more floor space to accommodate the stair than the previous type.

The stair planning and design have again been detailed, allowing for two quarter-space landings with three flights each 975 mm wide shown in the plan. The flight commences with a half-round step and all the strings are close strings. Sheet 79 shows the construction of the rounded step.

The trimmers of the first landing are framed into the newel post which runs down to the floor. The second landing newel finishes just below the string and is finished with a moulded drop to match the newel caps. Central carriages and brackets are indicated in the plan by dotted lines. Sheet 77.14 shows the detail at the landing with the handrail, capping and newel details shown in Sheet 77, figs. 15 and 16.

Tapered Steps

A quarter-space of winders with a splayed bottom step about a newel is shown in Sheet 78.17. Before setting out the tapered steps, reference should be made to the notes on page 136 and the details Sheet 74.12 and 13.

Support for the tapered steps is obtained from 100 mm by 50 mm bearers placed immediately below each of the risers of the third and fourth steps. These are housed into the newel post and the wall string. A 100 mm by 75 mm trimmer is placed immediately below the fifth riser, Sheet 78.18. This is notched and fixed to the newel and built into the wall to secure the lower end of the carriage which is birds-mouthed over it. Triangular blocks nailed to the carriage may be used as an alternative to brackets for additional support to the treads at the centre.

The tapered steps, because of their width, have to be jointed, preferably with cross tongues. The wall strings also have to be increased in width to accommodate the tapered steps with the top edge shaped with an easing as shown, these are tongued and grooved and nailed together at the internal angle.

Sheet 78.19 shows an alternative finish to tapered steps increasing the size of the newel. The faces of the newel are shown developed in Sheet 78.20. The side D is housed and mortised for the string; the sides C and B are housed for the tapered steps 13 mm deep, and the side A housed for the shaped bottom step.

Sheet 78.21 shows a further alternate finish to tapered steps keeping straight nosings in plan. This requires a short string to house the ends of the steps which fall outside the newel when meeting the conditions of the regulations on angle of taper and minimum going at the narrowest point of the tapered step.

Step Construction

The bottom step of the lowest flight in a staircase is often specially shaped and made to project beyond the newel. Several of these finishes are illustrated in Sheet 79.

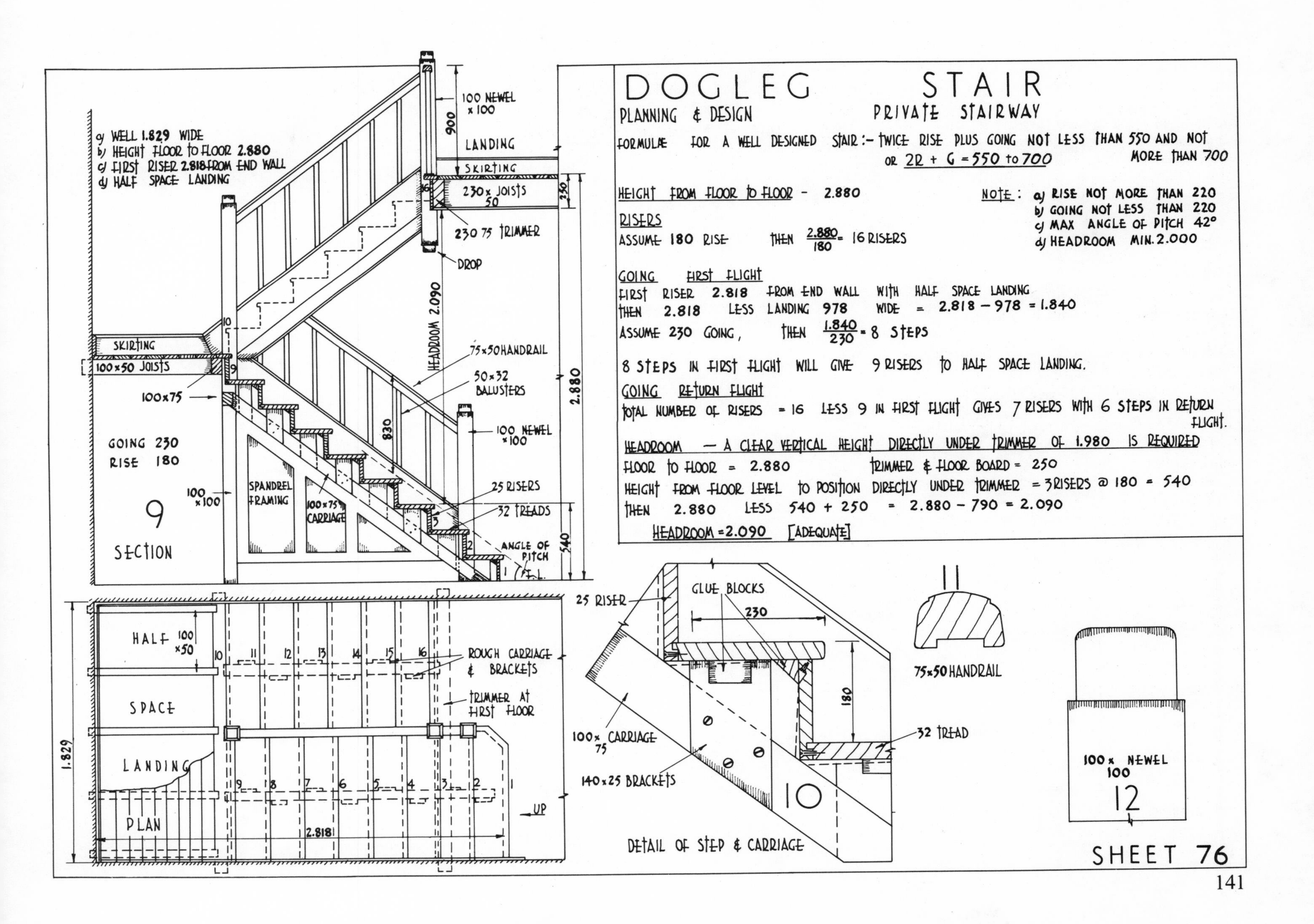
DOGLEG STAIR
PLANNING & DESIGN
PRIVATE STAIRWAY
FORMULAE FOR A WELL DESIGNED STAIR :- TWICE RISE PLUS GOING NOT LESS THAN 550 AND NOT MORE THAN 700
OR 2R + G = 550 to 700
HEIGHT FROM FLOOR TO FLOOR - 2.880
NOTE : a) RISE NOT MORE THAN 220
b) GOING NOT LESS THAN 220
c) MAX ANGLE OF PITCH 42°
d) HEADROOM MIN. 2.000
RISERS
ASSUME 180 RISE THEN 2.880/180 = 16 RISERS
GOING FIRST FLIGHT
FIRST RISER 2.818 FROM END WALL WITH HALF SPACE LANDING
THEN 2.818 LESS LANDING 978 WIDE = 2.818 – 978 = 1.840
ASSUME 230 GOING, THEN 1.840/230 = 8 STEPS
8 STEPS IN FIRST FLIGHT WILL GIVE 9 RISERS TO HALF SPACE LANDING.
GOING RETURN FLIGHT
TOTAL NUMBER OF RISERS = 16 LESS 9 IN FIRST FLIGHT GIVES 7 RISERS WITH 6 STEPS IN RETURN FLIGHT.
HEADROOM — A CLEAR VERTICAL HEIGHT DIRECTLY UNDER TRIMMER OF 1.980 IS REQUIRED
FLOOR TO FLOOR = 2.880
TRIMMER & FLOOR BOARD = 250
HEIGHT FROM FLOOR LEVEL TO POSITION DIRECTLY UNDER TRIMMER = 3 RISERS @ 180 = 540
THEN 2.880 LESS 540 + 250 = 2.880 – 790 = 2.090
HEADROOM = 2.090 [ADEQUATE]
a) WELL 1.829 WIDE
b) HEIGHT FLOOR TO FLOOR 2.880
c) FIRST RISER 2.818 FROM END WALL
d) HALF SPACE LANDING
100 NEWEL x 100
900
LANDING
SKIRTING
230 x 50 JOISTS
230
230 75 TRIMMER
DROP
HEADROOM 2.090
75x50 HANDRAIL
50x32 BALUSTERS
830
100 NEWEL x100
2.880
SKIRTING
100x50 JOISTS
100x75
GOING 230
RISE 180
100 x100
SPANDREL FRAMING
100x75 CARRIAGE
25 RISERS
32 TREADS
540
ANGLE OF PITCH
9
SECTION
HALF SPACE LANDING
100 x50
ROUGH CARRIAGE & BRACKETS
TRIMMER AT FIRST FLOOR
1.829
PLAN
2.818
UP
25 RISER
GLUE BLOCKS
230
180
32 TREAD
100x75 CARRIAGE
140x25 BRACKETS
10
DETAIL OF STEP & CARRIAGE
11
75x50 HANDRAIL
100 x 100 NEWEL
12
SHEET 76

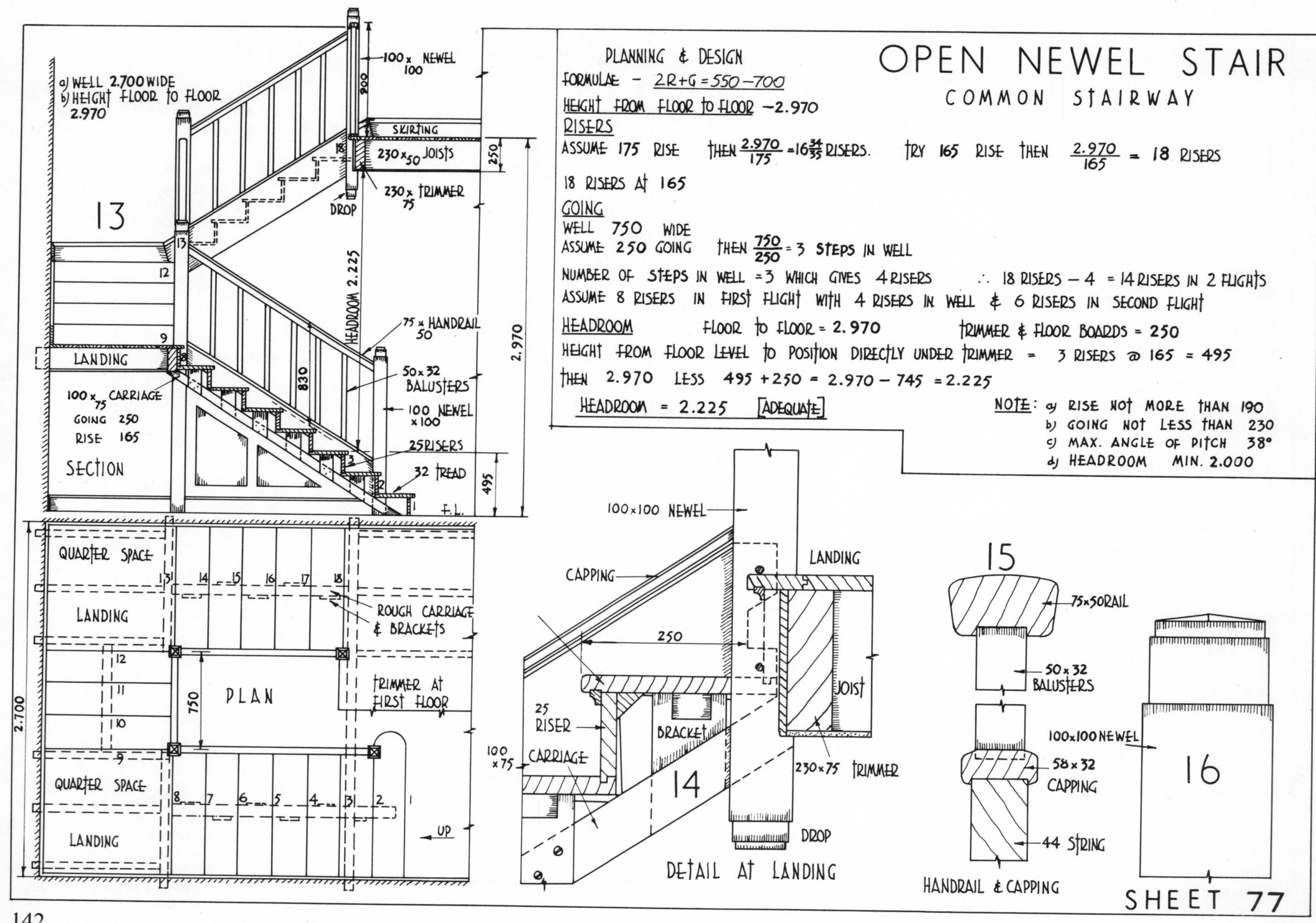

OPEN NEWEL STAIR
COMMON STAIRWAY
PLANNING & DESIGN
FORMULAE - 2R+G = 550-700
HEIGHT FROM FLOOR TO FLOOR - 2.970
RISERS
ASSUME 175 RISE THEN 2.970/175 = 16 34/35 RISERS. TRY 165 RISE THEN 2.970/165 = 18 RISERS
18 RISERS AT 165
GOING
WELL 750 WIDE
ASSUME 250 GOING THEN 750/250 = 3 STEPS IN WELL
NUMBER OF STEPS IN WELL = 3 WHICH GIVES 4 RISERS ∴ 18 RISERS - 4 = 14 RISERS IN 2 FLIGHTS
ASSUME 8 RISERS IN FIRST FLIGHT WITH 4 RISERS IN WELL & 6 RISERS IN SECOND FLIGHT
HEADROOM
FLOOR TO FLOOR = 2.970
TRIMMER & FLOOR BOARDS = 250
HEIGHT FROM FLOOR LEVEL TO POSITION DIRECTLY UNDER TRIMMER = 3 RISERS @ 165 = 495
THEN 2.970 LESS 495 + 250 = 2.970 - 745 = 2.225
HEADROOM = 2.225 [ADEQUATE]
NOTE: a) RISE NOT MORE THAN 190
b) GOING NOT LESS THAN 230
c) MAX. ANGLE OF PITCH 38°
d) HEADROOM MIN. 2.000
a) WELL 2.700 WIDE
b) HEIGHT FLOOR TO FLOOR 2.970
13
100 x 100 NEWEL
900
SKIRTING
230 x 50 JOISTS
250
230 x 75 TRIMMER
DROP
HEADROOM 2.225
75 x 50 HANDRAIL
2.970
830
50 x 32 BALUSTERS
100 NEWEL x 100
25 RISERS
32 TREAD
495
F.L.
LANDING
100 x 75 CARRIAGE
GOING 250
RISE 165
SECTION
QUARTER SPACE
LANDING
ROUGH CARRIAGE & BRACKETS
PLAN
750
TRIMMER AT FIRST FLOOR
2.700
UP
100 x 100 NEWEL
CAPPING
LANDING
250
JOIST
25 RISER
BRACKET
100 x 75 CARRIAGE
14
230 x 75 TRIMMER
DROP
DETAIL AT LANDING
15
75 x 50 RAIL
50 x 32 BALUSTERS
100 x 100 NEWEL
58 x 32 CAPPING
44 STRING
HANDRAIL & CAPPING
16
SHEET 77

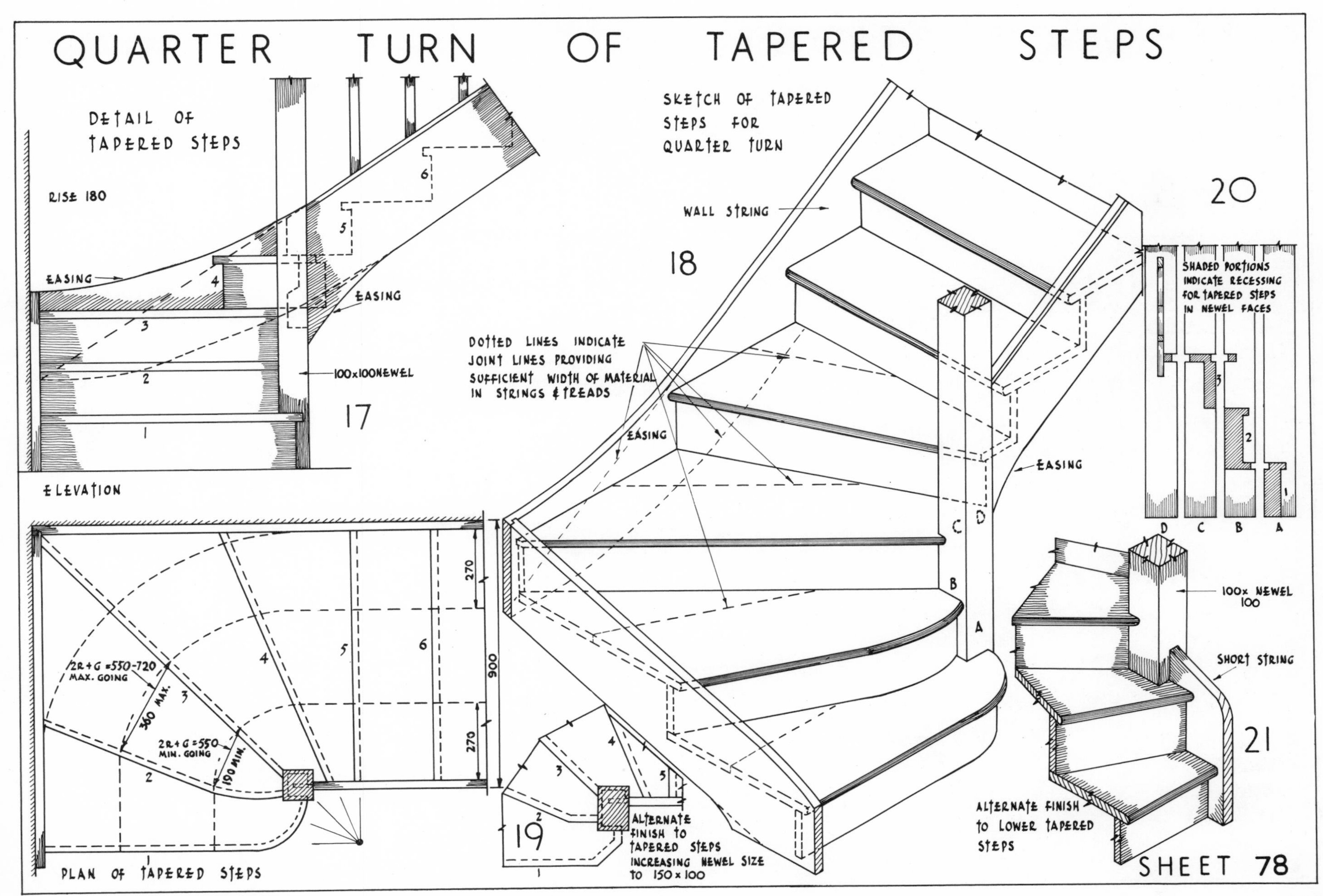
QUARTER TURN OF TAPERED STEPS
DETAIL OF TAPERED STEPS
RISE 180
EASING
EASING
100x100NEWEL
17
ELEVATION
SKETCH OF TAPERED STEPS FOR QUARTER TURN
WALL STRING
18
DOTTED LINES INDICATE JOINT LINES PROVIDING SUFFICIENT WIDTH OF MATERIAL IN STRINGS & TREADS
EASING
EASING
20
SHADED PORTIONS INDICATE RECESSING FOR TAPERED STEPS IN NEWEL FACES
D C B A
100x NEWEL 100
SHORT STRING
21
2R+G =550-720 MAX. GOING
360 MAX.
2R+G =550 MIN. GOING
190 MIN.
270
900
270
PLAN OF TAPERED STEPS
19
ALTERNATE FINISH TO TAPERED STEPS INCREASING NEWEL SIZE TO 150 x 100
ALTERNATE FINISH TO LOWER TAPERED STEPS
SHEET 78

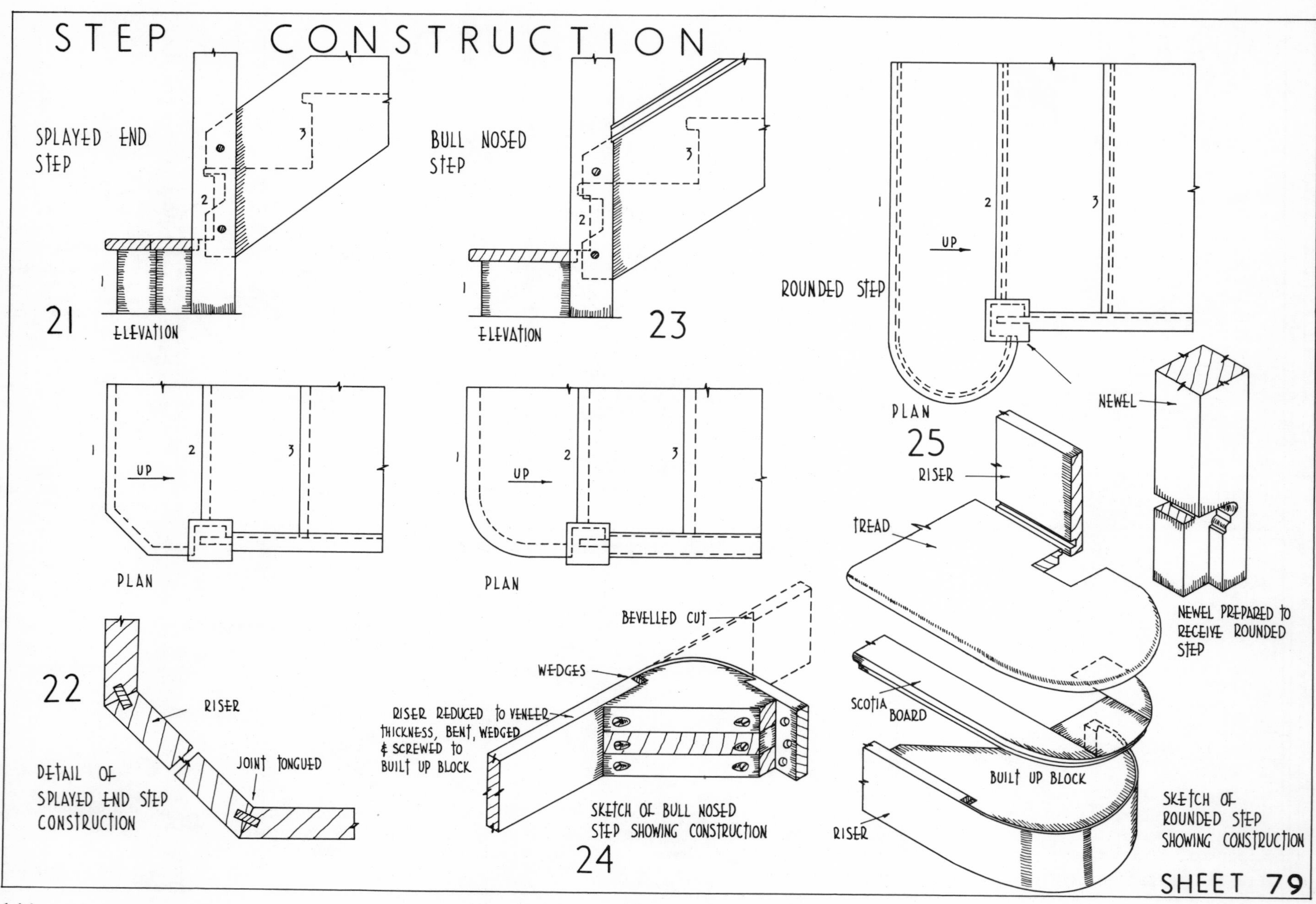
STEP CONSTRUCTION
SPLAYED END STEP
BULL NOSED STEP
1
2
3
21
ELEVATION
ELEVATION
23
ROUNDED STEP
UP
PLAN
25
NEWEL
RISER
TREAD
PLAN
PLAN
NEWEL PREPARED TO RECEIVE ROUNDED STEP
BEVELLED CUT
WEDGES
22
RISER
RISER REDUCED TO VENEER THICKNESS, BENT, WEDGED & SCREWED TO BUILT UP BLOCK
SCOTIA BOARD
JOINT TONGUED
DETAIL OF SPLAYED END STEP CONSTRUCTION
BUILT UP BLOCK
SKETCH OF BULL NOSED STEP SHOWING CONSTRUCTION
SKETCH OF ROUNDED STEP SHOWING CONSTRUCTION
24
SHEET 79

Splayed Step

Sheet 79.21 shows the part plan and elevation of this step with the detail showing the construction of the splayed riser in Sheet 79.22. It will be seen that the riser is in three pieces, being mitred and tongued at the joints.

Bullnosed Step

Sheet 79.23 shows the part plan and elevation of the quarter round or bull-nose step which consists of a tread and shaped riser. This is in one piece and is reduced to veneer thickness for bending to the required curve, as shown in Sheet 79.24. It also serves as a covering to the block which is built up in three laminations with the pieces arranged cross grained. The block strengthens the riser which is pocket screwed as shown. The newel is notched out at the foot and the riser is screwed to it. It will be seen that the face of the riser is made to coincide with the centre of the newel into which it is housed and screwed.

Rounded Step

The construction of this semicircular-ended step is similar to that described above. Where a scotia is used the curve of the scotia mould is worked on a piece of solid stuff which is fixed between the tread and the riser. Sheet 79.25 should make the construction clear. The newel prepared to receive the rounded step is also shown.

Geometrical Stairs

Sheet 80.26 shows the plan and sectional elevation of a geometrical stair having a cut and bracketed string with a half-space landing.

Geometrical stairs may be constructed with continuous cut and mitred strings or cut and bracketed strings. Both types are shown and will be described.

A Cut and Mitred String

A cut and mitred string suitable for the above stairs is shown in different stages of construction in Sheet 80.27. At the upper end one tread and riser is shown finished with the balusters and return nosing in position. The two other steps show the preparation of the string to receive the risers. Here, the return nosing is removed to show the fixing of the balusters which are dovetailed and screwed to the tread. The return nosing and scotia similar to Sheet 80.28 are cut to mitre with the tread nosing in front and mitred to a short return piece of itself at the back. The nosing pieces are fixed either by slot-screwing as shown, or by tongue and groove.

A Cut and Bracketed String

This is shown in the sketch, Sheet 80.29, and a plan and section in Sheet 80.30. These show the difference in treatment required when ornamental brackets are planted on the face of the string under the return ends of the treads. The bracket, cut from plywood, is mitred with the end of the riser which runs over the face of the string as shown, and the string is rebated to receive the riser to which it is nailed before the brackets are fixed. It will be seen that the tread in this case projects over the bracket equal to its thickness and when the balusters are fixed, the return nosing and scotia, similar to the previous example, are cut to fit over the bracket and mitre with the tread. The balusters are fixed at the top in various ways. They may be inserted in a groove in the handrail or screwed to an iron core as shown in the sketch, Sheet 80.31, the core being then screwed in the groove to the rail.

Wreathed Strings

In a geometrical stair the outside stringboards are wreathed, that is as they rise to suit the stairs they are curved in the plan at the change of direction. This is formed by reducing a short piece of string to a veneer between the springings and bending it upon a cylinder made to fit the plan in the manner shown in Sheet 81.32. When it is secured in position, the back of the veneer is filled up with staves glued across and a piece of canvas finally glued over the whole to give additional strength.

The string is set out before bending and positioned round the cylinder with the springing lines of the veneer placed exactly over the springing lines marked on the cylinder. The veneer portion will bend more easily if it is wetted or steamed.

The jointing of the wreathed portion to the straight strings is shown in Sheet 81.33. The joints are made square to the pitch, one riser distance from the springing with the two strings grooved and a loose tongue fitted. A counter cramp of hardwood strips mortised for wedges is used to pull the joint up tight.

To set out the string before bending, it is necessary to find the correct shape of the veneer. Sheet 81.34 shows the development of this for a stair having a quarter space of tapered steps. The plan is first drawn and the risers set out as shown. The positions of the faces of risers 4 and 7 are in the springing. A line

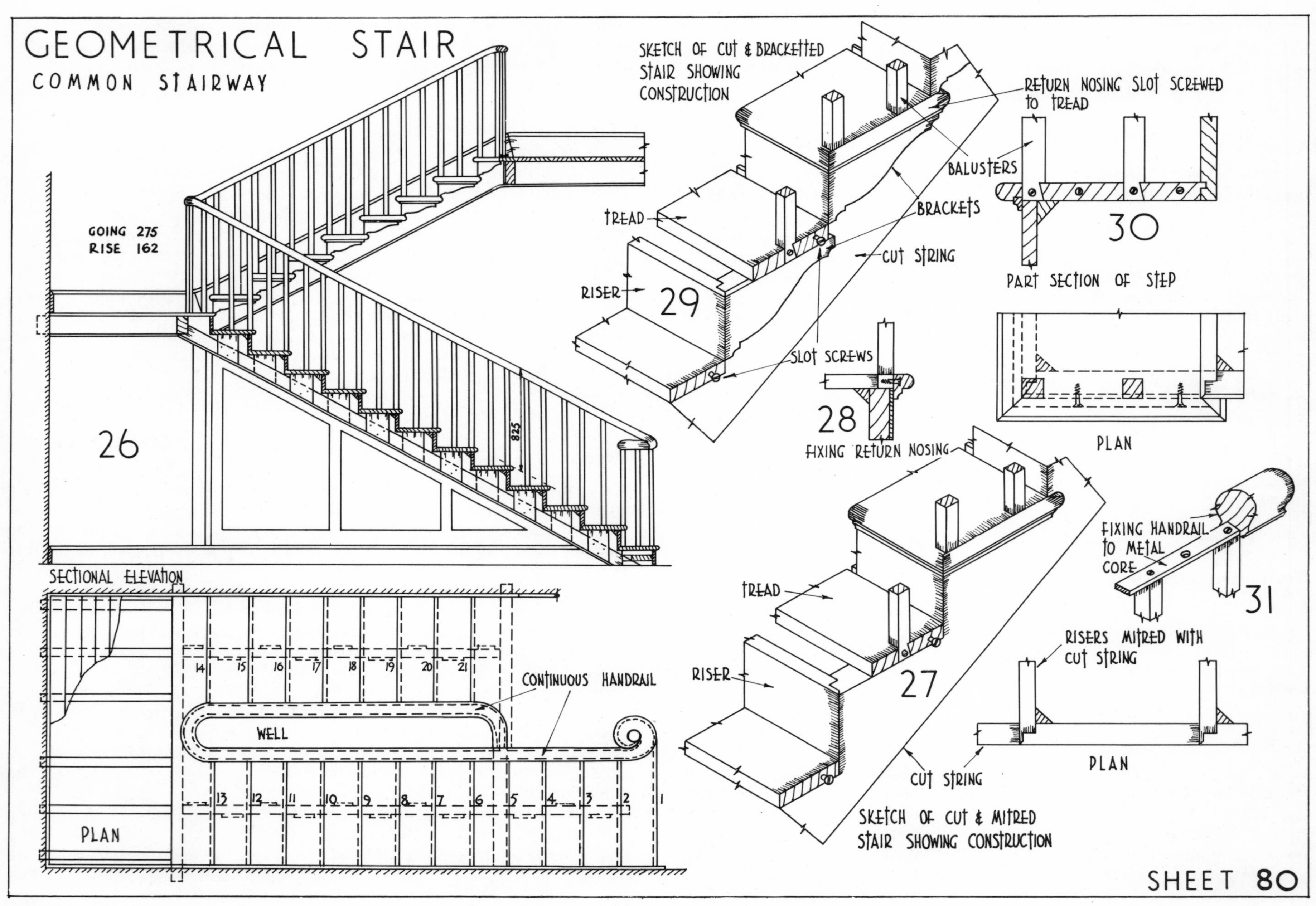
GEOMETRICAL STAIR
COMMON STAIRWAY
GOING 275
RISE 162
26
825
SECTIONAL ELEVATION
SKETCH OF CUT & BRACKETTED STAIR SHOWING CONSTRUCTION
RETURN NOSING SLOT SCREWED TO TREAD
BALUSTERS
BRACKETS
TREAD
CUT STRING
RISER
29
30
PART SECTION OF STEP
SLOT SCREWS
28
FIXING RETURN NOSING
PLAN
14
15
16
17
18
19
20
21
CONTINUOUS HANDRAIL
WELL
13
12
11
10
9
8
7
6
5
4
3
2
1
PLAN
TREAD
RISER
27
CUT STRING
SKETCH OF CUT & MITRED STAIR SHOWING CONSTRUCTION
FIXING HANDRAIL TO METAL CORE
31
RISERS MITRED WITH CUT STRING
PLAN
SHEET 80

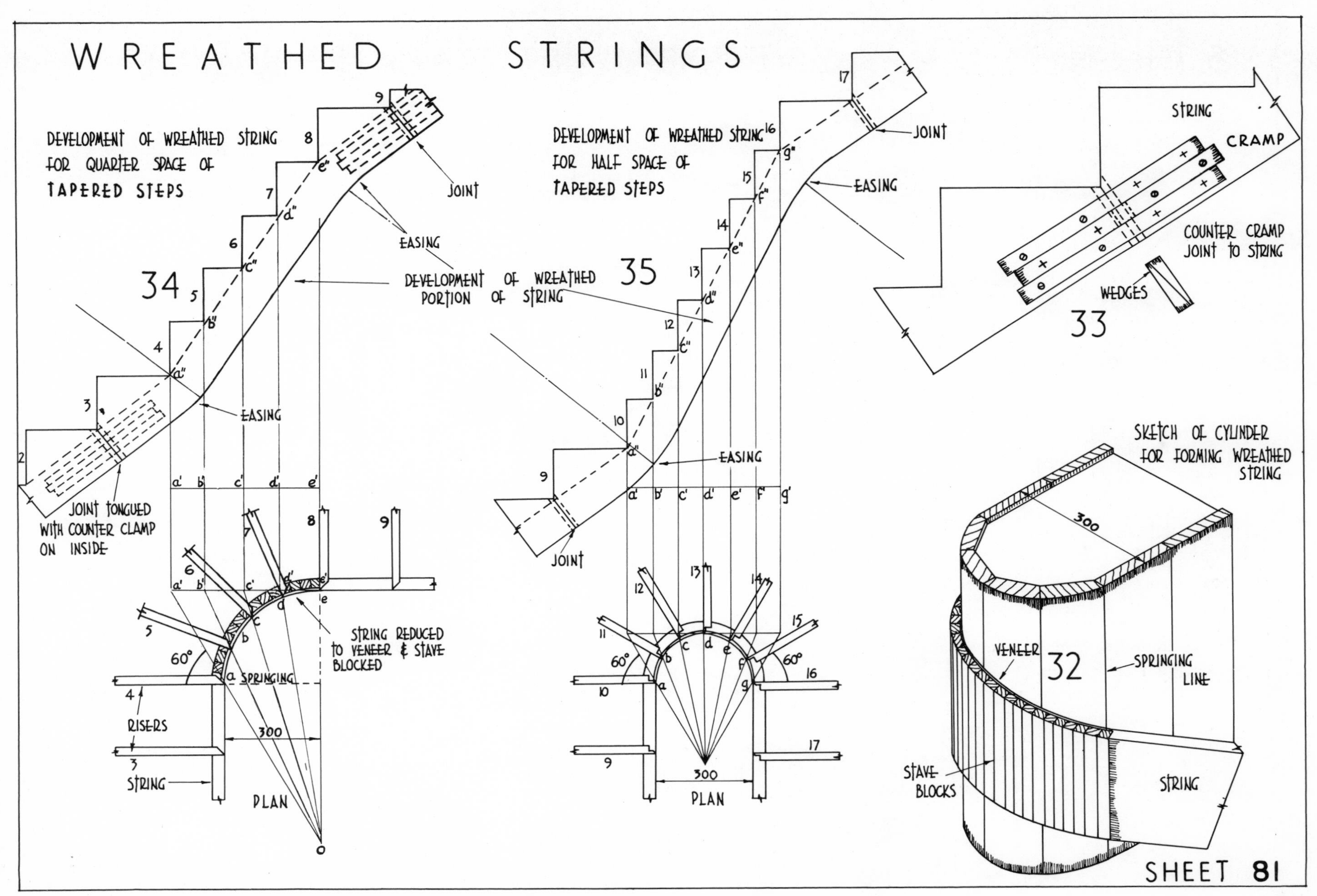
WREATHED STRINGS
DEVELOPMENT OF WREATHED STRING FOR QUARTER SPACE OF TAPERED STEPS
34
JOINT
EASING
DEVELOPMENT OF WREATHED PORTION OF STRING
JOINT TONGUED WITH COUNTER CLAMP ON INSIDE
STRING REDUCED TO VENEER & STAVE BLOCKED
SPRINGING
60°
300
RISERS
STRING
PLAN
DEVELOPMENT OF WREATHED STRING FOR HALF SPACE OF TAPERED STEPS
35
JOINT
EASING
PLAN
STRING
CRAMP
COUNTER CRAMP JOINT TO STRING
WEDGES
33
SKETCH OF CYLINDER FOR FORMING WREATHED STRING
VENEER
32
SPRINGING LINE
STAVE BLOCKS
SHEET 81

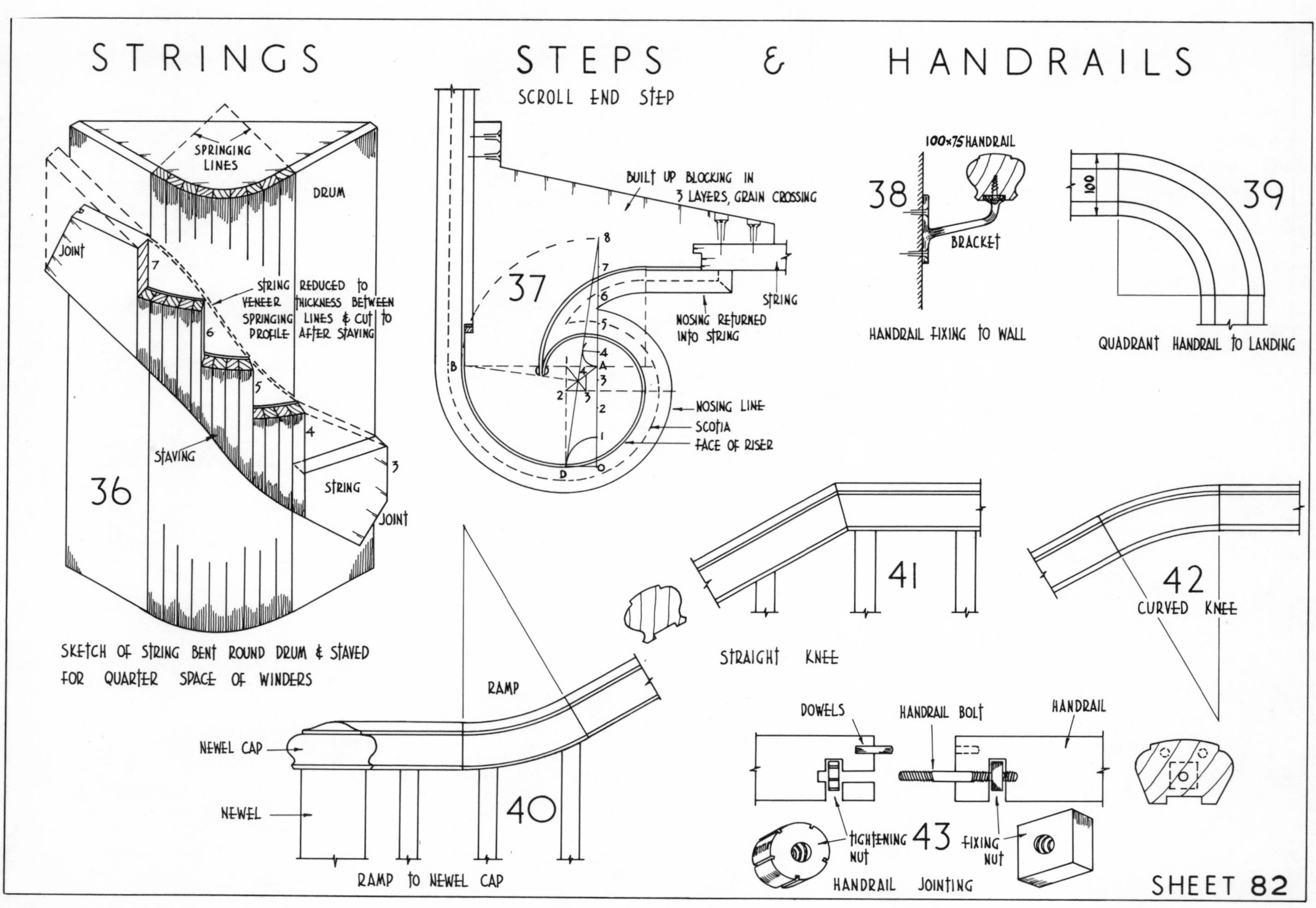
STRINGS STEPS & HANDRAILS
SCROLL END STEP
SPRINGING LINES
DRUM
JOINT
STRING REDUCED TO VENEER THICKNESS BETWEEN SPRINGING LINES & CUT TO PROFILE AFTER STAVING
STAVING
STRING
JOINT
36
SKETCH OF STRING BENT ROUND DRUM & STAVED FOR QUARTER SPACE OF WINDERS
BUILT UP BLOCKING IN 3 LAYERS, GRAIN CROSSING
37
STRING
NOSING RETURNED INTO STRING
NOSING LINE
SCOTIA
FACE OF RISER
100×75 HANDRAIL
38
BRACKET
HANDRAIL FIXING TO WALL
100
39
QUADRANT HANDRAIL TO LANDING
41
STRAIGHT KNEE
42
CURVED KNEE
RAMP
NEWEL CAP
NEWEL
40
RAMP TO NEWEL CAP
DOWELS
HANDRAIL BOLT
HANDRAIL
TIGHTENING NUT
43
FIXING NUT
HANDRAIL JOINTING
SHEET 82

drawn through the springing at point a, inclined at 60 degrees to intersect with a horizontal line brought out from d gives a′d′ which is the horizontal distance or stretch out of the string ad. Lines drawn through points b and c radiating to point o give the positions of risers 5 and 6 in the stretch out at b′ and c′.

At any convenient point above the plan, set out the elevation of the steps from points a′, b′, c′, d′. The lower edge of the string is set out parallel to the dotted line shown, touching the bottom corners of the steps with a margin of about 100 mm. This distance is necessary where a carriage is used and the underneath is to be plaster finish. Easings are required at the changes in direction to carry the string round in a sweeping curve.

The development of the wreathed portion of the string for a geometrical stair having a half-space of tapered steps is shown in Sheet 81.35 and is dealt with in a similar manner to the previous example.

Sheet 82.36 shows a sketch of the drum with the wreathed portion of the string previously developed in Sheet 81.34, prepared ready for jointing to the straight strings.

Curtail Step

The details of the shaped end of the step are shown in Sheet 82.37. As the outline of the step follows the outline of the handrail immediately above it, the geometrical setting out will be similar in both cases. 0–7 is the going of the step which is divided into seven equal parts. A further division is added giving 0–8. Set off one division at 90 degrees to 0–8 and join 8–D. Take the middle point of 0–8 which is 4 and draw an arc tangential to the diameter at A. Draw a line at right-angles from A and cut this at point 1 with a parallel to the diameter from D. The quadrant shown dotted is drawn from A as centre to cut a horizontal line from A in point B. Draw B at right-angles to 8–D and through the intersection draw lines from A and 1. The centre 2 will be found on the line from A and the centre 3 on the line from 1. A vertical line drawn from 3 to cut the diagonal from 2 gives the centre 4. The step outline is next drawn in from the centres, composed of a series of quadrants with their radii diminishing. The first quadrant shown dotted is continued, being drawn in from the centres obtained. The handrail scroll is set out from the centre line of the straight rail maintaining a parallel margin from the edge of the nosing.

The block is built up in laminations with the grain crossing to prevent warping, and cut to shape with sufficient material being allowed for fixing to the riser and string. The riser is then reduced to veneer thickness, well wetted and fixed at the end by wedging in the groove. The block is well glued while laying the veneer into position and the second pair of wedges driven in to pull the veneer tight and the screws put in. It will be seen that the string is tongued to the block and this also veneered on the internal curve.

The step is completed by screwing the scotia board to the block, and the riser and the tread through the scotia board and block from the underside.

Handrails

The handrails for newel stairs are usually in straight lengths housed and tenoned into the faces of the newel posts. In cheaper work they are tenoned but not housed. The balusters are housed into the underside of the handrail, although in ordinary work they are usually butted and nailed.

Sheet 82.38 shows the fixing of a handrail to a wall using brackets welded to the metal core.

A quadrant handrail to a landing is shown in the plan, Sheet 82.39.

Sheet 82.40 shows the finish of the handrail with the newel post which is fitted with a moulded capping the same section as the handrail. This allows a proper intersection between the rail and cap at the joint. To bring the two members level at the intersection a ramp is used which is an easing jointed between the short straight rail and the inclined one.

The jointing between the straight and inclined rails at the top of a flight where no newel is used is shown in Sheet 82.41. This is called a straight knee. An alternative finish is the curved knee shown in Sheet 82.42.

The jointing together of sections of handrails is made using a handrail bolt and dowels shown in Sheet 82.43.

Scrolls

In geometrical stairs the handrail usually terminates in the form of a scroll. These scrolls are either vertical, when they are sometimes known as monkey's tails, or horizontal as shown at the foot of the stairs in Sheet 80.26. As has already been stated, scrolls are usually composed of a series of quadrants with their radii diminishing and to set these out, various methods are used.

In Sheet 83, figs. 44 and 45 the scrolls are set out in a similar way, these have been detailed for comparison of the finished outline. In Sheet 83.44 the overall width of the scroll is divided into nine equal parts, whereas in Sheet 83.45, eight divisions are used. The conclusion between the two examples is that if the angle 90b is altered by dividing 09 into more or less parts than eight, and the same setting out used, the volute approaches the centre at quicker or slower rate, according to the proportion of 9b to 09 in the figure. If 9b is less than one-eighth the diameter, the approach is slower, and vice versa.

Sheet 83.46 shows a further method of setting out a scroll where various

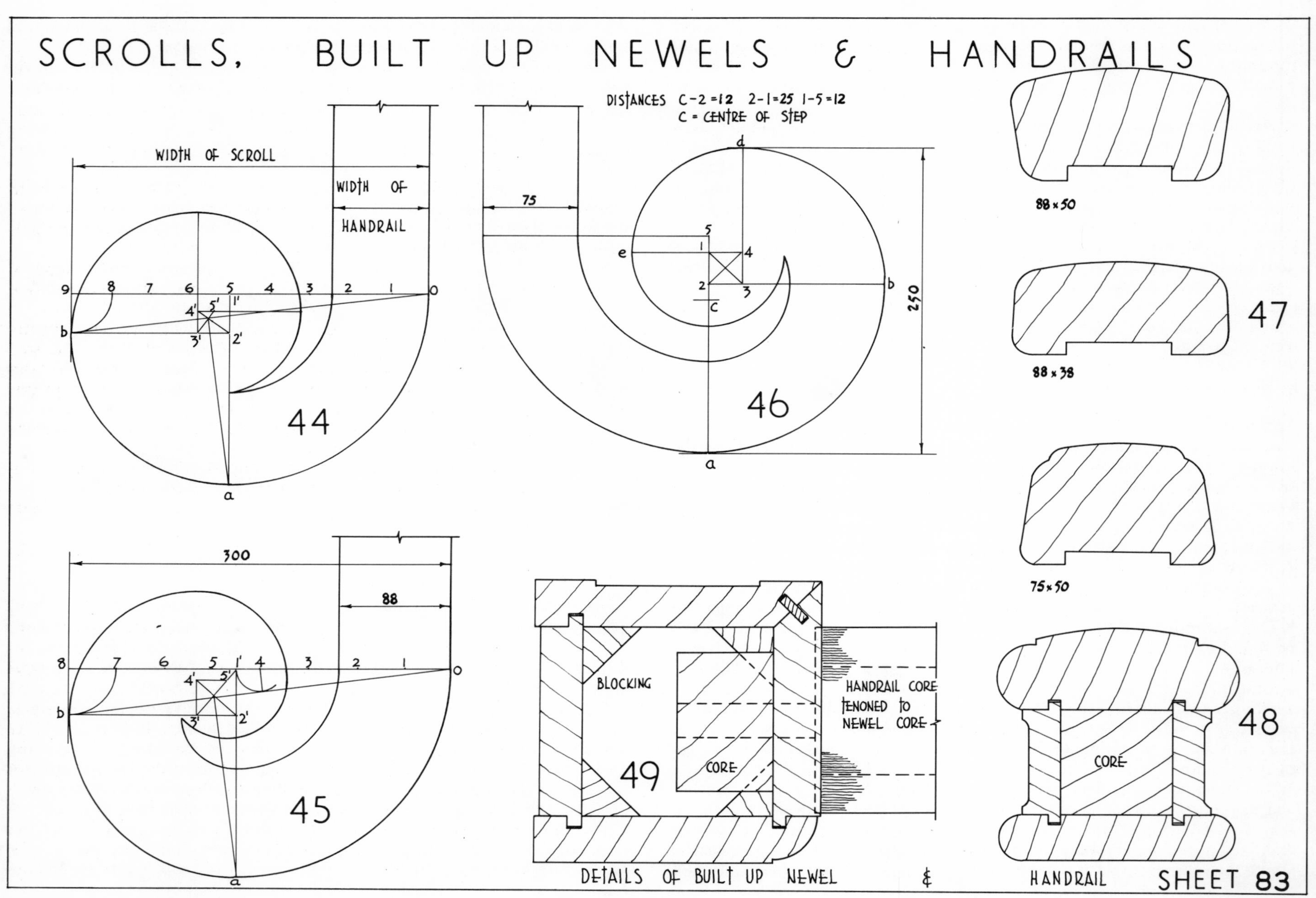
SCROLLS, BUILT UP NEWELS & HANDRAILS
WIDTH OF SCROLL
WIDTH OF HANDRAIL
9 8 7 6 5 4 3 2 1 0
4' 5' 1'
3' 2'
b
a
44
75
DISTANCES C-2 = 12 2-1 = 25 1-5 = 12
C = CENTRE OF STEP
d
5
1
e
4
2
3
c
b
250
46
a
88 × 50
88 × 38
47
75 × 50
300
88
8 7 6 5 1' 4 3 2 1 0
4' 5'
b
3' 2'
45
a
BLOCKING
CORE
49
HANDRAIL CORE TENONED TO NEWEL CORE
CORE
48
DETAILS OF BUILT UP NEWEL & HANDRAIL
SHEET 83

proportions are adopted to determine the centres. C is the centre of the step, distances C–2 equal to 12 mm, 2–1 equal to 25 mm, and 1–5 equal to 12 mm. The centres 3 and 4 lie on diagonals drawn from 1 and 2 as shown.

Sheet 83.47 shows three sections of moulded handrails with a built-up handrail section shown in Sheet 83.48. A softwood core is faced with moulded top, sides and bottom in hardwood tongued and grooved together as shown. As with the built-up newel, Sheet 83.49, this method conserves a certain amount of hardwood, and in work where newels larger than 100 mm square are required this method is the best, on account of the possibility of splitting and warping in large solid members. Alternative face finishings and corner jointing to the newel are shown.

The jointing between the built-up newel and rail is by mortising the newel core to receive the tenon worked on the core of the handrail. Similar jointing is used between the string and the newel.

Open Tread Stairs

In modern buildings such as shops, multi-stores, offices and showrooms, stairs are much more contemporary in design than those which have been dealt with so far. The whole treatment is simple and effective, where elaborately moulded nosings, cappings, newels and handrails are avoided. In many cases the closed strings, employed in newel stairs to receive the steps, give way to an open design of stair where risers too are often left out with the support provided in various other ways.

Examples of this modern treatment are to be found where a combination of materials is used, i.e., concrete stair with timber finishings; open timber treads with steel stringer supports; open timber treads with laminated beam supports, metal balusters, and either timber or extruded plastic handrails.

In most cases the work of the joiner is much simpler than hitherto, although the planning is the same for whatever form the stair may take. Various finishings to balustrades include timber panelling, open metal balusters, etched plate glass, wrought iron, aluminium alloy castings and woven ribbon wirework.

The stairs in large stores and showrooms are usually wide to allow people to pass in both directions, these may be divided down the centre and up to 1.800 m or more in width, and for this reason the treads of hardwood are at least 38 mm thick. Iroko, agba, afrormosia, oak, teak and meranti are suitable timbers for this type of work.

A suitable finish to a concrete stair is shown in Sheet 84.50. The outer faces of the stair are faced with a hardwood string cut to the shape of the steps and left projecting 19 mm. This is fixed by screwing through the face and pelleting. Non-slip stairtreads are fitted between the strings and screwed to the concrete treads, the riser and the space between the back of the stairtread and the riser filled in with linoleum or cork, using impact adhesive.

The stair shown in Sheet 84.51 has laminated stringers cut to receive the wide 38 mm thick treads and inclined risers which are fixed by screwing and pelleting to the stringers. It will be seen in the part elevation of the same figure that the risers finish flush to the stringer with the treads projecting some 100 mm with the corners rounded. The bases of the metal balusters are housed and fixed to the treads as shown.

Sheet 84.52 shows a portion of a stair having steel channel stringers with support for the treads obtained from 50 mm by 6 mm mild steel straps. These are welded to the channel and housed and fixed to the underside of the treads as shown in the detail, Sheet 84.52.

Sheet 84.53 shows a stair with a laminated stringer, the tread supports, which are from 200 mm by 50 mm hardwood, are formed with the lamination of the stringer.

Sheet 84.54 shows a part section of a stair with steel channel stringers and welded steel angle brackets, faced with timber treads and risers. The step detail, Sheet 84.55, shows the hardwood riser tongued into the treads on both edges with fixing by screwing through the angle bracket. Blocking pieces, coach screwed to the channel, provide for the fixing of the hardwood cut string which is screwed and pelleted from the face. The treads project over the strings with metal balusters fixed to them.

A further type of modern laminated stair is designed using shaped laminated beams, built up of 19 mm softwood lamina as central supports to laminated treads, fixed to mild steel brackets which are welded to a steel plate housed in the top edge of the central beams.

Handrail Wreaths

Geometrical or continuous stairs are so called because the setting out of the strings and handrails is based upon geometrical principles. The handrails and one or both strings run continuously from top to bottom of the successive flights.

In the making of a continuous handrail the chief difficulty is in the wreaths, where the rail is of double curvature. This requires the joiner to produce a templet, or face mould, along with bevels which may be applied to the plank so that the wreath may be cut out, squared, and jointed correctly with straight rails. Many arrangements embracing handrails of double curvature are met, a selection of the more common forms only are dealt with here.

There are several systems of handrailing, all giving satisfactory results, but each requires a knowledge of solid geometry. The system described in this chapter is that known as the Square Cut Tangent System. In this method the

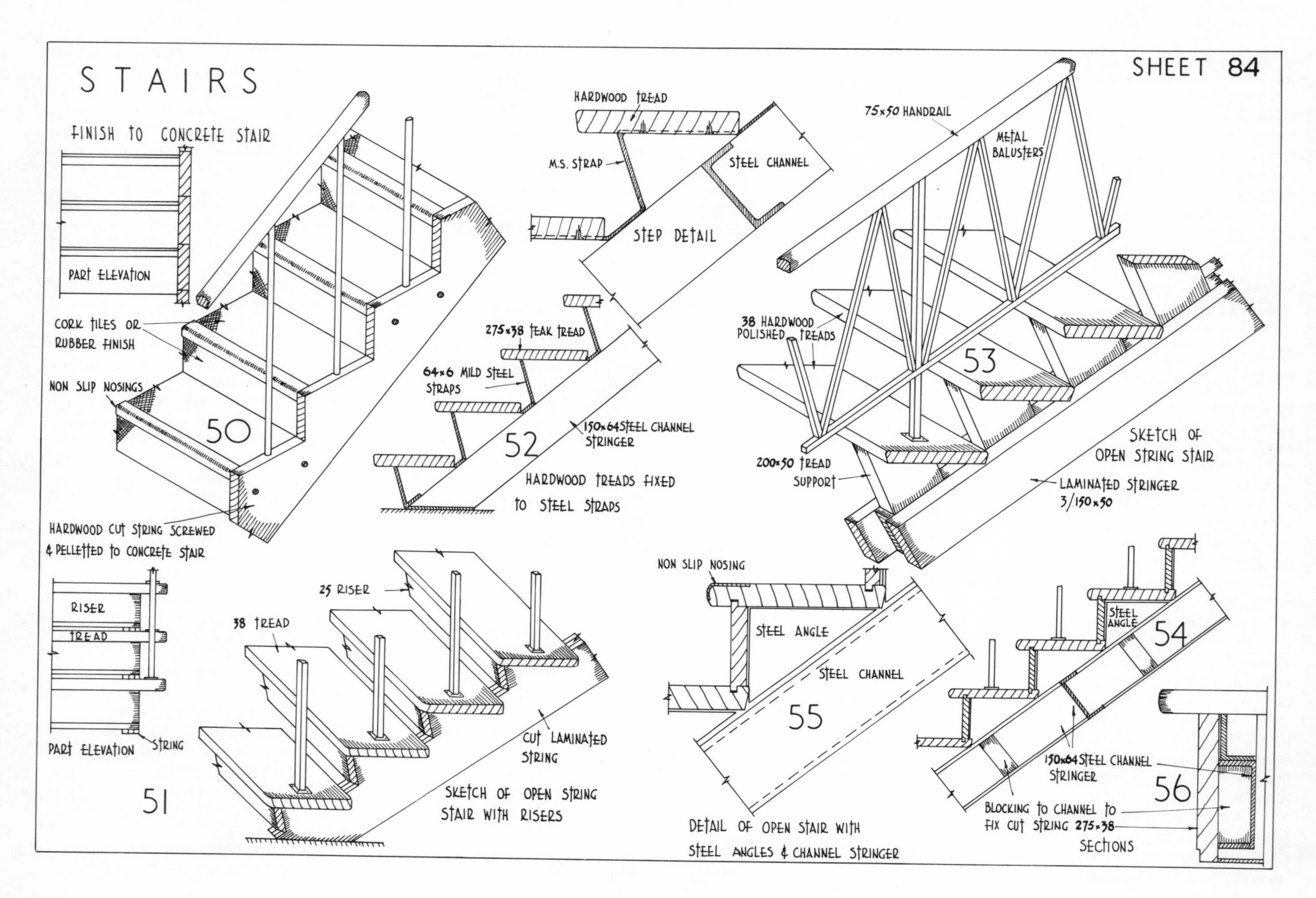

STAIRS
SHEET 84
FINISH TO CONCRETE STAIR
PART ELEVATION
CORK TILES OR RUBBER FINISH
NON SLIP NOSINGS
50
HARDWOOD CUT STRING SCREWED & PELLETTED TO CONCRETE STAIR
HARDWOOD TREAD
M.S. STRAP
STEEL CHANNEL
STEP DETAIL
275×38 TEAK TREAD
64×6 MILD STEEL STRAPS
150×64 STEEL CHANNEL STRINGER
52
HARDWOOD TREADS FIXED TO STEEL STRAPS
75×50 HANDRAIL
METAL BALUSTERS
38 HARDWOOD POLISHED TREADS
53
200×50 TREAD SUPPORT
SKETCH OF OPEN STRING STAIR
LAMINATED STRINGER 3/150×50
RISER
TREAD
PART ELEVATION
STRING
51
25 RISER
38 TREAD
CUT LAMINATED STRING
SKETCH OF OPEN STRING STAIR WITH RISERS
NON SLIP NOSING
STEEL ANGLE
STEEL CHANNEL
55
DETAIL OF OPEN STAIR WITH STEEL ANGLES & CHANNEL STRINGER
STEEL ANGLE
54
150×64 STEEL CHANNEL STRINGER
56
BLOCKING TO CHANNEL TO FIX CUT STRING 275×38
SECTIONS

rail may be treated as a part of a hollow cylinder whose base is the plan of the rail, the edges of the squared wreath forming parts of the vertical surfaces of the cylinder.

The plan of the centre line of the curved portion of the rail starts at one corner of the plan of a four-sided prism and finishes at the opposite corner as shown in Sheet 85.1. The inclined plane, which is assumed to cut the cylinder, is determined from the plan and vertical stretch out of the stair. The section of the cylinder made by the inclined plane is elliptical and forms the pattern for cutting out the wreathed portion of the rail. This pattern is called the face mould. In the sketch of the block, the plan of the centre line of the wreath and its intersection with the portions of straight rails is shown. The upper rail is level and the lower one inclined. The tangent planes are represented by the vertical surfaces AB and BD of the block.

Sheet 85.2 shows the plan of the wreath connecting a straight flight and a level landing with the top riser placed at the springing. The plan of the rail and prism A–B–D–C is drawn, the tangents AB and BD being the centre lines of the two rails produced. The prism represents the block shown in the sketch, Sheet 85.1. Next, project the riser lines upwards vertically and construct one or two steps to give the pitch of the stair. This is also the inclination of the top surface of the block or prism. Project the plan and centre line of the straight rail upwards to cut the pitch line in point E. The rectangular section of the rail is drawn from E with the lower surface horizontal. A line drawn parallel to the nosing line passing through point F is the top face of the plank. The bottom face of the plank is taken as the nosing line. The thickness of the plank is the distance between the two surfaces and the bevel is equal to the pitch of the straight flight.

Draw the VT and at points from where the lines projected from the plan cut this line, draw out at right-angles. From A′ and B′ set off the distances AC and BD to give points C′D′ in the development. A′B′D′C′ is the development of the inclined surface of the block, C′D′ is the semi-major axis and C′A′ the semi-minor axis. The width of the face mould at D′ is obtained from the sides of the rectangular rail section projected on to the pitch line. The width of the face mould at A′ is equal to the width of the rail shown in plan.

Draw the curves of the face mould with the trammel, the shank may be made any convenient length.

The sketch of a prism, Sheet 85.3, cut by an oblique plane shows the bevels or angles which the cutting plane makes with the vertical faces of the prism. The upper pitch is produced to meet the ground line at K, through this point and B draw the horizontal trace to find the angle between the plane and this face. With A as centre, draw an arc tangential to the pitch line to cut the perpendicular AE, draw a line from the intersection to B to give the first bevel. The top face of the prism is contained in a second plane. Produce B1 to meet edge FE produced, then with E as centre, draw an arc tangential to B1 produced cutting the perpendicular AE. The dotted line drawn from this point to H gives the second required bevel.

Sheet 85.4 shows the above prism drawn in plan and elevation with the quadrant BD the centre line of the rail. AB and AD are the tangent lines. The bevels are obtained as already described for Sheet 85.2. The line X′–Y′ is drawn at right-angles to the horizontal trace through the centre C so that a line which shows the true inclination will also contain the major axis of the elliptical section. Lines from A, C and D are drawn parallel to HT, and on a line from D set up the height taken from elevation. Draw 1, 3′ which is the true inclination of the plane. From 1, 2′, C′, 3′ draw lines at right-angles and step off distances 1B′ equal to 1B, 2′A′ equal to 2A, C′ minor to C minor, 3′D′ to 3D. Join A′B′C′D′, this is the section of the prism and to be correct A′B′ must equal B″L and A′D′ equal to HL. The section of the cylinder is drawn by continuing the centre line of the rail in the plan to meet X′Y′ at 4. Draw 4–4′ parallel to 3–3′ then from C′ to 4′ is the semi-major axis, the semi-minor being drawn, the semi-ellipse is completed to pass through B′ and D′.

Sheet 85, figs. 5, 6, 7 and 8 show the various stages of marking and shaping the wreath.

The wreath is cut from a plank the required thickness to the outline of the face mould, with additional material left on at the edges if required. The tangent lines and springing lines are marked on both faces of the face mould. After truing up the faces of the wreath the face mould is applied and the tangent lines marked as shown in Sheet 85.5 along with the joint lines square to the tangents and to the face of the plank. The positions for the handrail bolts are next accurately marked.

Sheet 85.6 shows the marking of the wreath for bevelling, which requires the bevel to be marked at the end, and the face mould slid along so that the tangent lines lie exactly on the new tangent lines drawn from the twist bevel applied at the centre of the joint. The mould is then marked on the face of the plank and repeated on the lower face. The material may be cut away with a bow saw, gouge and chisel and finished by spokeshave. The bevelling of the wreath may also be done on the bandsaw.

The top and bottom surfaces are next prepared, being marked as in Sheet 85.7 with free-hand falling lines on the edge of the material kept to the thickness of the handrail.

Either the top or bottom of the wreath may be finished first, Sheet 85.8, but where a metal core is fitted to the lower face, this must be finished first. The wreath should be bolted to the straight rails before finishing.

Handrail Wreath for a Half-space Landing

Sheet 86.9 shows the plan of a staircase consisting of two straight flights connected by a half-space landing. The straight flights do not finish at the

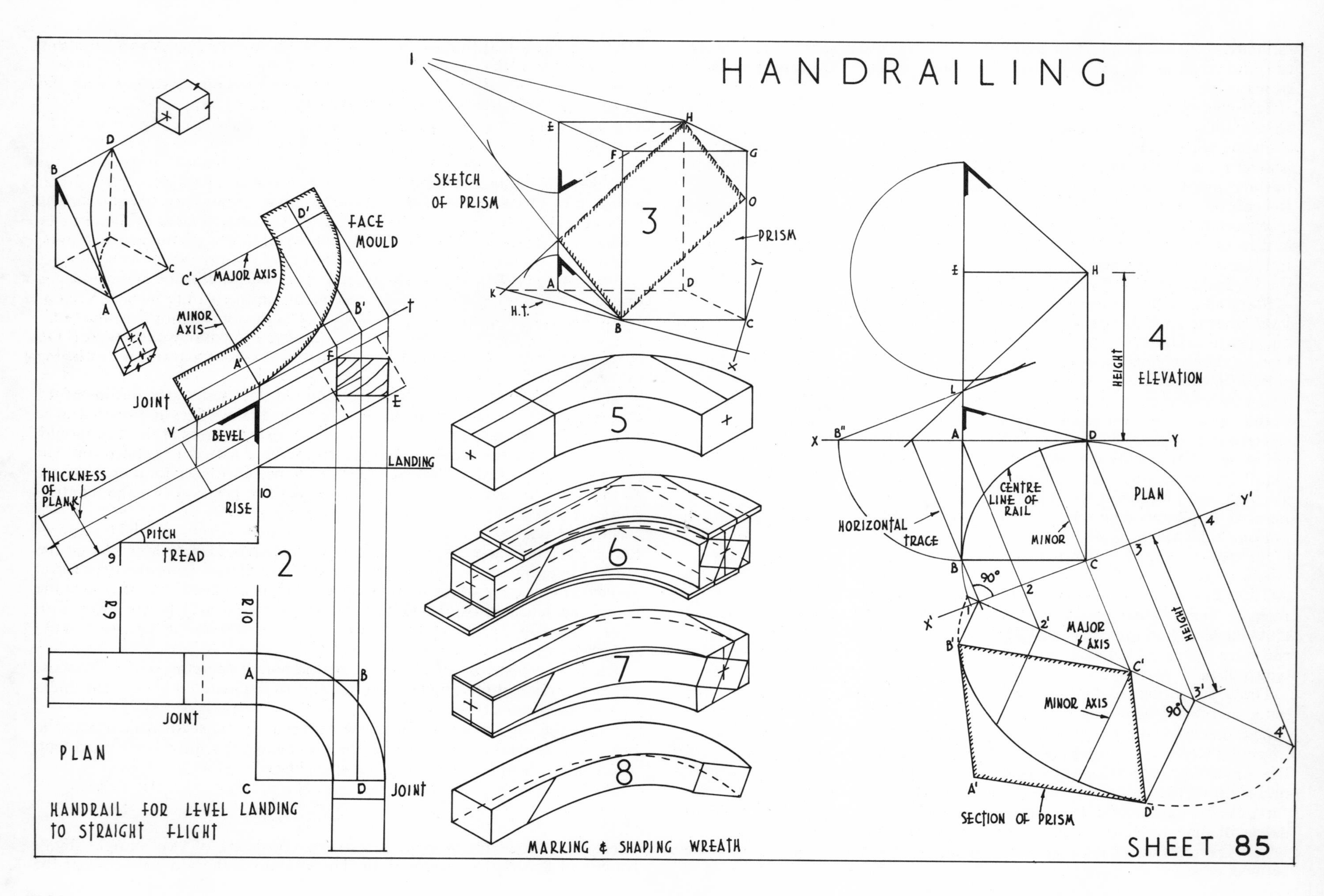
HANDRAILING
SKETCH OF PRISM
PRISM
H.T.
FACE MOULD
MAJOR AXIS
MINOR AXIS
JOINT
BEVEL
THICKNESS OF PLANK
PITCH
RISE
TREAD
LANDING
PLAN
JOINT
HANDRAIL FOR LEVEL LANDING TO STRAIGHT FLIGHT
MARKING & SHAPING WREATH
HEIGHT
ELEVATION
HORIZONTAL TRACE
CENTRE LINE OF RAIL
MINOR
PLAN
MAJOR AXIS
MINOR AXIS
SECTION OF PRISM
SHEET 85

springings, the position of the landing risers 8 and 9 should be noted, being placed at half a going from the tangents B and D. The centre line of the wreath is enclosed by the tangents A, B, C, D, E, found by producing the centre lines of the straight rails and connecting them by a line at right-angles, touching the wreath at its centre and the centre of the well.

The wreath should be made in two parts, with the joint made at C to avoid cross grain and to economize in material, both parts being alike. The springing lines are drawn through the centre O at points A and E to complete the plan of the containing prisms. Risers 6, 7, 10 and 11 are drawn on each side to obtain the going and the development of the tangents in the plan, making the points A, B, C, D, E on a line B, C, D produced. A horizontal line representing the landing is next drawn in any convenient position, and the risers 8 and 9 at the same distance from the lines A and E that they are from the corresponding marked points in the plan. Draw in the treads and risers at 175 mm rise and 230 mm going to determine the pitch of the stair. The nosing lines and the depth of the handrail are drawn with the centre lines of the rails cutting the tangents B and D in points F and G. These are the elevations of the tangents of the section of the prism that contains the elliptical section of the cylinder represented by the plan of the wreath. The joint lines for the straight rails are drawn square to the centre line and about 75 mm from the springing.

The face mould shown in Sheet 86.10 is obtained in a similar way to Sheet 85.2. The true length of the diagonal AC must first be found. With centre C in the plan and radius CA draw the arc to cut the horizontal line drawn out from C. A perpendicular to cut a horizontal brought out from the intersection of the springing line with the centre line of the rail at H gives A″. Join A″ to C″ which is the true length of the diagonal AC.

The diagonal A′C′ is drawn and A′B′ is made equal to HF and C′B′ equal to FC″. The resulting triangle is the true shape of the section of the prism A, B, C, A in the plan, and the lines A′B′ and B′C′ are the tangents of the face mould which are square to each other. Point O, which is the centre of the elliptical curves, is found by drawing from A′ and C′ parallel to the two tangents.

The width of the rail in the plan is marked off equally on each side of the tangent at A′ in the face mould, and the shank is drawn with the joint made at right-angles to the tangent. The width of the face mould at the other end is found by stepping off the width of the handrail on each side of the tangent HF. Lines drawn parallel to the tangent to cut the perpendicular from B give the width of the mould at C′. The semi-major and semi-minor axes of the curves being determined, the curves are now drawn in by trammel. As both halves of the wreath are alike, only one face mould is required.

The twist bevel is given by the angle which the tangent HF makes with the vertical shown.

The thickness of the plank from which to cut the wreath is shown in Sheet 86.11. The twist bevel is set out and the rectangle circumscribing the section of the rail is drawn. The thickness is shown between the parallel lines through the top and bottom edges of the squared wreath.

Handrail Wreath for a Quarter-space of Tapered Steps

Sheet 86.12 shows the plan of a staircase consisting of two straight flights connected by a quarter-space of tapered steps. The centre line of the wreath is enclosed by the tangents AB, and BC found by producing the centre lines of the straight rails as in the previous examples.

The positions of the risers from the plan are drawn into the stretch-out by swinging the points of intersection with the tangents in the plan as shown. The pitch of the upper flight gives the inclination of the edge of the prism BC. The lower rail is drawn to rest on the nosings. The lower tangent A′B″ is drawn from B″ to intersect the centre line of the lower rail so that the handrail is kept slightly higher over the tapered steps than over the straight stair. A handrail which is lower at the tapered steps is of great danger to the users of the stairs and cannot be permitted. The lower rail must therefore be ramped. An easing is used on the straight rail to join the different pitches with a graceful curve. The curves employed in easings are purely a matter of taste and in the workshop can be drawn in, using a flexible lath. The lines C″B″ and B″A′ are the lengths and inclinations of the two sides of the prism ABC of the section containing the centre line of the wreath. These will also provide the tangents of the face mould.

The length of the diagonal AC is required, and found by turning AC in the plan with C as centre to cut BC produced. This point is projected vertically to intersect a horizontal line brought out from A′ in A″. Join A″C″ which is the true length of the diagonal AC.

AC, Sheet 86.13, is drawn equal to A″C″ and AB and BC equal to A′B″ and B″C″, and through A and C draw lines parallel to AB and BC which will intersect at O, the centre of the elliptical curves. The tangents are next produced beyond the springings with the joints drawn square to them. The average length for the shank is 75 mm worked on the wreath for convenience of preparing for the handrail bolt in straight wood, and also because the change of curve to straight is easier to work in the solid than on separate pieces.

The widths at the ends of the face mould and the two bevels are shown in Sheet 86.12. Both ends in this case are wider than the handrail as shown in the plan.

The top bevel is found by producing the tangent BC to A″, Sheet 86.13. A line drawn at right-angles from A gives AA″. This distance is taken and struck

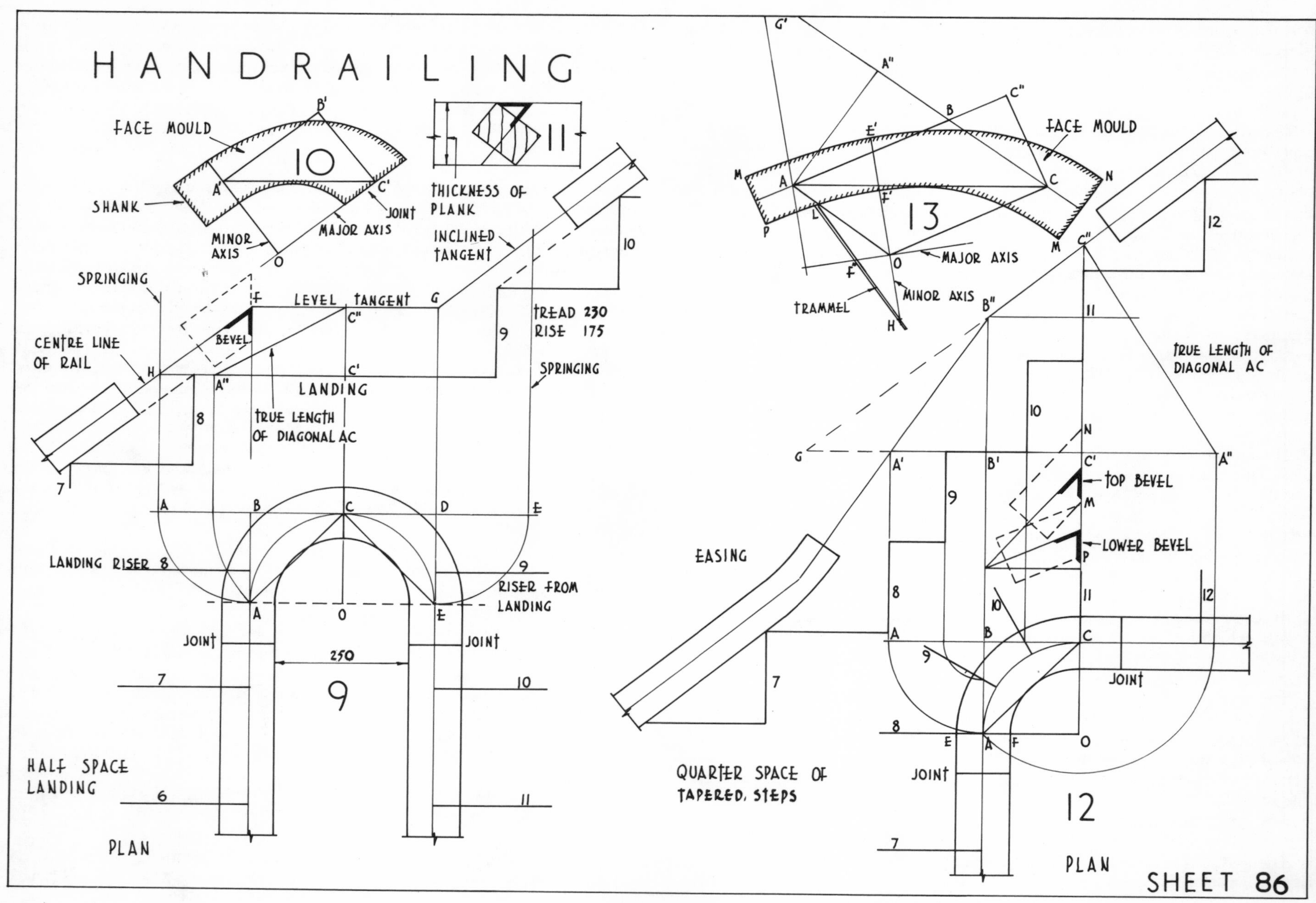
HANDRAILING
FACE MOULD
B'
10
A'
C'
SHANK
JOINT
MINOR AXIS
MAJOR AXIS
O
11
THICKNESS OF PLANK
INCLINED TANGENT
SPRINGING
F
LEVEL TANGENT
G
C''
BEVEL
CENTRE LINE OF RAIL
H
A''
C'
LANDING
TRUE LENGTH OF DIAGONAL AC
TREAD 230
RISE 175
SPRINGING
LANDING RISER 8
RISER FROM LANDING
JOINT
250
9
HALF SPACE LANDING
PLAN
G'
A''
C''
B
E'
FACE MOULD
M
A
C
N
F'
P
L
13
MAJOR AXIS
TRAMMEL
MINOR AXIS
H
B''
TRUE LENGTH OF DIAGONAL AC
TOP BEVEL
LOWER BEVEL
EASING
QUARTER SPACE OF TAPERED STEPS
12
PLAN

SHEET 86

between the elevation of the tangents BC to give the top bevel. The width of the handrail drawn on each side and parallel to the tangent gives the width of the face mould NM. The lower bevel and width of mould is obtained in a similar way.

As the axes of the elliptical curves are not in line with the springing lines due to section having two pitches, the direction of these is to be found. The tangent B″C″, Sheet 86.12, is produced to a horizontal taken out from A′ in point G. The length B″G is added to the tangent BC in Fig. 13, to give point G′. A line from G′ through A is the horizontal trace, and a line drawn parallel to this from O will be the direction of the minor axis with the major axis drawn at right-angles to it. The length of the minor axis OE′ is taken from the centre O in plan to the inside and outside faces of the rail, OF and OE. The trammel rod is laid across the axis from one of the given points with the point F″ resting in the direction of the major axis; then a mark made at point H where the rod passes the minor axis, will be the length of the major axis to strike that particular curve of the face mould.

The wreath, having been marked and squared, Sheet 85, figs. 5, 6, 7 and 8, now requires moulding. For this purpose a variety of tools are used depending on the finished section or shape of the rail. These will include straight and bent chisels and gouges, quirk and bead routers, thumb planes of various sections, shaped scrapers, bent files and shaped cork rubbers. Not all of these tools will be required for wreaths to match the modern sections of handrails shown in Sheet 83.47, but for more traditional moulded wreaths, the bulk of the waste is removed with relief grooves worked with a quirk router followed by gouges or chisels, the surfaces are then worked up with thumb planes, scrapers and various glass-papers. Cross-grain may be shaped up using the files and before final papering up to finish off the joints, the wreath should be bolted to the straight rail.

The handrail bolt used in jointing handrails is shown in Sheet 82.43. This is a bolt threaded for a nut at each end, one nut being square the other circular; the latter has grooves spaced round it to allow tightening with a handrail punch. Both nuts are mortised into the underside of the rail, the slot for the square nut being cut small enough to prevent the nut turning while the bolt is being turned into it. The slot for the circular nut is cut large enough to take a washer and the handrail punch for tightening up.

To prevent the rails turning about the bolt at the joint, two short dowels are used positioned as shown.

After the joint has been made the two slots are pelleted with material of the same kind as that from which the rail is made, the grain running with the grain of the rail.

Chapter Nine: Joinery

Wall Panelling—1

The traditional type of panelling for the covering of walls in large houses and public buildings consists of mortised and tenoned framing prepared to receive well-figured and finely toned panels, combined with solid mouldings. The height varies considerably. Panelling which is about window bottom or chairback height is termed dado panelling. Panelling may be taken to the height of the doors in a room, or the whole of the walls may be covered to cornice height.

As this type of panelling still exists in older buildings, work on restoration, or in additions and extensions, requires to be done, and for this reason has been included here, followed by more modern designs and treatments.

Sheet 87.1 shows the part elevation and section of the lower portion of traditional wall panelling. The height of the moulded dado is 900 mm and represents the lower portion of the panelling in the key elevation, Sheet 87.2. This shows the arrangement of panelling suitable for a board-room where almost the full height of the wall is covered.

The framing of this panelling would be prepared in the shop and transported to the site for fixing. Special considerations should be given to work of this size from the point of view of handling, transporting and getting the framing upstairs and through doors. This particular design lends itself to the dividing up of the frames. It will be seen in the section that at dado height the framing has been split, the dado rail breaking the joint between the two sections. At this point, softwood lining-out pieces are shown where they are covered by the dado rail. This conserves a certain amount of hardwood and reduces the widths of the rails, thereby counteracting warping.

The construction of the panelling is similar to that for doors. Solid moulded framing is shown grooved for solid raised panels, the work is only finished on the face side, the back being left from the saw. Work of this type is secret-fixed wherever possible, but first it is necessary to fix a system of rough grounds. These are 19 mm thick, being the thickness of the plaster and placed behind the frame members so as to give a solid bearing to the work. Sheet 87.3 shows a typical arrangement of the groundwork where framed grounds are used at the doorway. These are fixed to plugs in the walls which must be plumbed and squared most carefully, particularly at openings and angles.

The fixing of the work to the grounds, whether by nailing or screwing, is done through the face where the fixing is covered by the skirting or dado. The fixing of the skirting is best done by first screwing a moulded fillet to the floor to receive the bottom edge, the top edge being pinned through the square or screwed through the face to skirting blocks or soldiers. Where screws cannot be hidden, the heads must be let into the face of the panelling and pelleted. Other forms of secret fixing will be dealt with later.

Sheet 87.4 shows the method of fixing panelled framing at the internal angles. The edge of one stile is tongued to fit a groove made in the stile of the other. The jointing of the skirting at all internal angles, Sheet 87.5, shows the skirtings stop grooved and tongued with the top mouldings scribed. All external angles are mitred.

Sheet 88.6 shows the finish at the cornice. This is the horizontal member fixed to the framing and is usually built up, with provision for ventilation at the back of the panelling as shown.

If concealed lighting is used at cornice level, metal brackets screwed to the moulded member and groundwork are used, Sheet 88.7. As a considerable amount of heat is given off by this type of lighting, care should be taken with the mitres and other jointing. Canvas glued on the back side of the members helps to strengthen the joints and prevents light from showing through in the event of shrinkage.

Two forms of pilasters used for joining up sections of wall panelling and as a decorative feature are shown in Sheet 88.8. The concealed fixing of the second form of pilaster is clearly shown in the sketch, Sheet 88.9, and requires no further explanation.

An alternative secret fixing is shown in Sheet 88.10, again illustrated from the rear and secret fixing, using slot screws shown in Sheet 88.11.

Wall Panelling—2

The introduction of new materials and new applications has resulted in a new conception of joinery finishings. The modern trend is for solid mouldings with veneered plywood or laminboards, from 10 mm thick upwards. Sheet 89.12 shows an example of this type where veneered panels are used with bolection mouldings and finished with a moulded capping.

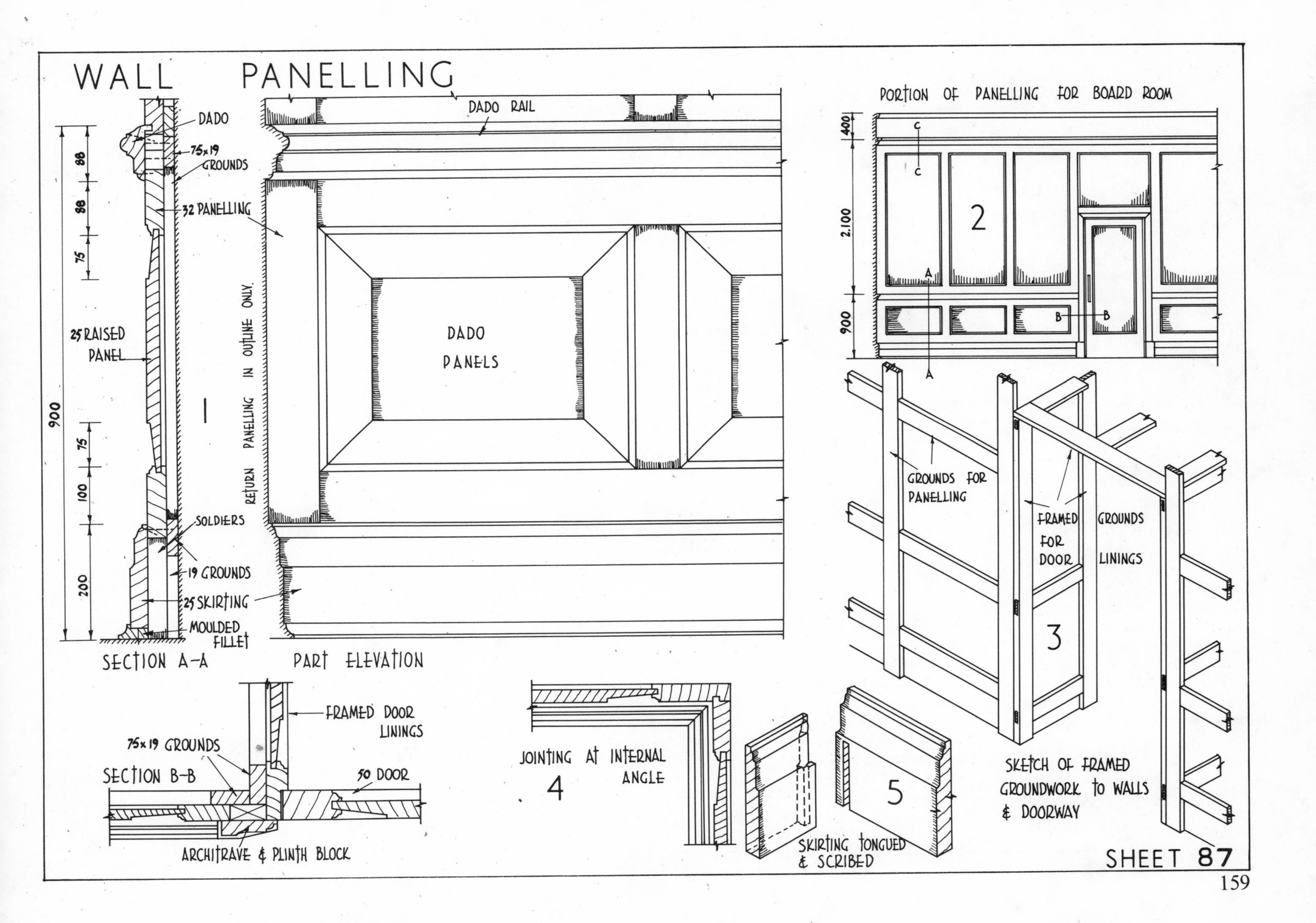
WALL PANELLING
DADO
75×19 GROUNDS
32 PANELLING
25 RAISED PANEL
SOLDIERS
19 GROUNDS
25 SKIRTING
MOULDED FILLET
88
88
75
900
75
100
200
1
RETURN PANELLING IN OUTLINE ONLY.
DADO RAIL
DADO PANELS
SECTION A–A
PART ELEVATION
PORTION OF PANELLING FOR BOARD ROOM
400
2.100
900
2
GROUNDS FOR PANELLING
FRAMED GROUNDS FOR DOOR LININGS
3
SKETCH OF FRAMED GROUNDWORK TO WALLS & DOORWAY
FRAMED DOOR LININGS
75×19 GROUNDS
SECTION B–B
50 DOOR
ARCHITRAVE & PLINTH BLOCK
JOINTING AT INTERNAL ANGLE
4
5
SKIRTING TONGUED & SCRIBED
SHEET 87

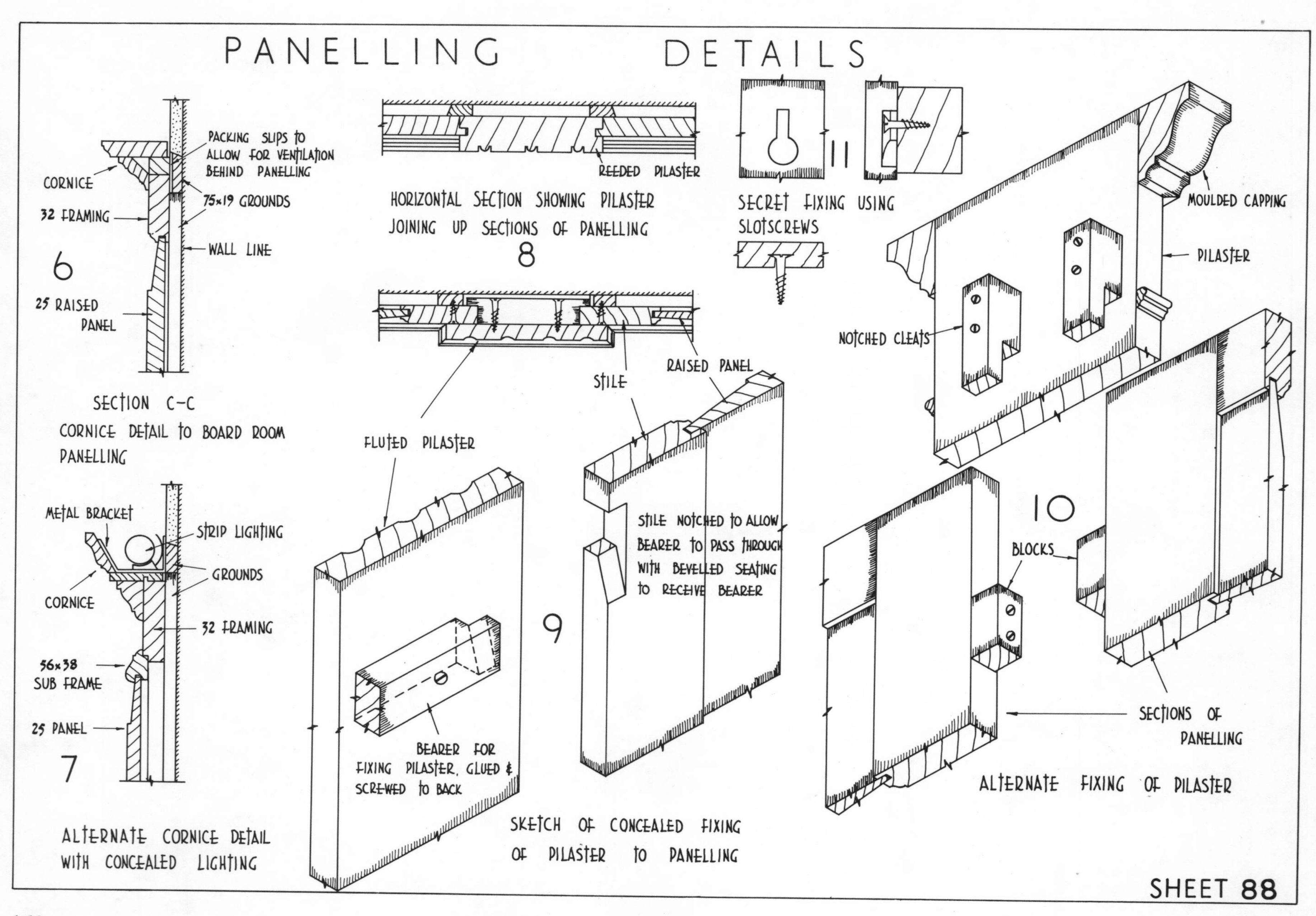
PANELLING DETAILS
PACKING SLIPS TO ALLOW FOR VENTILATION BEHIND PANELLING
CORNICE
75×19 GROUNDS
32 FRAMING
WALL LINE
6
25 RAISED PANEL
SECTION C–C
CORNICE DETAIL TO BOARD ROOM PANELLING
METAL BRACKET
STRIP LIGHTING
GROUNDS
CORNICE
32 FRAMING
56×38 SUB FRAME
25 PANEL
7
ALTERNATE CORNICE DETAIL WITH CONCEALED LIGHTING
REEDED PILASTER
HORIZONTAL SECTION SHOWING PILASTER JOINING UP SECTIONS OF PANELLING
8
STILE
RAISED PANEL
FLUTED PILASTER
STILE NOTCHED TO ALLOW BEARER TO PASS THROUGH WITH BEVELLED SEATING TO RECEIVE BEARER
9
BEARER FOR FIXING PILASTER, GLUED & SCREWED TO BACK
SKETCH OF CONCEALED FIXING OF PILASTER TO PANELLING
11
SECRET FIXING USING SLOTSCREWS
MOULDED CAPPING
PILASTER
NOTCHED CLEATS
10
BLOCKS
SECTIONS OF PANELLING
ALTERNATE FIXING OF PILASTER
SHEET 88

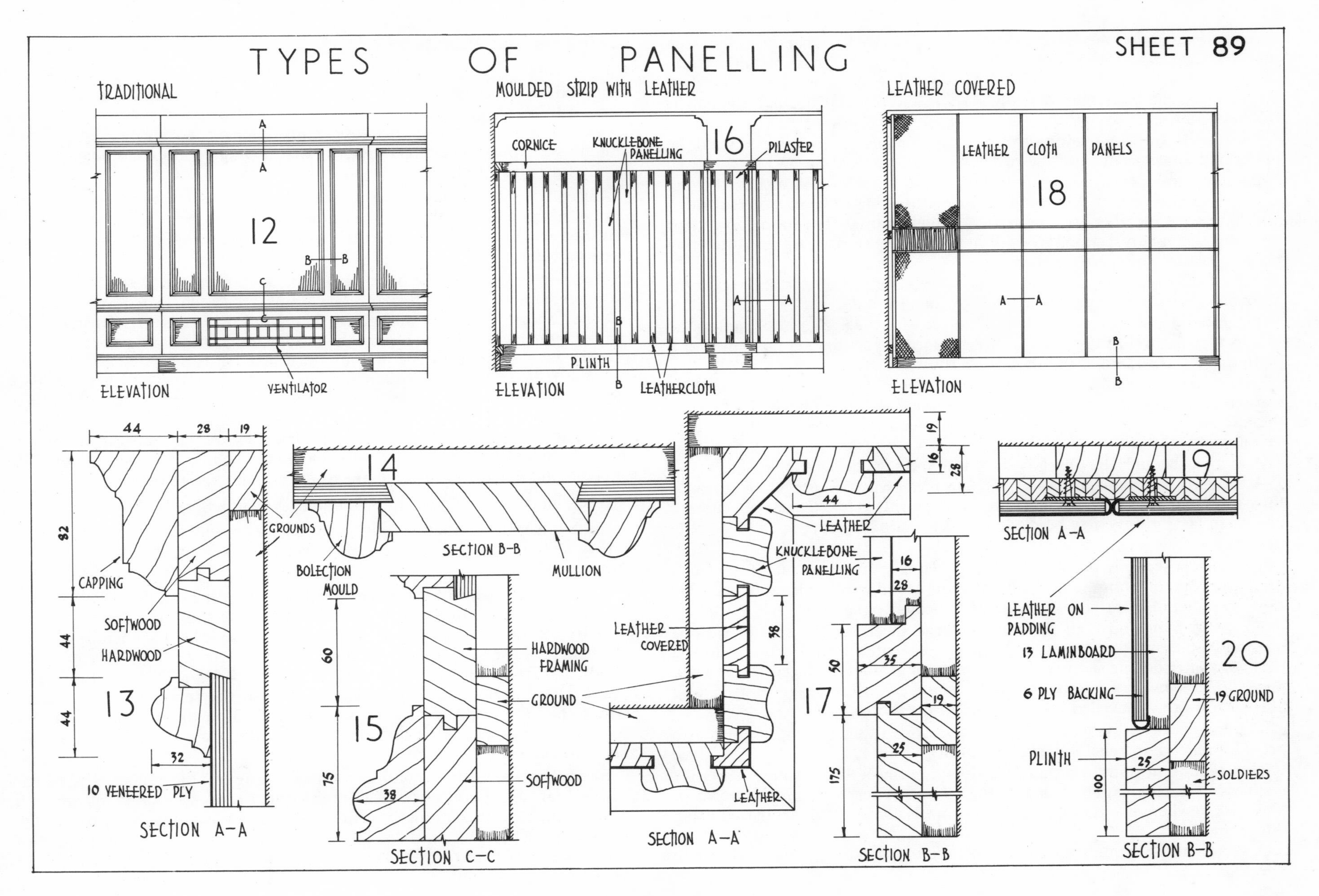
TYPES OF PANELLING
SHEET 89
TRADITIONAL
MOULDED STRIP WITH LEATHER
LEATHER COVERED
12
16
18
CORNICE
KNUCKLEBONE PANELLING
PILASTER
LEATHER
CLOTH
PANELS
PLINTH
LEATHERCLOTH
ELEVATION
VENTILATOR
ELEVATION
ELEVATION
14
19
SECTION B-B
SECTION A-A
CAPPING
SOFTWOOD
HARDWOOD
GROUNDS
BOLECTION MOULD
MULLION
LEATHER
KNUCKLEBONE PANELLING
LEATHER COVERED
LEATHER ON PADDING
13 LAMINBOARD
6 PLY BACKING
19 GROUND
PLINTH
SOLDIERS
13
15
17
20
HARDWOOD FRAMING
GROUND
SOFTWOOD
10 VENEERED PLY
LEATHER
SECTION A-A
SECTION C-C
SECTION A-A
SECTION B-B
SECTION B-B

A section through the cornice is shown in Sheet 89.13, where it will be seen that the edges of the veneered ply panels are splayed and flush with the framing on the back side. This allows for a bolder bolection moulding to be used on the face. These are mitred at the corners and screwed through the panels from the rear.

Sheet 89.14 shows a section through a mullion and the jointing between the upper and lower sections of the panelling is shown in Sheet 89.15.

Another example of modern panelling suitable for a bank or board-room is shown in Sheet 89.16. A combination of polished hardwood moulded strips alternating with leather-covered narrow panels is used. The hardwood strips are knucklebone in section, grooved in both edges to receive the contrasting leather panels, which are tongued.

A part section through a pilaster showing the panelling and the treatment at an external and internal angle are shown in Sheet 89.17.

The groundwork is prepared in a similar way to the previous examples, the moulded strips and leather panels being pinned at the top and bottom with a plain rebated cornice and plinth fixed as shown.

A further example, Sheet 89.18, shows a complete wall area lined with leathercloth. This has a padded backing applied to 6 mm plywood panels by the upholsterer. A section through the panels is shown in Sheet 89.19, where secret fixing is used.

A section through the plinth is shown in Sheet 89.20.

A modern treatment to dado panelling is shown in Sheet 90.21, giving a flush finish with little projection from the main walls. Here the use of veneered boards in large sheets may be made with the dado rail, capping and skirting of contrasting woods.

The groundwork is prepared as for previous examples, but the fixing is done working up from the skirting, the moulded fillet being fixed first. Again care is needed in fixing so that the appearance is not spoiled.

To avoid fixing through the face of large sheets at the centre, various methods are employed. Sheet 90.22 shows a rebated bearer screwed to the back and interlocking with rebated horizontal grounds plugged to the wall. Sheet 90.23 shows an alternative where metal hook brackets are used, and Sheet 90.24 shows bevelled grounds used at the centre.

The joining up of panel sections may be done using a muntin which also provides a break or relief in large flat surfaces.

The jointing of thin flush panels is shown in Sheet 90.25 using alternative methods. A further method of jointing 19 mm veneered board is shown in Sheet 90.26 where the gluing area at the joint is considerably increased using this method.

A sketch of the groundwork and panelling to a square column is shown in Sheet 90.27.

Columns and Pilasters

In panelled work, where sections are to be joined together or stanchions screened, engaged columns and pilasters are used. They also serve as a decorative feature, providing a break in large flat surfaces. Full columns are often placed round steel stanchions supporting floors in public buildings. Both columns and pilasters are made and framed-up in the shop, as is the panelling, ready for fixing on the site. They may be left plain or fluted.

Sheet 91.1 shows a part elevation at the base of a half column, engaged with and joining up two sections of wall panelling. To the left of the centre line the finished work is shown and, to the right, the section showing the built-up base and shaft.

The base, Sheet 91.2, is built up of four members arranged with cross or staggered joints shown in plan. The joints are made with loose tongues, stopped short of the face and glued. The plinth is formed by mitring at the angles, glue blocks being used on the inside, and screws used where they will not be seen on the face. The first moulded member is prepared and fitted on the plinth, being jointed on the centre line in two sections. The other members are prepared in the same way, glued and screwed as before. It will be seen that the top moulded member is rebated to receive the shaft. This is built up as shown to the left of the centre line in plan, the jointing between the staves being loose tongues, well glued with glue blocks behind. The staves should be tapered and fluted to suit the shape of the column before being glued.

The plan at the base of a full column is shown in Sheet 91.3.

Columns usually taper in their height with a gradual swelling at the centre. This swelling is called the 'entasis', and improves the appearance of the finished column.

Sheet 91, figs. 4 and 5 shows two methods of setting out the column shafts to give the entasis. In Sheet 91.4 the height of the shaft and points AB and CD are set out. With centre O and radius OC describe a semicircle. Divide the height of the shaft into a number of equal parts, 1, 2, 3, 4. Drop a perpendicular from A to cut the semicircle in point 4. Divide C4 into four equal parts. Draw horizontal lines at right-angles to the centre line. The points where the vertical projectors from points 1, 2, 3 on the semicircle intersect horizontal lines 1, 2, 3 on the shaft give points through which to draw the curve of entasis.

Sheet 91.5 shows the second method of setting out the entasis. The height of the shaft and points AB, CD are set out. From A describe an arc with radius equal to CO, cutting the centre line at E. Draw a line from A through E to meet CD produced in M. Divide the centre axis EO into a number of equal parts numbered 1, 2, 3. Through these points draw lines radiating to M as shown. From points 1, 2, 3 with radius CO mark off the points 1′, 2′ and

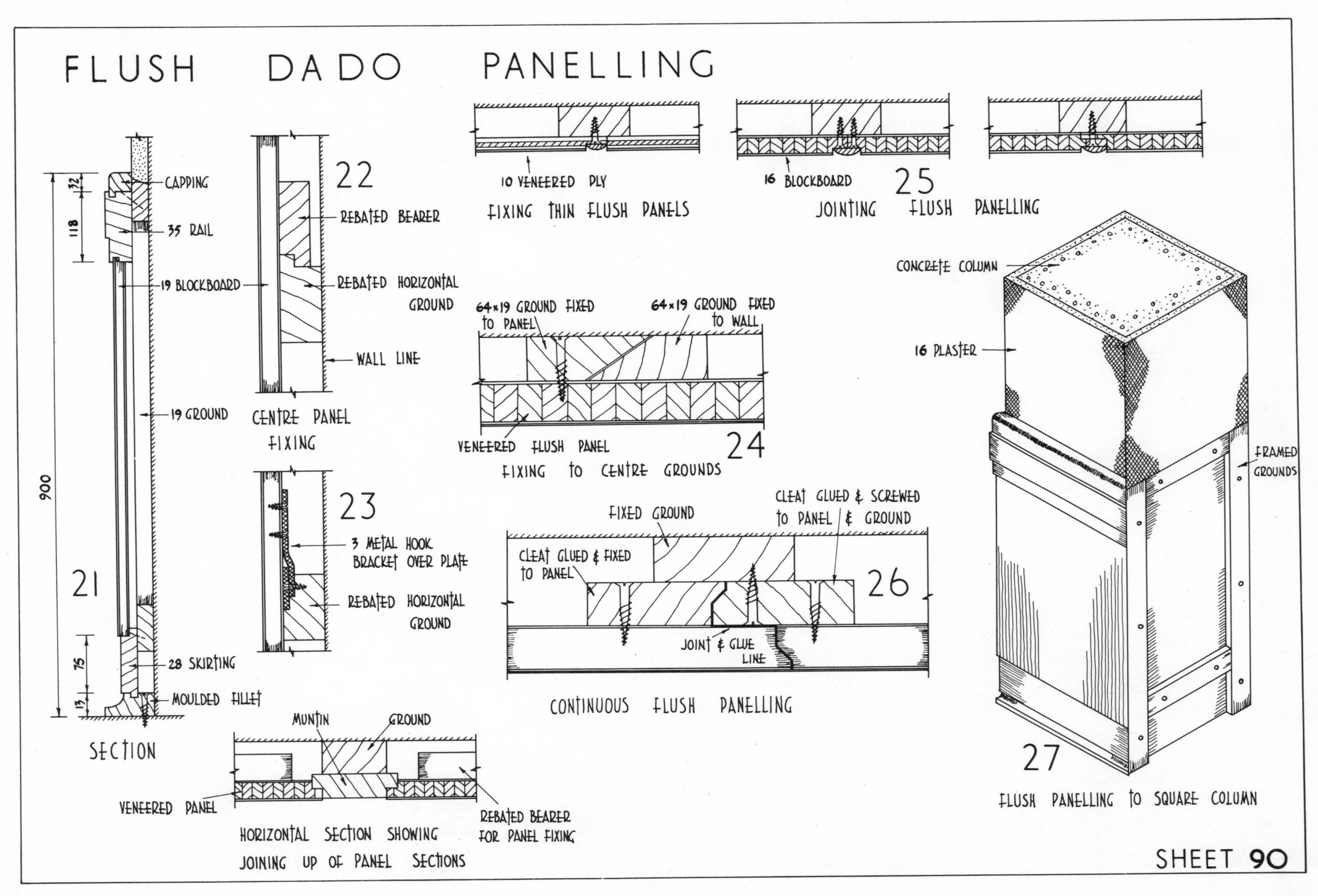
FLUSH DADO PANELLING
CAPPING
35 RAIL
19 BLOCKBOARD
19 GROUND
28 SKIRTING
MOULDED FILLET
32
118
900
75
13
21
SECTION
22
REBATED BEARER
REBATED HORIZONTAL GROUND
WALL LINE
CENTRE PANEL FIXING
23
3 METAL HOOK BRACKET OVER PLATE
REBATED HORIZONTAL GROUND
10 VENEERED PLY
FIXING THIN FLUSH PANELS
16 BLOCKBOARD
25
JOINTING FLUSH PANELLING
64×19 GROUND FIXED TO PANEL
64×19 GROUND FIXED TO WALL
VENEERED FLUSH PANEL
24
FIXING TO CENTRE GROUNDS
FIXED GROUND
CLEAT GLUED & SCREWED TO PANEL & GROUND
CLEAT GLUED & FIXED TO PANEL
26
JOINT & GLUE LINE
CONTINUOUS FLUSH PANELLING
MUNTIN
GROUND
VENEERED PANEL
REBATED BEARER FOR PANEL FIXING
HORIZONTAL SECTION SHOWING JOINING UP OF PANEL SECTIONS
CONCRETE COLUMN
16 PLASTER
FRAMED GROUNDS
27
FLUSH PANELLING TO SQUARE COLUMN
SHEET 90

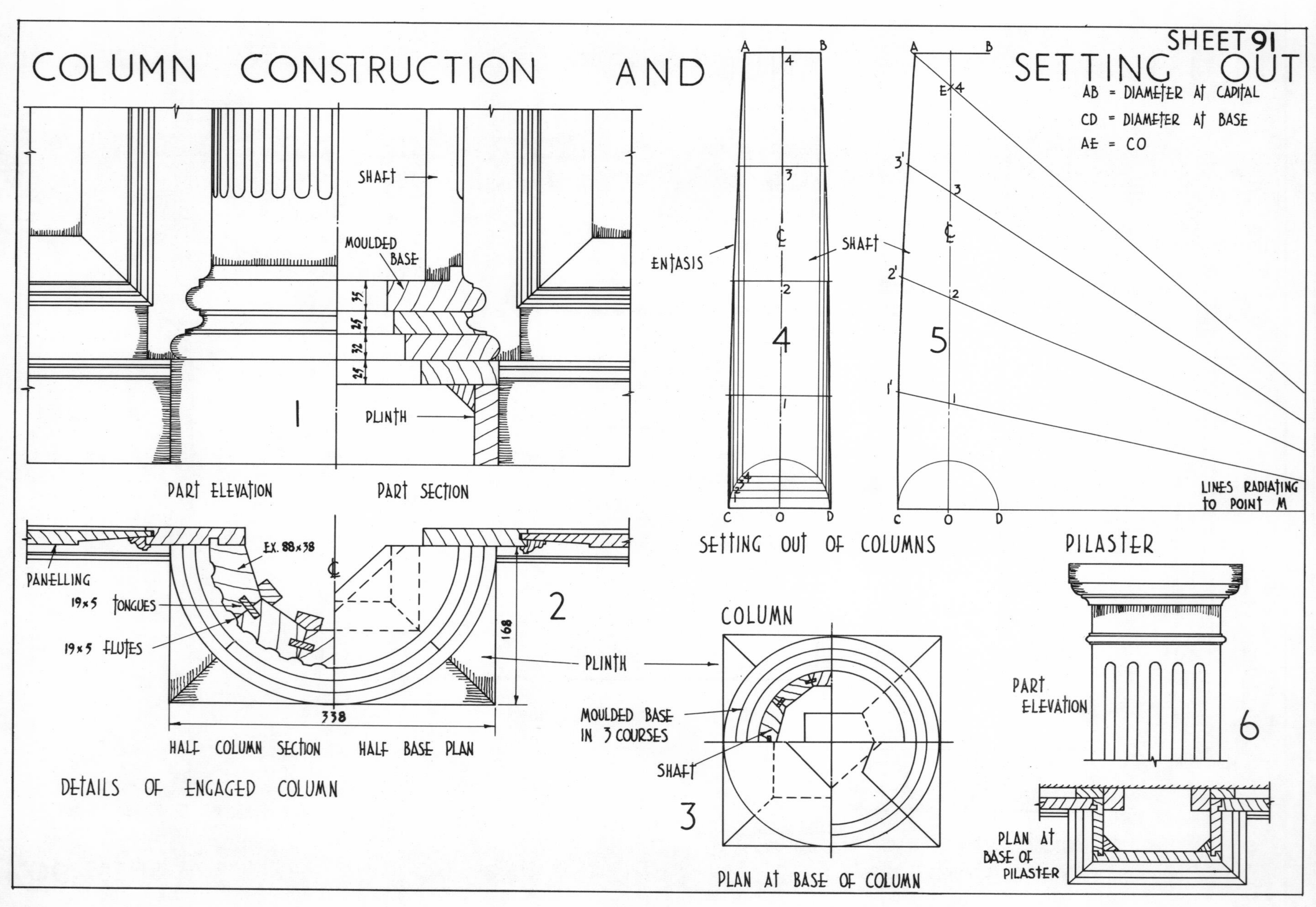
SHEET 91
COLUMN CONSTRUCTION AND SETTING OUT
SHAFT
MOULDED BASE
35
25
32
25
PLINTH
1
PART ELEVATION
PART SECTION
PANELLING
EX. 88×38
19×5 TONGUES
19×5 FLUTES
168
2
338
HALF COLUMN SECTION
HALF BASE PLAN
DETAILS OF ENGAGED COLUMN
ENTASIS
SHAFT
4
5
AB = DIAMETER AT CAPITAL
CD = DIAMETER AT BASE
AE = CO
LINES RADIATING TO POINT M
SETTING OUT OF COLUMNS
COLUMN
PLINTH
MOULDED BASE IN 3 COURSES
SHAFT
3
PLAN AT BASE OF COLUMN
PILASTER
PART ELEVATION
6
PLAN AT BASE OF PILASTER

3′. A fair curve drawn through these points gives the entasis which is repeated for the other side of the shaft.

If the column is to encase a steel stanchion, the base would be built up in position, having been first prepared in the shop. The shaft too is made in the shop and left with two dry joints so that it may be split into two parts for fixing. Special clamps with a sprung steel band are used for clamping the two parts together after gluing.

Sheet 91.6 shows a part elevation and plan of a pilaster showing the finish at the top or capital and the method of fixing.

Raking Mouldings

Before dealing with the problems on this subject, it is necessary to mention certain points. Moulds may be required to suit varying conditions as follows: (a) two level moulds and one raking over a square plan; (b) two raking moulds over a square plan; (c) one level mould and one raking over an obtuse-angled plan; (d) two level moulds and one raking over obtuse- and acute-angled plans; (e) two raking moulds over an obtuse-angled plan. Any mould in the above conditions may be taken as the given mould, from which the others may be developed.

The following points should be noted:

(1) The true shape at the intersection of two moulds with a common intersection on a vertical mitre is the same for both moulds.
(2) The section of one mould must always be known or given.
(3) The inclination of the raking moulds and the plan angles will be known.
(4) The thickness of raking moulds with common intersections is the same for each pair of moulds.
(5) The widths of raking moulds differ in accordance with the pitch of the moulds.

Sheet 92.1 shows the key plan for condition (c) above over an obtuse-angled plan, Sheet 92.2 the geometrical construction with the level mould being the given mould. The plan is first drawn and the section of the given mould set out as shown, giving the points 0, 1, 2, 3, 4, 5, 6, 7, 8. Ordinates projected from the section of this mould on to the mitre line in plan are projected up into elevation to intersect with the given mould ordinates to determine the elevation of the mitre. The elevation of the mitre is reversed and the ordinates projected on to this. These points are drawn to the pitch of the raking mould to determine the required section as shown.

Sheet 92.3 shows the method of determining the bevels for the moulds. The bevel for the inclined mitre is shown at AB″ while the bevel for the level mould is shown in plan AB. The bevel at A′ in the elevation is the side bevel for the back of the moulding.

Mouldings to Pediments

Sheet 92.4 shows the elevation of an inclined mould with a returned level mould at the top and bottom. When the section of the bottom mould is given, it is necessary to find the sections of the other two so that they will properly intersect.

Dealing with the given mould. On the outline of the mould select points a, b, c, d, e, f, g and erect perpendiculars on to a horizontal datum line in points 0, 1, 2, 3, 4, 5, 6, cutting the pediment at H. With point H as centre and radius H–6, etc., describe arcs as shown. From the points thus determined, drop perpendiculars from the points 0, 1, 2, 3, 4, 5, 6, cutting lines from corresponding letters a, b, c, d, e, f, g, giving the true section of the inclined mould. At the top the same construction is employed to give the section of the returned level mould.

Moulding to Lantern

In a lantern light having its members moulded, in order that all the mouldings may properly intersect, it is necessary to determine the true shape of each mould. Draw in the glazing bar in section, Sheet 92.5, pitched at 30 degrees, divide the moulded part 0, 1, 2, 3, 4. Drop perpendiculars from the pitch through these points and where they cut through the mould draw in the lines a, b, c, d, e, f parallel to the pitch of the roof. Produce these both ways until they cut the ridge and bed mould. On the ridge set off the distances 0 to 4 as shown, equal to those on the bar, draw perpendiculars until they cut the corresponding points on the lines a, b, c, d, e, f. Where these lines intersect complete the section of the ridge as shown. It should be noted that the ridge is thicker than the glazing bar and is grooved for glass. The bottom edge is usually finished as shown.

The true shape of the bed mould is found in the same way as for the ridge.

To find the true shape of the hip rafter the dihedral angle (see roof bevels p. 66) is required. The distances a, b, c, d, e, f, g, h are taken from the glazing bar section and the geometrical construction to determine the true section, the same as for the ridge.

Hoppers

A hopper is an open box having inclined sides with the top edges made parallel to the horizontal or left square.

The plan and section of a square hopper are shown in Sheet 93.1. The section is drawn first and the plan projected from it. The side to the left is shown lapping over the other two, and that on the right is mitred at the corners. To develop the side D, with centre O and radius OA describe an arc to give A″ on a horizontal line brought out from A′ in plan. Join A″ to the

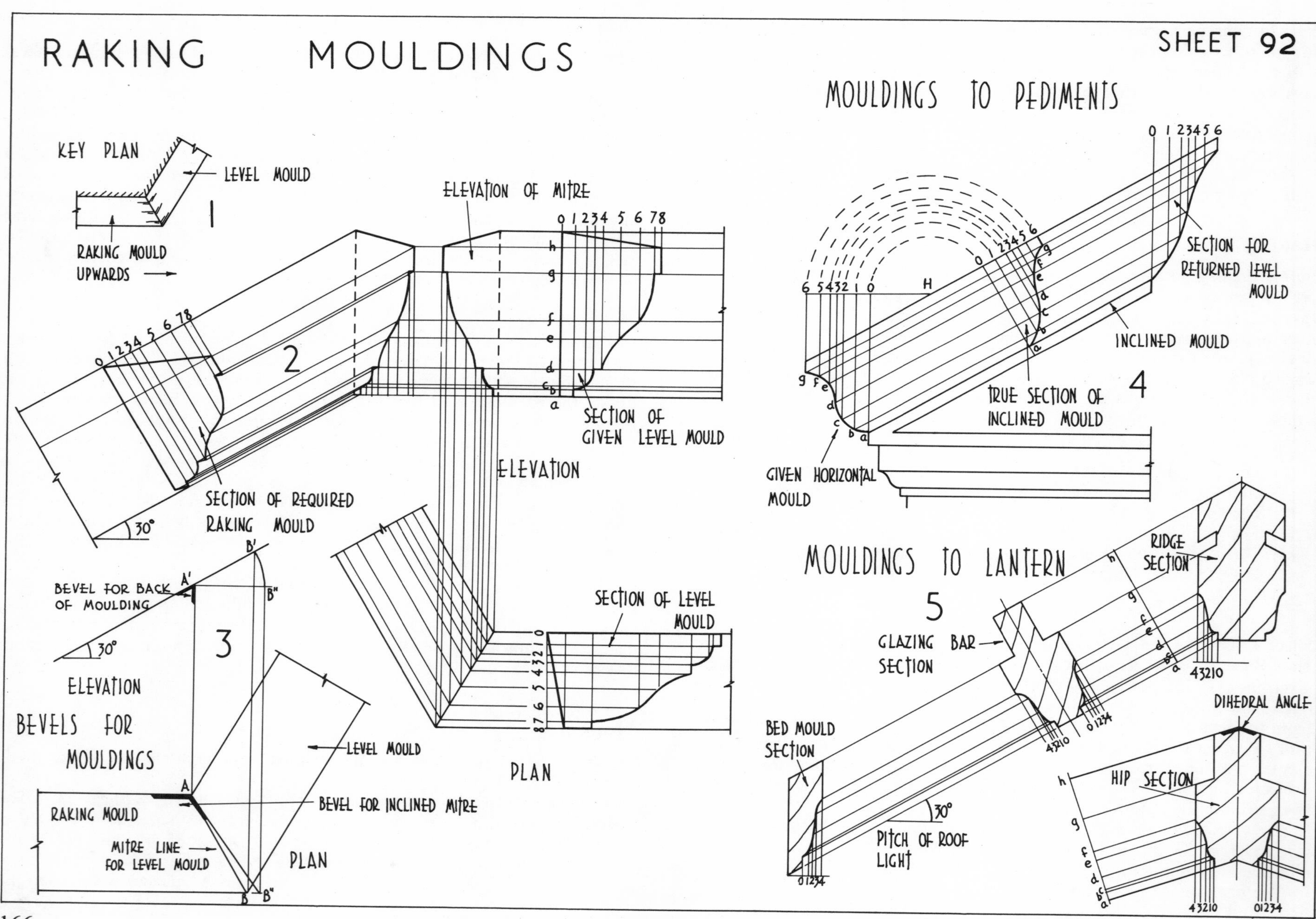

RAKING MOULDINGS
SHEET 92
KEY PLAN
LEVEL MOULD
1
RAKING MOULD UPWARDS
ELEVATION OF MITRE
0 1 2 3 4 5 6 7 8
h
g
f
e
d
c
b
a
2
SECTION OF GIVEN LEVEL MOULD
ELEVATION
SECTION OF REQUIRED RAKING MOULD
30°
BEVEL FOR BACK OF MOULDING
A'
B'
B"
30°
3
ELEVATION
BEVELS FOR MOULDINGS
SECTION OF LEVEL MOULD
LEVEL MOULD
PLAN
A
RAKING MOULD
BEVEL FOR INCLINED MITRE
MITRE LINE FOR LEVEL MOULD
PLAN
B
B"
MOULDINGS TO PEDIMENTS
0 1 2 3 4 5 6
SECTION FOR RETURNED LEVEL MOULD
H
INCLINED MOULD
TRUE SECTION OF INCLINED MOULD
4
GIVEN HORIZONTAL MOULD
MOULDINGS TO LANTERN
RIDGE SECTION
5
GLAZING BAR SECTION
43210
01234
BED MOULD SECTION
30°
PITCH OF ROOF LIGHT
DIHEDRAL ANGLE
HIP SECTION

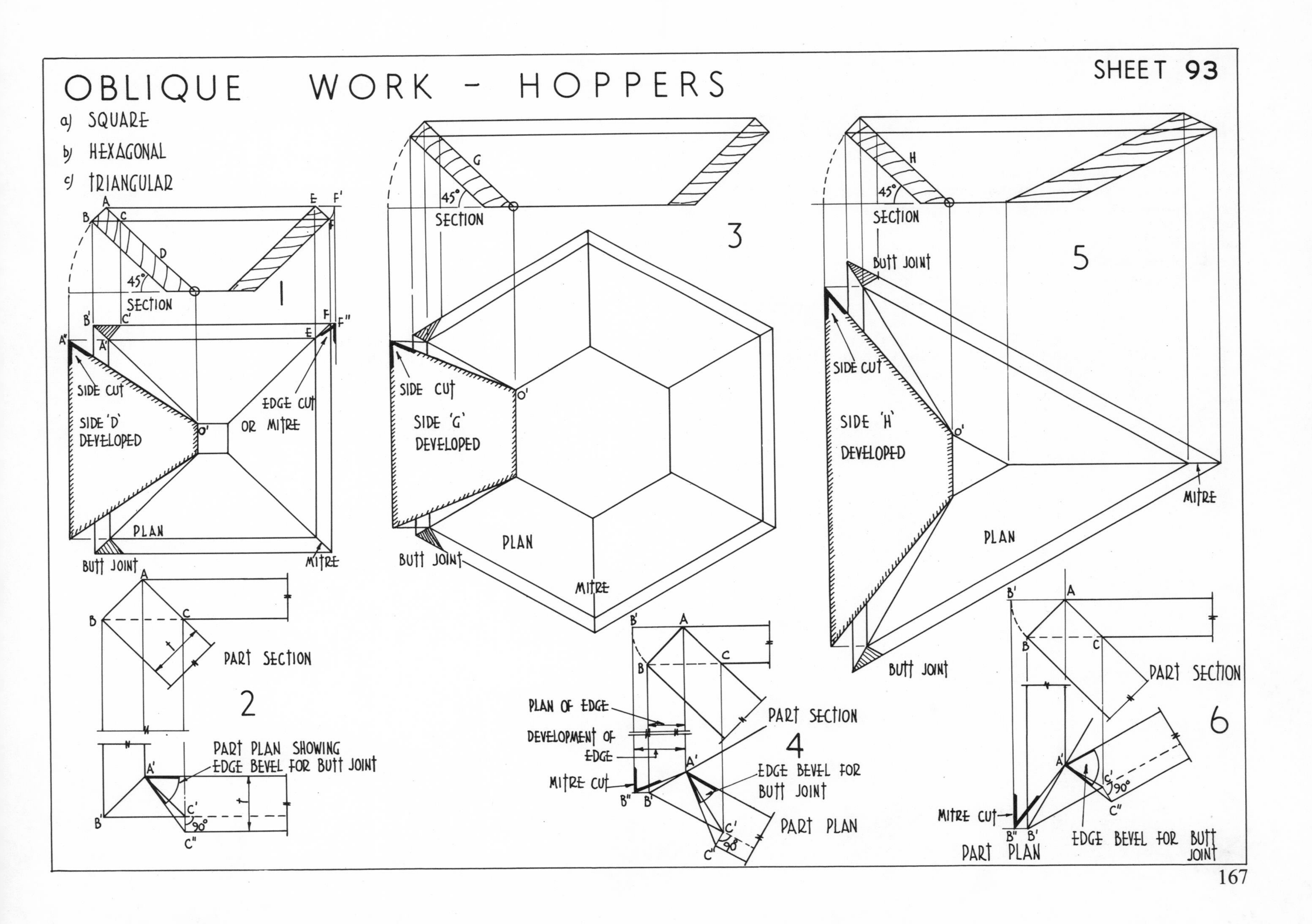
OBLIQUE WORK - HOPPERS
SHEET 93
a) SQUARE
b) HEXAGONAL
c) TRIANGULAR
SECTION
45°
SIDE CUT
SIDE 'D' DEVELOPED
EDGE CUT OR MITRE
PLAN
BUTT JOINT
MITRE
1
PART SECTION
2
PART PLAN SHOWING EDGE BEVEL FOR BUTT JOINT
90°
SIDE 'G' DEVELOPED
3
PLAN OF EDGE
DEVELOPMENT OF EDGE
MITRE CUT
EDGE BEVEL FOR BUTT JOINT
PART PLAN
4
SIDE 'H' DEVELOPED
5
PART PLAN
6

inside bottom edge O′ to show the side cut and when repeated at the bottom corner, this gives the true shape of the inside face of the side.

The mitre cut is found by developing the edge EF. With centre E and radius EF, describe an arc to give F′ on a horizontal line brought out from E. Draw a line from F′ in section to F″ in plan to intersect with a horizontal line brought over from F in plan. Join E to F″ to give the edge or mitre cut to be applied to the sides on the right.

To obtain the edge cut, so that one side may lap over the other two as shown to the left in plan, draw the enlarged section through one side and part plan of the corner, Sheet 93.2. Draw a horizontal line from B in the section and from the point where this cuts out at C, drop a perpendicular to a corresponding point C′ in plan; join C′ to the edge on the mitre line A′, the resulting triangle A′B′C′, represented by the etched portion in plan, shows the cross grain. Develop the edge, by drawing from C′ to C″ as shown and the required bevel is shown after joining A′C″.

An alternative method of finding the mitre cut is to obtain the dihedral angle between the two surfaces. Half the dihedral angle is then taken as the necessary bevel to apply to the two sides to be mitred.

Sheet 93.3 shows the plan and section of an hexagonal hopper one side lapped over two of the others to the left, the others mitred. The mitre cut and the bevel for sides butting are shown in Sheet 93.4.

The plan and section of a triangular hopper are shown in Sheet 93.5. The cuts and bevels are obtained in the same way as the previous examples. The mitre cut and the bevel for the sides butting in this hopper are shown enlarged in Sheet 93.6.

An alternative method for determining the bevel when the corners of the hoppers are to be butt jointed, is to obtain the dihedral angle between the two sides in a similar way to that shown on the work in splayed linings on Sheet 58. The supplement of the dihedral angle between the sides is the required bevel.

The supplement of the dihedral angle is also the angle between the sides of the groove and the face of the grooved side, if a tongued and grooved joint is to be used at the corner, instead of being mitred together. The grooved side is usually allowed to run on a short distance beyond the outer face of the tongued side to give additional support to the tongue when this particular jointing is used.

Chapter Ten: Joinery

Library Furniture

This comprises issue counters, bookcases, wall fitments, shelves, card index cabinets, magazine racks, tables and newspaper stands. Again this type of work provides the highest class of joiner's work, done in hardwood and polished all to match. The following illustrations are typical of those to be found in most parts of this country.

Counters

Sheet 94.1 shows the half front elevation and half sectional elevation of an issue counter. A half plan is shown in Sheet 94.2. The counter is normally sited within the entrance to the library so that borrowers pass along one side of the counter on entering and along the other when leaving. The main dimension in this type of counter is the height, the width in this case is unimportant and may be anything up to 675 mm wide. The shape of the counter depends upon the area of floor space available.

The counter illustrated has splayed sides to the front, with a flush door 600 mm wide giving access to the counter at the rear. The veneered plywood front is set in 100 mm at the bottom with an inset skirting to allow foot-room.

The design of the counter provides open shelves for book storage and drawers on the inside. The construction consists of a series of frames or partitions, at positions indicated in plan, and shelves along the two sides. The frames are mortised and tenoned together and connected by shelves running through them. The front is glued and pinned to the front edge of the frames and to the shelves and carcassing.

In the sections where the nests of drawers are placed, additional rails are housed into the framed ends to receive the drawer runners and guides. A drawer handle is detailed in Sheet 94.3.

The solid jointed top is buttoned on to the framing, the mitres being cross-tongued and glued.

The counter is raised above the normal floor level on 75 mm by 50 mm boarded joists as shown in Sheet 94.4, and screwed down. The 100 mm skirting is fixed last covering the sub-floor. The door is framed and covered on both sides and hung to a light rebated jamb with a planted stop on the other side of the opening.

Bookcase

Sheet 94.5 shows the elevation of a bookcase, the upper portion is glazed and fitted with shelves, the lower portion provided with a cupboard in the centre and drawers on each side.

A section through the cupboard is shown in Sheet 94.6, where it will be seen that the lower portion is wider than the glazed top portion. In the manufacture of this type of fitting two separate units screwed together are often used.

The construction consists of solid or framed ends covered with plywood or veneered blockboard. The horizontal rails are dovetailed and housed into the ends, with the vertical members housed into the rails. The tops of both units are slot screwed to the rails. Both units may have independent plywood backs nailed on, or a single back rebated in as shown.

Veneered blockboard doors with all the edges lipped and rebated at the centre are fitted to the lower cupboard, with the plywood bottom forming the rebate and a planted stop at the top.

The shelves in the top portion are loose and are adjustable on patent strip fittings shown in Sheet 94.8 for varying sizes of books. The strip fittings are housed and screwed into the ends and divisions, and for this reason these members should not be framed but either solid timber or blockboard used. Cleats screwed to the lower division members support the shelf as shown.

The upper doors are glazed with beads on the inside. The centre pair are rebated at the centre with planted stops.

Sheet 94.7 shows a detail of the plinth.

Wall Units

The elevation of an open wall unit for the display and storage of library books is shown in Sheet 95.8. It consists of two similar units 900 mm high and 1.950 m wide on each side of a centre unit, which is 1.130 m high and 1.050 m wide. This centre unit projects beyond the outer units as shown in Sheet 95.9 and has two fixed shelves. The outer units have a centre division and a fixed shelf in the centre.

The construction of the fitting is similar to that for the bookcase described above to which reference should be made.

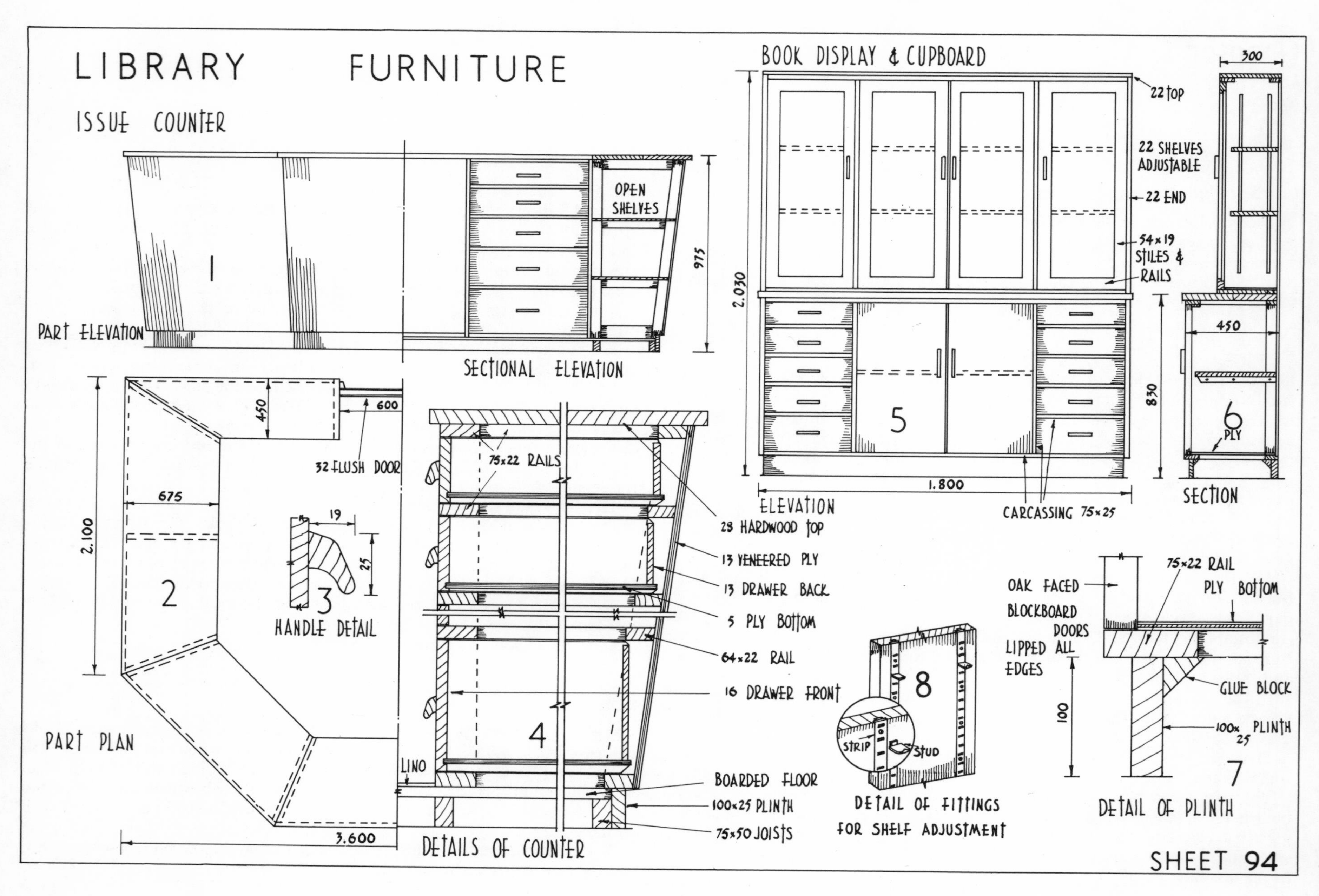
LIBRARY FURNITURE
ISSUE COUNTER
OPEN SHELVES
1
975
PART ELEVATION
SECTIONAL ELEVATION
450
600
32 FLUSH DOOR
675
2.100
2
19
25
3
HANDLE DETAIL
PART PLAN
3.600
75×22 RAILS
4
LINO
DETAILS OF COUNTER
28 HARDWOOD TOP
13 VENEERED PLY
13 DRAWER BACK
5 PLY BOTTOM
64×22 RAIL
16 DRAWER FRONT
BOARDED FLOOR
100×25 PLINTH
75×50 JOISTS
BOOK DISPLAY & CUPBOARD
22 TOP
22 SHELVES ADJUSTABLE
22 END
54×19 STILES & RAILS
2.030
5
1.800
ELEVATION
CARCASSING 75×25
300
450
830
6
PLY
SECTION
STRIP
STUD
8
DETAIL OF FITTINGS FOR SHELF ADJUSTMENT
OAK FACED BLOCKBOARD DOORS LIPPED ALL EDGES
75×22 RAIL
PLY BOTTOM
GLUE BLOCK
100× 25 PLINTH
100
7
DETAIL OF PLINTH
SHEET 94

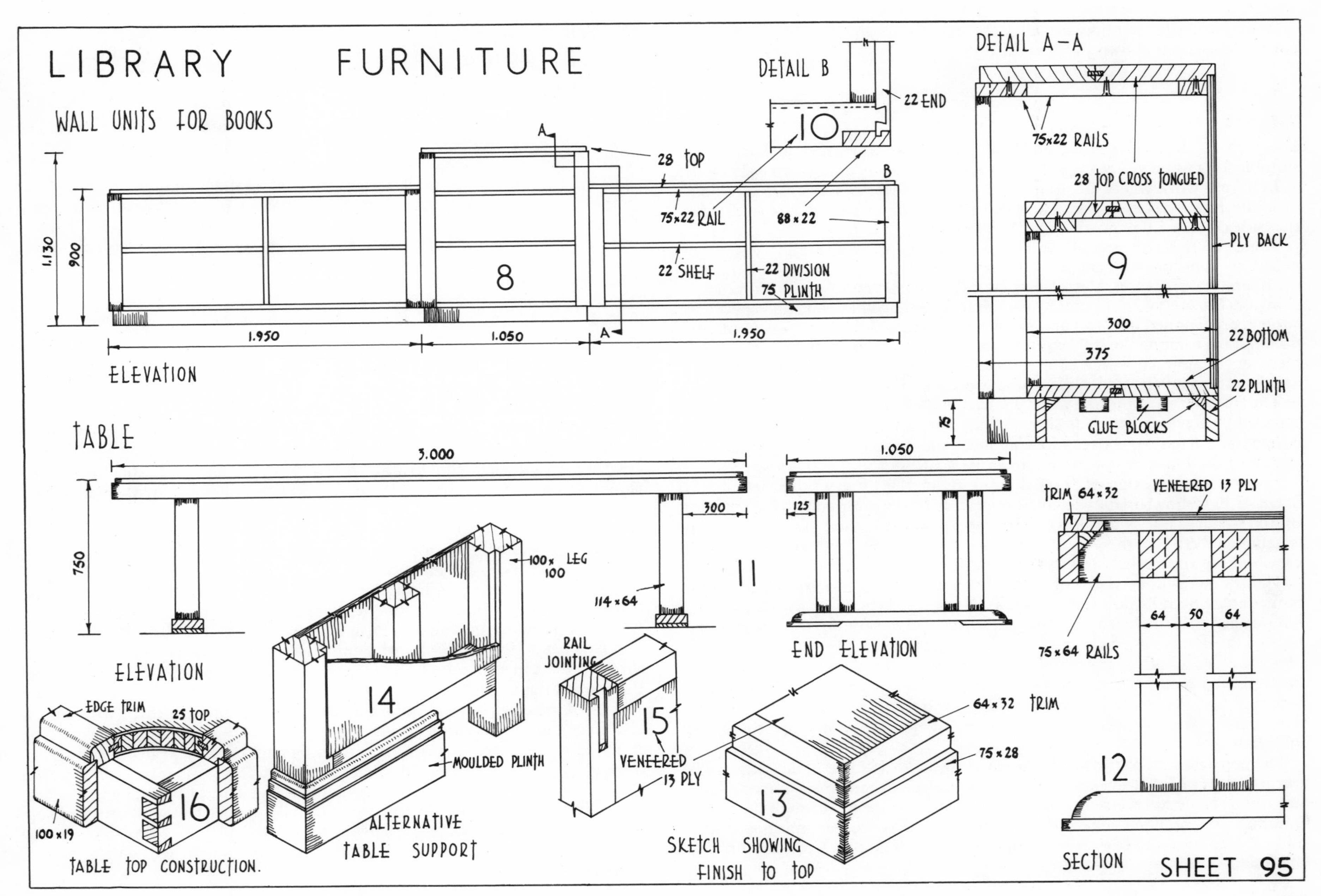
LIBRARY FURNITURE
WALL UNITS FOR BOOKS
DETAIL B
22 END
10
28 TOP
75×22 RAIL
88×22
22 SHELF
22 DIVISION
75 PLINTH
8
1.130
900
1.950
1.050
1.950
ELEVATION
DETAIL A-A
75×22 RAILS
28 TOP CROSS TONGUED
PLY BACK
9
300
375
22 BOTTOM
22 PLINTH
GLUE BLOCKS
75
TABLE
3.000
300
750
114×64
11
1.050
125
END ELEVATION
ELEVATION
TRIM 64×32
VENEERED 13 PLY
75×64 RAILS
64
50
64
12
SECTION
100× 100 LEG
14
MOULDED PLINTH
ALTERNATIVE TABLE SUPPORT
RAIL JOINTING
15
VENEERED 13 PLY
64×32 TRIM
75×28
13
SKETCH SHOWING FINISH TO TOP
EDGE TRIM
25 TOP
16
100×19
TABLE TOP CONSTRUCTION.
SHEET 95

A detail of the jointing between the carcassing and the end at the top corner of a unit is shown in Sheet 95.10.

Tables

Sheet 95.11 shows the front and end elevations of a table in hardwood, suitable for a library. The important dimension is the height, which is 750 mm.

In library work where tables are in use by a number of readers at one time, absolute rigidity is essential. The table shown allows for readers to use the two sides, being 1.050 m wide and 3.000 m long.

Tables of this size present particular problems in construction and technique, as the construction must be such as to prevent movement in any members which causes loss of rigidity. The construction of the framing is shown in Sheet 95.12, 114 mm by 64 mm legs, arranged in pairs and spaced as shown, are tenoned into 150 mm by 48 mm runners at the base and 75 mm by 64 mm rails running the full length of the table at the top. Into these rails cross members are jointed with a 28 mm outside rail completing the under framing and fixed so that the underside of the top rests directly upon them.

The framed top is finished with veneered blockboard or plywood, glued and screwed to a rebated trim as shown. The trim is mitred and tongued at the corners and fixed by pocket screwing and blocking to the under framing. A sketch of the finish to the top is shown in Sheet 95.13.

An alternative design of framed support for the library table detailed in Sheet 95.11 is shown in the sketch, Sheet 95.14. The framing consists of 100 mm by 100 mm legs into which top and bottom rails are tenoned. A centre muntin is tenoned between these rails and the whole flushed on both sides with veneered plywood rebated in the legs and rails.

The framed supports are connected by means of three long 174 mm by 64 mm connecting rails dovetail housed as shown in Sheet 95.15 with cross rails half-lapped over these, so that the underside of the made-up top rests on the tops of the connecting rails and the framed ends.

A moulded plinth is mitred round the supports as shown, which may be protected by metal kicking strips. These are often used round the bases of furniture and fittings in public buildings, not only to protect the woodwork from damage by kicking, but from staining by floor cleaning appliances and materials.

The construction of the top is shown in Sheet 95.16. The under framing is through dovetailed at the corners with a moulded edge member mitred round to form a rebate to receive the top and is fixed by screwing through the under framing. The moulded edge trim is tongued to the veneered laminboard top, and rebated over the edge member and secured by pocket screwing through the under framing.

Church Work

Modern church work, perhaps more than any other branch of joinery, is based on and closely follows traditional methods, both in construction and in design. Design is mostly of a gothic character with Austrian and American oak the principal timbers used. The characteristics of church or ecclesiastical work are heavy mouldings, mouldings stuck on the solid and finished with mason's mitres or stops, carved figures and tracery work.

Church fittings comprise lecterns, pews, choir stalls, litany desks, screens, pulpits, communion rails and font covers. Space will not allow all of these to be detailed, but students will find many textbooks devoted solely to illustrations of such work in our churches and cathedrals. This work, although chiefly carried out by specialist firms, is still much a part of the joiner's work, with ornamentation executed by carvers.

Pews

Sheet 96.1 shows a section through a modern pew or bench. The shaped vertical member which is housed to receive the seat and back is called a pew end and may be worked on the edges of both sides with a chamfer or scotia.

The essential dimensions of the pew have been detailed and, as these are most important, they should be remembered. Some further points concerning design are that seats should not be less than 350 mm wide, the back may slope between 25 mm and 100 mm and 500 mm allowed for the seating of each person.

The pew end may be carved or finished with some form of decorative treatment on one or both sides, although on modern work much of this decoration has been dispensed with.

Sheet 96.2 shows a part back elevation of the pew. The back is framed and filled between muntins with tongued and grooved boards, tongued top and bottom to the rails and finished flush on the face side. The top rail is moulded and finished on the top edge with a moulded capping slot screwed or screwed through the top and pelleted. The capping serves as a handrail and may be used to aid rising, after kneeling, or from the seat.

A book-board is shown and a hat-rack in the form of a shelf, which may also store kneeling cushions below the seat.

The seat, which is tongued to the back is pocket-screwed from below and through the bottom rail and slopes 25 mm from front to back.

Sheet 96.3 shows a further example of an older type of pew having a wider seat. This has a framing acting as additional support below the seat, along with intermediate supports as shown. It will be seen that the wider book-board is tongued to the top rail of the back and supported with moulded brackets in this design.

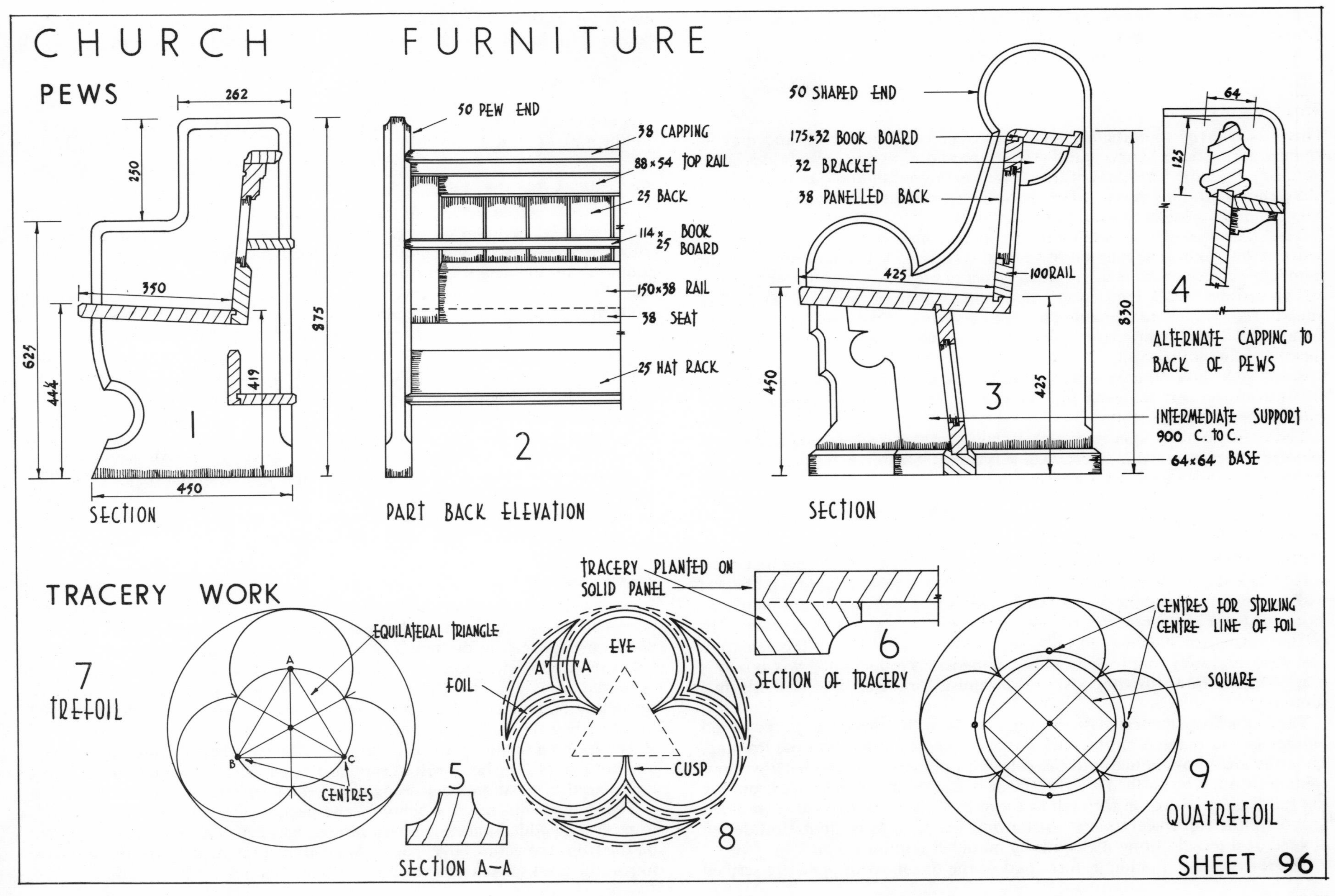
CHURCH FURNITURE
PEWS
262
250
350
625
444
419
875
450
1
SECTION
50 PEW END
38 CAPPING
88×54 TOP RAIL
25 BACK
114×25 BOOK BOARD
150×38 RAIL
38 SEAT
25 HAT RACK
2
PART BACK ELEVATION
50 SHAPED END
175×32 BOOK BOARD
32 BRACKET
38 PANELLED BACK
425
100 RAIL
450
425
830
3
INTERMEDIATE SUPPORT
900 C. TO C.
64×64 BASE
SECTION
64
125
4
ALTERNATE CAPPING TO
BACK OF PEWS
TRACERY WORK
7
TREFOIL
EQUILATERAL TRIANGLE
A
B
C
CENTRES
5
SECTION A–A
TRACERY PLANTED ON
SOLID PANEL
FOIL
EYE
A
A
CUSP
8
6
SECTION OF TRACERY
CENTRES FOR STRIKING
CENTRE LINE OF FOIL
SQUARE
9
QUATREFOIL
SHEET 96

The shaped pew end is tenoned and mortised to a moulded base, which is housed to receive the end, and also the lower framework.

An alternative capping finish is detailed in Sheet 96.4.

Tracery Work

This was referred to earlier as a characteristic of church work and may be defined as a series of curved and straight mouldings which require to be set out geometrically. The basis of the designs is the trefoil and quatrefoil.

Tracery work may be cut from the solid or pierced and moulded on both sides, giving a section as shown in Sheet 96.5, or it may be planted on a solid panel, Sheet 96.6. The moulding used may be a chamfer or a scotia.

Sheet 96.7 shows the method of setting out a trefoil. The term refers to the number of part circles enclosed in the larger circle.

The setting out is constructed upon an equilateral triangle ABC. The angles of the triangle are bisected to give the centre of the circle enclosing the triangle, and the larger circle. Points A, B and C are the centres for striking the centre lines of the foil.

Sheet 96.8 shows the completed tracery based on the above construction. The leaf shape between each foil is called a 'cusp' and the piercing in the tracery the 'eye'.

The setting out of a quatrefoil, which has four part circles contained within a larger circle, is shown in Sheet 96.9. It is constructed upon a square with the setting out similar to that for the trefoil.

Book-Rest

Sheet 97.10 shows a part front and the end elevation of a book-rest to a choir stall. The choir stalls are usually placed on each side of the church chancel with a book-rest along the front of each side. A fitting of similar design placed along the front of the first row of pews in the body of the church is often used.

The seating for the choir has not been shown and although the construction follows closely that for pews, their dimensions differ. The book board is much higher, and so are the ends, to allow music for the choir to be read while standing.

The book-rest illustrated has framed ends with tracery work prepared separately and planted on the back panel, which is housed into the framing. The rails are stub-tenoned into the stiles and pinned from the back without coming through on to the face. The front is framed in a similar way, but the top rail is allowed to run through to the stile. This is moulded and prepared to receive the top tracery panel as shown. This panel is of quatrefoil design divided at intervals along its length by moulded mullions.

The lower panel mouldings form part of the tracery work, and are scribed on to the chamfer of the bottom rail. A moulded capping is fixed by slot-screwing to the top rail and framed ends, being mitred at the corners. Shaped buttresses, moulded from the solid, are slot-screwed to the framed ends on the face providing additional decoration.

Sheet 97.11 shows a part back elevation of the choir stall with the book-rest housed into the framed end. A detail of the top rail and capping is shown in Sheet 97.12, and the bottom rail in Sheet 97.13.

Communion Rail

The fitting is placed in front of the altar table and usually spans the full width of the chancel between walls. Sheet 97.14 shows a part elevation of the rail which is normally heavily moulded. As the congregation kneel at the communion rail, the height is most important, usually between 600 mm and 675 mm. Fixing is done at each end to posts plugged to the walls and through the bottom rail into the floor. The centre bay is hinged to fold to each side, giving access to the altar as it is only in position during the Communion Service.

The tracery work is pierced or open tracery, moulded on both sides and housed in the framework as shown in the detail, Sheet 97.15. A detail of the top rail with wide capping and tongued bed moulds is shown in Sheet 97.16. The built-up base detail is shown in Sheet 97.17.

In some churches a heavily moulded rail, usually about 125 mm by 75 mm in section, is supported on pillars carrying ornamental wrought iron brackets. The pillars are generally octagonal and spaced about 900 mm apart. The centre section directly in front of the altar table is either hinged, so that it lies on the fixed rail when it is open, or made to be removed altogether with pins at each end to drop in slots cut in the fixed rails.

Litany Desk

There are many variations in design of this small desk, which is used by the reader in a kneeling position during a church service. Sheet 98.18 shows the elevation and section of such a desk. Again, before attempting to design a litany desk, the dimensions, of which the height is the most important, should be noted. This church fitting consists of a framed base with two ends or uprights containing a decorative front panel or tracery, a book-rest and kneeling board.

The framing for the panel is housed into the two shaped ends and the base and housed into the book-rest as shown. The panel is housed into the framing and placed in position when wedging up the framing.

The shaped ends are stub-tenoned and pegged into the base which extends to receive a padded kneeling board housed between. The book-rest, moulded on the front projecting edge, has a moulded fillet tongued on the bottom edge to rest the books against as the board slopes 32 mm.

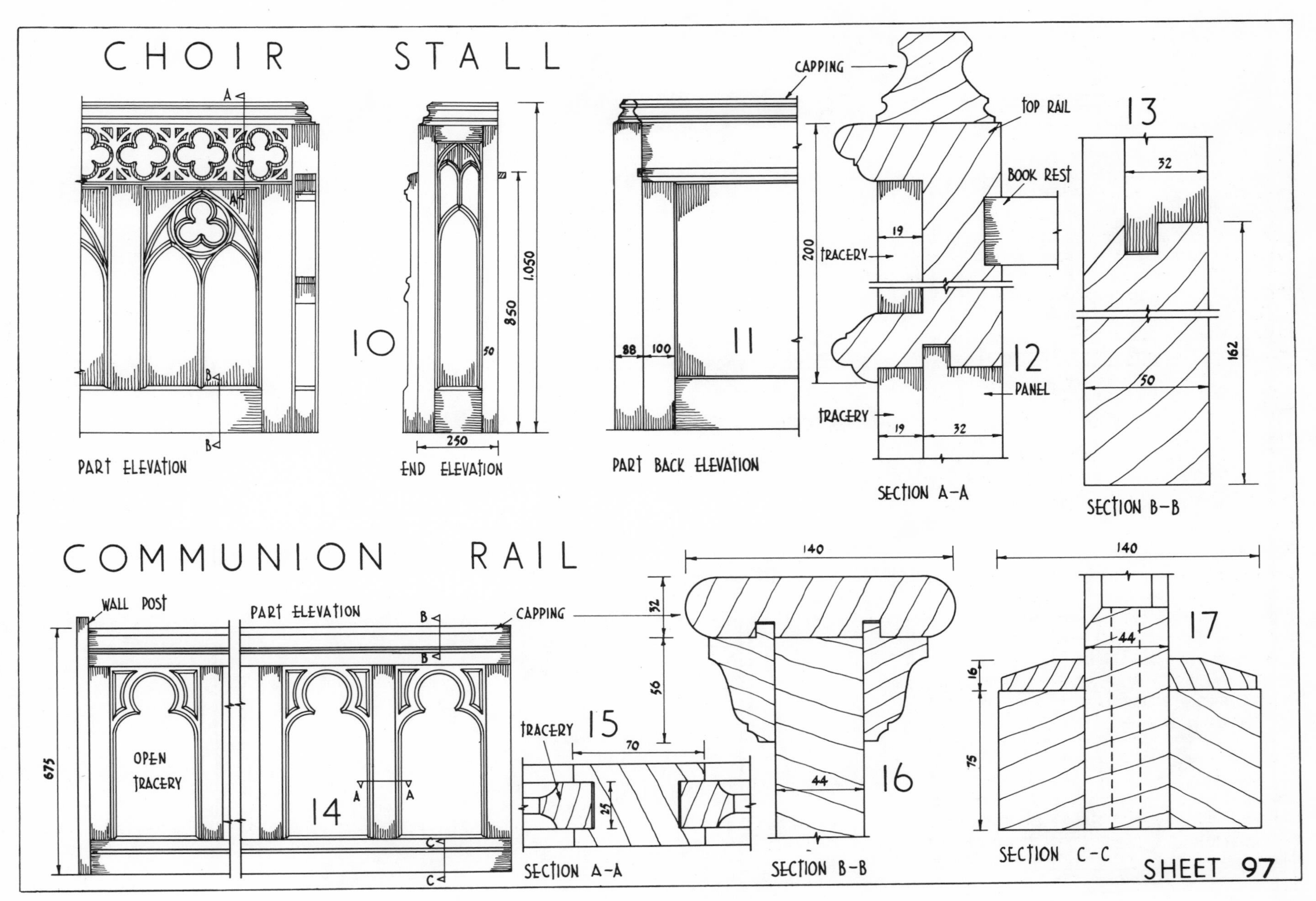
CHOIR STALL
CAPPING
TOP RAIL
BOOK REST
13
32
TRACERY
19
200
1,050
850
50
10
11
88
100
12
PANEL
TRACERY
19
32
162
50
250
PART ELEVATION
END ELEVATION
PART BACK ELEVATION
SECTION A–A
SECTION B–B
COMMUNION RAIL
140
140
WALL POST
PART ELEVATION
CAPPING
32
17
44
56
16
TRACERY
15
70
OPEN TRACERY
675
14
25
44
16
75
SECTION C–C
SECTION A–A
SECTION B–B
SHEET 97

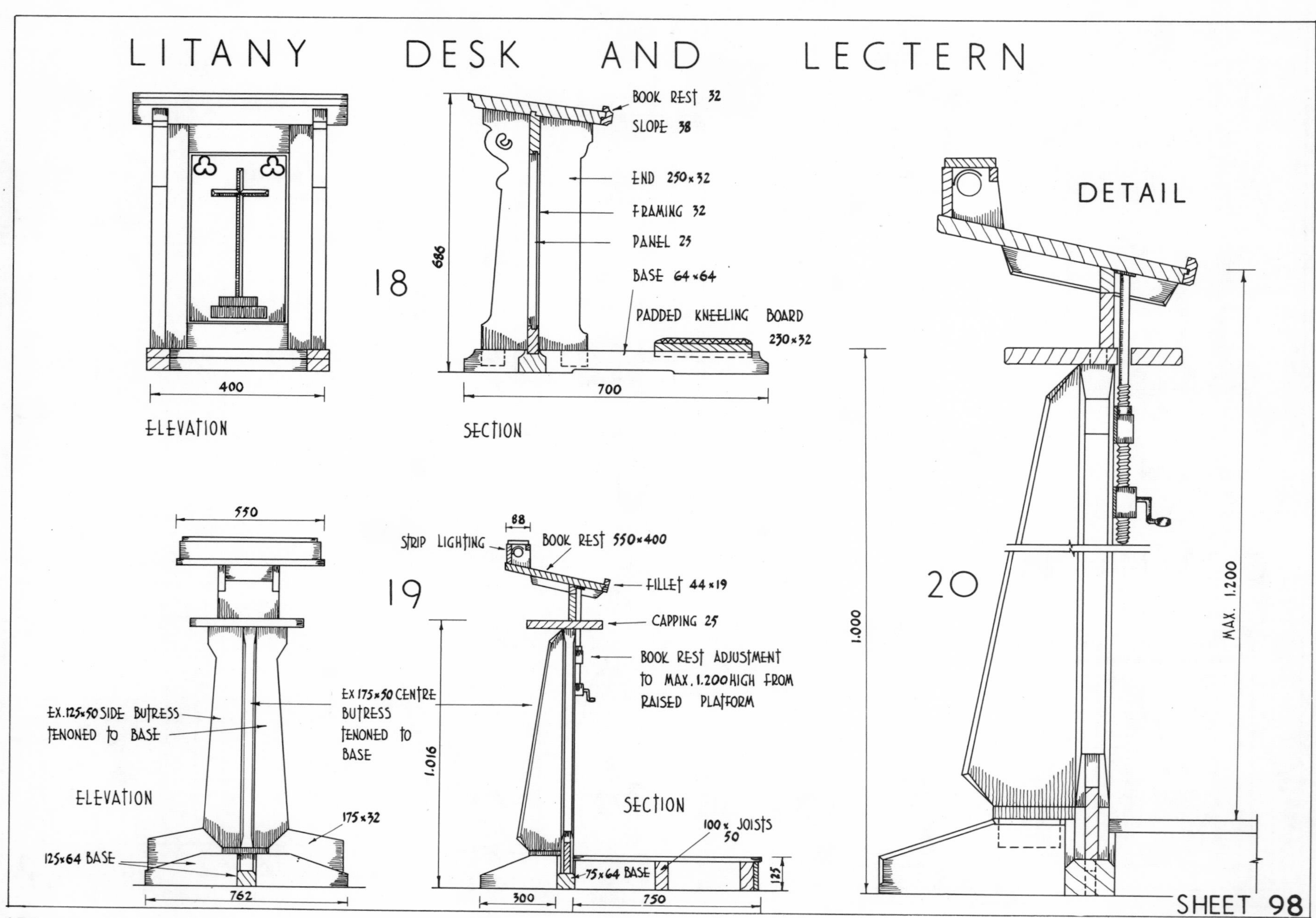
LITANY DESK AND LECTERN
BOOK REST 32
SLOPE 38
END 250×32
FRAMING 32
PANEL 25
BASE 64×64
PADDED KNEELING BOARD
230×32
686
18
400
700
ELEVATION
SECTION
DETAIL
550
88
STRIP LIGHTING
BOOK REST 550×400
FILLET 44×19
CAPPING 25
BOOK REST ADJUSTMENT
TO MAX. 1.200 HIGH FROM
RAISED PLATFORM
19
EX.125×50 SIDE BUTRESS
TENONED TO BASE
EX 175×50 CENTRE
BUTRESS
TENONED TO
BASE
1.016
ELEVATION
175×32
125×64 BASE
762
SECTION
100× JOISTS
50
75×64 BASE
125
300
750
20
1.000
MAX. 1.200
SHEET 98

Lectern

This piece of furniture is a desk designed to carry the Bible from which the lessons are taken by the reader when standing. Again they vary in design and are often made both in metal and wood, with the book-rest adjustable to suit individual requirements. In older churches this fitting is usually found to be elaborate and rich in ornamentation with carved figures to the pedestals and intricate tracery patterns between.

Sheet 98.19 shows the elevation and section of a modern lectern much simpler in design. The base is made in a tee section cut out of 125 mm by 64 mm mortised and tenoned together. Shaped centre and side buttresses are screwed together and tenoned to the base to form a centre pedestal. This carries the book-rest adjusting mechanism on the inside, and is finished with a capping as shown. Strip-lighting is provided to the sloping book-rest which has a tongued fillet along the bottom edge. It is usual to elevate the reader on a platform above the normal floor level. A 25 mm boarded platform on 100 mm joists is shown, the reader's feet being masked by a 175 mm by 32 mm board fixed to the base and side buttresses. The maximum height of the book-rest from the platform need not exceed 1.200 m.

Sheet 98.20 shows an enlarged detail of the lectern.

A familiar lectern in older churches is the 'Golden Eagle' in brass mounted on a heavily moulded shaft octagonal in section and carrying the Bible on its extended wings. The base of the shaft is also heavily moulded to the same shape to give the lectern balance.

Lecterns may have double-sided book-boards, that is, two sloping boards back to back forming two of the sides of an equilateral triangle. These are mounted on a shaft, either hexagonal or octagonal in section, with cap mouldings below the book-rests. The shaft is taken out of 125 mm square timber with sunk panels at each face. Heavy base mouldings are built up round the shaft at the bottom for the purpose of balance.

The triangular shaped end panels of the book-board may be solid or pierced to add further decoration. Book-rests are tongued into the face of the book-boards and screwed from the back side.

Screens

These comprise parclose, chancel and organ screens. The parclose screen is provided to separate a portion of the church. Usually they are very heavily moulded with heavy pierced tracery work moulded on both sides in the upper portion of the screen and panels below. As the tracery work is open tracery, the divided part is only partially screened.

The chancel screen is of similar design to the parclose, being a partial screen, tracery work in the upper portion with solid panels or panels formed of tongued and grooved boards below. The priests' stalls are often formed against the lower portion in chancel screens, facing the altar.

Sheet 99.21 shows the elevation of a screen 1.800 m high suitable for screening an organ or the entrance to a vestry. It consists of solid panels in a number of bays tongued into 32 mm framing. This is chamfered and prepared to receive the tracery work in the upper portion as shown.

At the open end of the screen the framed panelling is tongued into a 100 mm by 70 mm end post which is stop chamfered. The moulded plinth runs into this with the moulded capping mitred round it to finish the screen.

A quatrefoil design has been used for the tracery work to the screen, but in the elevation the details of the tracery have been omitted because of the small scale of the drawing.

A detail of part of this panel is shown in Sheet 99.22. As the panel is only narrow, the tracery is cut from the solid and moulded on one side with the corner carvings completing the decoration.

Sheet 99.23 shows a section through the moulded cappings and top rail. Sheet 99, figs. 24 and 25, sections through the panelled framing, end post and muntin. Sheet 99.26 shows a part section through the plinth.

The examples of church fittings which have been covered in this book are of a typical and simple character, the object being to explain the construction and uses rather than depict elaborate designs. It should be mentioned that, so far as this type of work is concerned, the finish varies from church to church. Often the furniture is left in its natural state to mellow over the years. It is common also to treat the surface with wax or to have it stained or fumed and polished.

Much carved work is to be found in many churches. The reredos is an example. This is positioned as a background to the altar and is elaborate and rich in ornamentation with carved figures the highlight. This type of work, of course, is the work of specialists who are carvers or sculptors, not joiners.

Bank Fittings

These comprise counters for cashiers and clerks, desks and screens. Traditional designs in this type of work, invariably done in hardwood, provide the highest class of joiner's work. While basic constructions remain, the designs of these fittings have changed in the post-war years with the introduction of new materials and applications.

Counters

A part elevation and section of a traditional bank counter is shown in Sheet 100, figs. 1 and 2. The top is made strategically wide for the cashiers to work at during business with the general public. It may be fitted with a protective

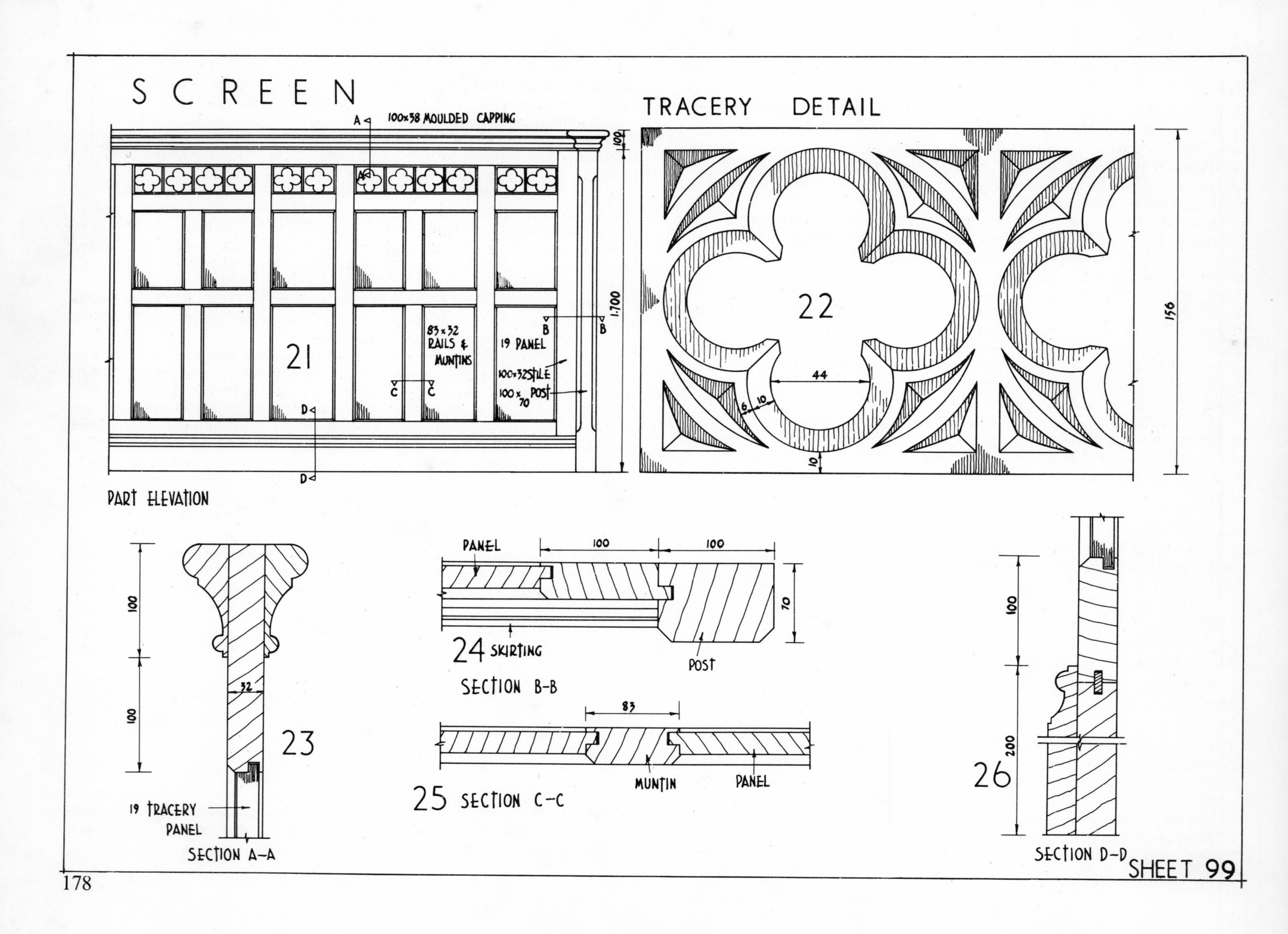

SCREEN
TRACERY DETAIL
100×38 MOULDED CAPPING
100
1.700
21
83×32 RAILS & MUNTINS
19 PANEL
100×32 STILE
100×70 POST
22
44
6
10
10
156
PART ELEVATION
100
100
32
23
19 TRACERY PANEL
SECTION A–A
PANEL
100
100
70
24 SKIRTING
POST
SECTION B–B
83
25 SECTION C–C
MUNTIN
PANEL
100
200
26
SECTION D–D
SHEET 99

grille to provide extra security, set in from the front and screwed to the top. There are generally no openings in bank counters, one end abuts against a wall, the other framed into a return counter or screen. The height of bank counters ranges from 900 mm to 1.050 mm.

The moulded counter top overhangs the front framing to allow the public to stand close without kicking the skirting. Supporting the overhanging top are shaped trusses 64 mm thick, reeded on the front edge to add further decoration, and fixed by screwing through the front framing from the rear.

The front panelling is made up in a series of frames with joints hidden by the moulded trusses. Bed moulds and moulded skirtings are cut between the trusses and fixed by screwing through the framing.

In the underfitting cash drawers are fitted immediately below the top and below these a series of cupboards and shelves, with flush doors either side-hung or sliding. A feature of the underfitting is the construction of the cash drawers. Since the drawers may have to hold considerable weights in cash, they must be robust and run freely. They are made of hardwood with 100 mm diameter removable till bowls and note divisions. Drawer sides 22 mm thick are used with patent ball-bearing runners screwed to them, the guides being fixed to the sides of the pedestals.

A section through a modern cashiers' counter and desk is shown in Sheet 100.3. The clerks' counter shown in Sheet 100.4 is placed 1.350 m clear on the staff side of the banking counter.

Blockboard veneered with all exposed edges lipped in hardwood is used for the construction. The basic framework is built up out of 75 mm by 50 mm softwood for the front with 75 mm by 25 mm hardwood carcass rails dove-tailed into the ends. The pedestal ends, shelves, bottom and front are of block-board with the horizontal members housed and glued. The top is of veneered blockboard or finished with plastic, with the nosing and lipped edges in teak and fixed by gluing and screwing to the carcassing.

A detail of the nosing and finish to the front edge of the counter top is shown in Sheet 100.5. The finish to the front of the counter shows leather, padded on a plywood backing, with a mitred teak surround glued and screwed through the blockboard as shown in Sheet 100.6. Veneered blockboard sliding doors are fitted with lipped edges in hardwood and running on fibre track.

The clerks' counter, Sheet 100.4, is constructed in a similar manner. A plate glass screen with teak surround and shelf behind is fixed to the counter top.

Cash drawers are similar to those already described for traditional bank counters.

All heading joints in the counter top are made with a counter cramp. The tops are grooved and cross tongued and are pulled together and held firmly in place by the cramps. These are screwed across the joints on the underside of the top with folding wedges driven to pull the joint up tight.

The detail of the cashiers' desk is shown in Sheet 100.7.

Service Counters

Almost every shop requires a counter of one kind or another. For small general businesses the simple framed structure providing storage shelves and a cash drawer below the top may be the most suitable. Specialist shops and stores require counters to display goods in the front, with shelves and cup-boards behind, and with the top arranged to display trays with compartments for small articles.

A type of counter suitable for a ladies' gown shop or a gents' outfitters is shown in Sheet 101.8. The design of this particular counter shows a centre portion with glazed top and front. The two side wings are panelled at the front in leather, with veneered blockboard tops in mahogany.

The centre portion is fitted with movable trays for quick service and the side wings with drawers, which are closed to the customers' side. The counter being raised on legs gives a sense of continuity to the floor space, which is a feature of modern shop and store design.

The 100 mm by 38 mm hardwood legs are tapered on the two inside edges and have 64 mm by 38 mm rails tenoned into them to form the underframe. The carcassing for the closed ends consists of 50 mm by 38 mm framing mor-tised and tenoned together and covered on the face side and ends with 25 mm blockboard. Sheet 101.9 shows the section through this portion of the counter with the drawers of varying depths, in position. It will be noticed that the fronts of the drawers are kept narrower than the sides and require no handles. These run on hardwood runners with guides, and are easily taken out for quick service. The fixing of the counter to the base is by screwing through the framing into the rails of the base.

Sheet 101.10 shows a section through the glazed centre portion of the coun-ter. This has an overlay of 19 mm laminboard, veneered on the top side, to form a base over the under frame. Bronze metal sections are used for the glazing of the counter, these are welded at the corners with the members fitting against the ends of the side wings fixed to them. A 25 mm veneered laminboard centre division is fixed to the base and the back top rail to carry the drawer runners in the centre.

A sketch of a glazed counter showcase is shown in Sheet 101.14 which may be fitted with glass shelves or movable trays. Hardwood rims used in this fitting are jointed with the horizontal members dovetailed together and a stub tenon on the vertical member mortised into them as shown in the detail, Sheet 101.12, the whole are then well glued. The completed joint viewed from the inside is shown in the sketch in Sheet 101.11. A more common and simpler joint is shown in the sketch, Sheet 101.13, where the horizontal members are mitred together with a stub tenon on the vertical member. The detail at the base of the case is shown in Sheet 101.15. A 25 mm veneered blockboard base is lipped with a hardwood edging strip which is screwed to the plinth with glue

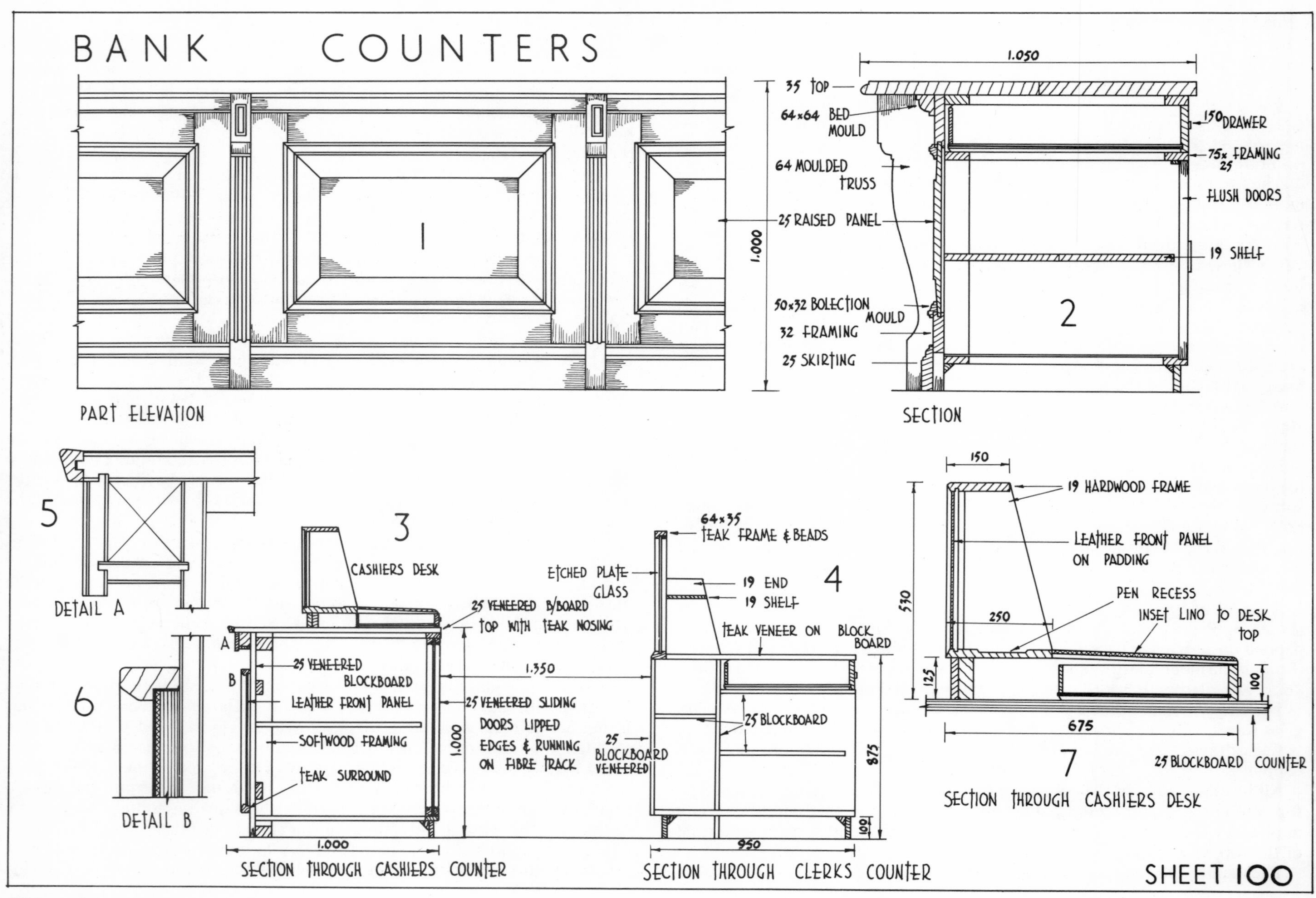
BANK COUNTERS
1
PART ELEVATION
1.050
35 TOP
64×64 BED MOULD
150 DRAWER
75×25 FRAMING
64 MOULDED TRUSS
FLUSH DOORS
25 RAISED PANEL
1.000
19 SHELF
50×32 BOLECTION MOULD
32 FRAMING
2
25 SKIRTING
SECTION
5
DETAIL A
6
DETAIL B
3
CASHIERS DESK
25 VENEERED B/BOARD TOP WITH TEAK NOSING
A
B
25 VENEERED BLOCKBOARD
LEATHER FRONT PANEL
SOFTWOOD FRAMING
TEAK SURROUND
25 VENEERED SLIDING DOORS LIPPED EDGES & RUNNING ON FIBRE TRACK
1.000
1.000
SECTION THROUGH CASHIERS COUNTER
64×35 TEAK FRAME & BEADS
ETCHED PLATE GLASS
19 END
19 SHELF
4
TEAK VENEER ON BLOCK BOARD
1.350
25 BLOCKBOARD VENEERED
25 BLOCKBOARD
875
100
950
SECTION THROUGH CLERKS COUNTER
150
19 HARDWOOD FRAME
LEATHER FRONT PANEL ON PADDING
530
PEN RECESS
250
INSET LINO TO DESK TOP
125
100
675
7
25 BLOCKBOARD COUNTER
SECTION THROUGH CASHIERS DESK
SHEET 100

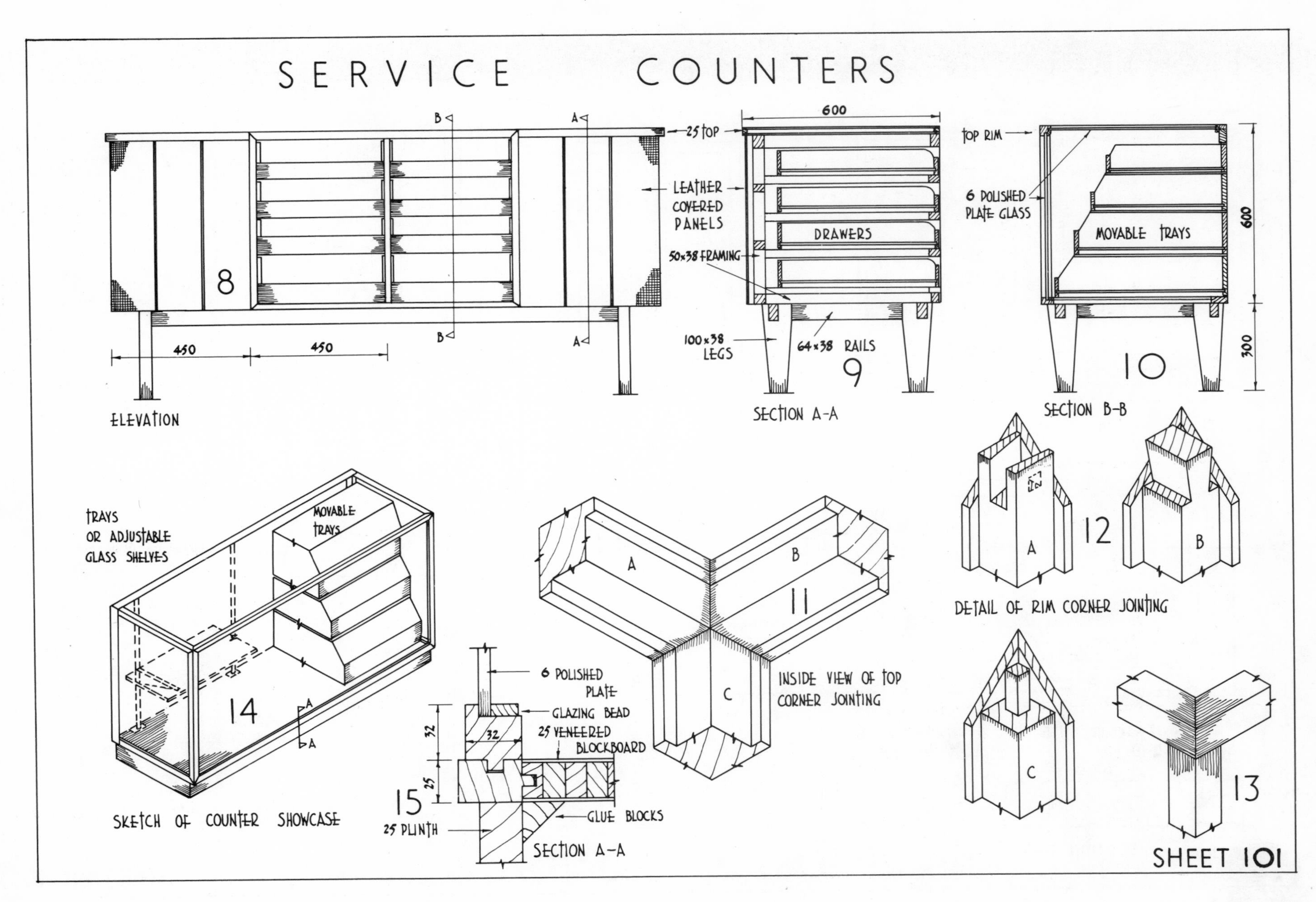

SERVICE COUNTERS
B
A
25 TOP
LEATHER COVERED PANELS
8
450
450
ELEVATION
600
50×38 FRAMING
DRAWERS
100×38 LEGS
64×38 RAILS
9
SECTION A-A
TOP RIM
6 POLISHED PLATE GLASS
MOVABLE TRAYS
600
300
10
SECTION B-B
A
12
B
DETAIL OF RIM CORNER JOINTING
TRAYS OR ADJUSTABLE GLASS SHELVES
MOVABLE TRAYS
14
SKETCH OF COUNTER SHOWCASE
A
B
11
C
INSIDE VIEW OF TOP CORNER JOINTING
6 POLISHED PLATE
GLAZING BEAD
25 VENEERED BLOCKBOARD
32
32
25
15
GLUE BLOCKS
25 PLINTH
SECTION A-A
C
13
SHEET 101

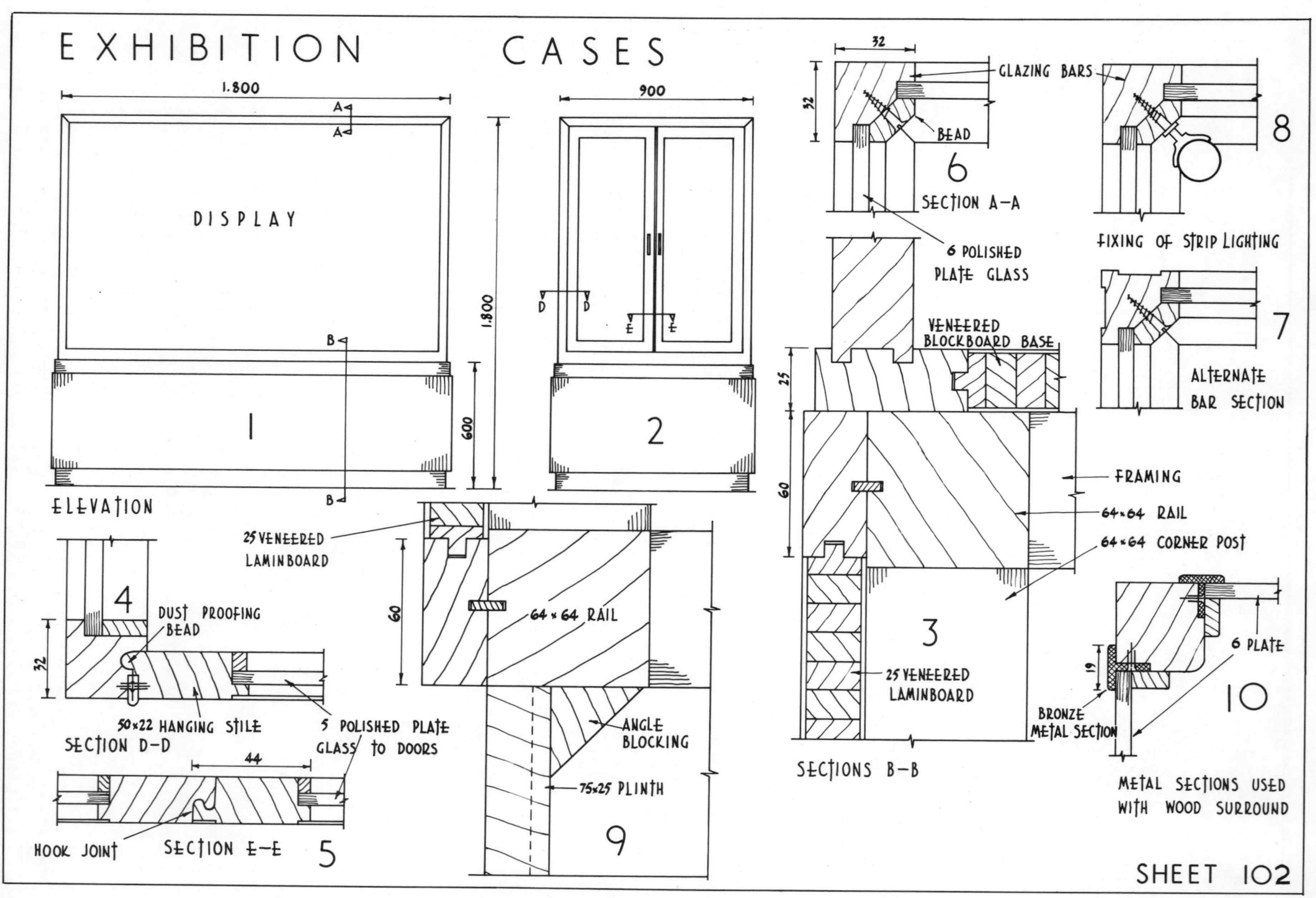
EXHIBITION CASES
1.800
900
DISPLAY
1
2
1.800
600
ELEVATION
A
B
D
E
32
GLAZING BARS
BEAD
6
SECTION A-A
8
FIXING OF STRIP LIGHTING
6 POLISHED PLATE GLASS
7
ALTERNATE BAR SECTION
VENEERED BLOCKBOARD BASE
25
60
FRAMING
64×64 RAIL
64×64 CORNER POST
25 VENEERED LAMINBOARD
3
SECTIONS B-B
6 PLATE
19
BRONZE METAL SECTION
10
METAL SECTIONS USED WITH WOOD SURROUND
4
DUST PROOFING BEAD
50×22 HANGING STILE
5 POLISHED PLATE GLASS TO DOORS
SECTION D-D
44
HOOK JOINT
SECTION E-E
5
25 VENEERED LAMINBOARD
64×64 RAIL
ANGLE BLOCKING
75×25 PLINTH
9
SHEET 102

blocking behind. The lower rim is tongued into the edging strip, as shown. The plinth is dovetailed at the corners with intermediate rails fitted to strengthen the base.

Exhibition Cases

Sheet 102.1 shows the front elevation and Sheet 102.2 the end elevation of a typical island display case suitable for the display of collections of works of art. The practice illustrated in this example is applicable to shop as well as museum cases. In this design, the case is mounted on a framed stand 600 mm high. This is a simple mortise and tenoned framework in softwood with corner posts which are connected by top and bottom rails tenoned into them and serves a double purpose, providing useful storage space as well as a base. The framework is then faced with 25 mm veneered laminboard edged as shown in Sheet 102.3, mitred at the vertical angles and tongued, glued and screwed to the framing. The detail of the recessed plinth is shown in Sheet 102.9. This is dovetailed at the corners and secured to the stand by pocket screwing and angle blocking.

Access to the case is provided by a pair of narrow door frames with hook-jointed meeting stiles. This arrangement is generally better than a single door frame as the plate glass panels in wood door frames impose a great strain on the mortise and tenon joints of the frames.

A method of forming the special joints, which are an essential feature of this class of work in order to prevent the entry of dust, is by the use of double rebates in the construction.

In airtight showcases the protection against the change of air is obtained by a system of beads and grooves. Sheet 102.4 shows a section through a vertical member of the frame which is grooved to receive the bead worked on the edge of the hanging stile of the door. The bead is arranged so that the hinge does not penetrate it when screwed in position. The jointing at the centre stiles of the doors is shown in Sheet 102.5, where a hook joint is used.

A further feature of this class of work is that the framework is reduced to the smallest possible dimensions that will keep the plate glass in position, so that an uninterrupted view of the contents may be had. This slenderness of frame requires that the joints at the corners between the rims shall be made with the greatest care and accuracy. Special methods are employed: one method is shown in Sheet 101.12, two members are jointed using a secret dovetail and the third stub-tenoned, with the surfaces mitred. The more common method is to mitre the two top rim members together and stub-tenon the corner vertical bar into these as shown in Sheet 101.13, the whole then being glued together.

Sheet 102.6 shows a section of a glazing member of the frame. This may be left square and rebated for the glass or prepared as Sheet 102.7 with a sinking on each face. The glass is bedded on putty, with a tight joint between the glass and the frame and held in position with beads screwed into the bars. The edges of the glass are usually painted black before glazing to stop any reflection from the cut edges. If tubular strip-lighting is to be fitted (Sheet 102.8), the beads will require grooving on the back side to accommodate the wiring for the lights.

A section through the base of the case is shown in Sheet 102.3. Veneered blockboard is tongued into an outer frame which is mitred and tongued at the corners and screwed from below. The lower member of the case is double tongued into the base frame as shown.

The detail of a corner bar with bronze metal tee sections forming the rebate for the polished plate glass is shown in Sheet 102.10. Metal now plays a large part in the construction of showcases, where bronze is the primary metal used. The angle sections are welded at the corners. Putty is used to bed the glass which is held in position by small metal blocks having a tap screwed into them.

Chapter Eleven: Electric Tools

Portable Electric Tools

The building industry has at its disposal a wide range of portable electric tools for use both in workshops and on the site. These tools have relieved the craftsman of much of the laborious work done by hand and speeded up the production of work, without affecting the skill of the craftsman.

Industry now demands greater productivity at the lowest possible cost and labour. This demands the full use of new techniques, mechanical aids and power tools, by all trades.

The tools available to the carpenter and joiner are electric drills, saws, planers, sanders, screwdrivers and routers.

Electric Drills

These are the most widely used of any power tools. They are made in three categories, light, general and heavy duty, in a variety of sizes and speed ranges. The 13 mm capacity drill is the most popular. It should be pointed out here that the capacity under which the drill is listed by the makers is half its maximum capacity in hardwood. The reason for this is that these tools were originally designed for use in the engineering trades and the rating based upon their capacity to drill steel. The alternative capacity of the drill in hardwood is usually included by the makers. Plate 1.1 shows a 16 mm capacity drill.

A wide range of accessories and attachments is available for use with the power drill. An ingenious attachment to a drill is the holesaw, which will cut clean round holes up to 100 mm diameter in any material a hacksaw will cut. A chisel mortising attachment is also available suitable for small mortises.

A drill stand is shown in Plate 1.2. The stand supports the full weight of the drill, freeing the hands for drilling accuracy and speed. The feed-arm leverage allows tremendous pressure to be exerted for high speed heavy-duty drilling.

Electric Saws

Plate 1.3. These are available in 150 mm to 300 mm diameter blade sizes. The most popular size for general use by the joiner are the 175 mm models with a cutting depth of 60 mm. They can be adapted by changing the blade to cut brick, tile, cement, asbestos, cast-iron guttering, stone, plastics and a wide variety of composite materials.

Saws are adjustable for angle cuts up to 45 degrees, such as occur at rafter ends, and for depth. They are fitted with a telescopic safety guard so designed as to cover the saw blade completely. As the guard is spring-loaded, it uncovers the blade as it enters the work and snaps back over the teeth on the completion of the cut. An outside fence works in the normal way and the saws may be used for ripping or cross-cutting as desired. Most of these saws are provided with a pointer to allow a marked cutting line to be easily followed.

Sabre Saws

An illustration of this versatile saw is shown in Plate 1.4. The chief use of the saw is in making internal cuts and intricate shapes in wood or metal. A variety of blades for different materials makes them adaptable for use by various trades.

When using the saw, the work should be cramped down as, by the cutting action, there is considerable tendency to vibration. This is due to the saw cutting on its up stroke which tends to pull the machine into the wood. Generally boring a hole through the wood is the most convenient way of starting the saw on internal panel cuts, although on thin material the saw can cut its own starting place.

Hardwood up to 50 mm thick can be cut, sheet metal, 13 mm diameter steel bars, aluminium, brass and asbestos sheeting or wall boards with the special blades.

Plate 1.5 shows a two-speed reciprocating saw. At high speeds timber up to 75 mm thick can be cut either cross-cutting or ripping. With a variety of blades available, composition materials, nails and screws can be cut. A wide range of metals, plastics, pipes, etc. can be cut at low speeds. The saw can be converted quickly into a jigsaw for shaped work, as well as for straight cutting and will cut flush up or down, left or right without additional attachments. The blades can be mounted in six different ways to operate in any cutting position.

Scrudrill

Plate 1.6. This machine can be used as a drill and a screwdriver by the adjustment of a collar. Where large numbers of screws have to be driven this

PORTABLE ELECTRIC POWER TOOLS

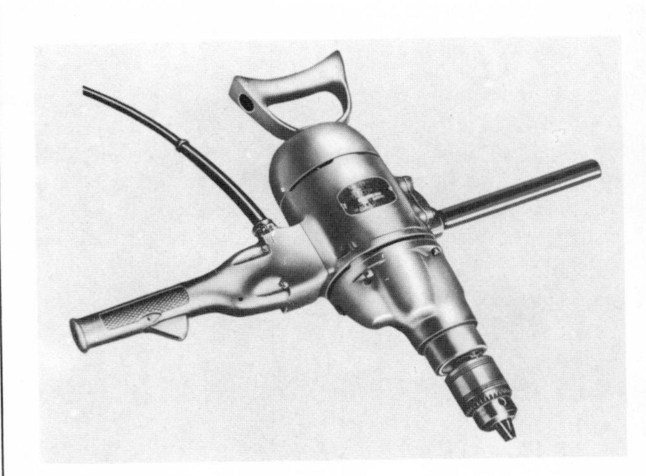

1 DRILL

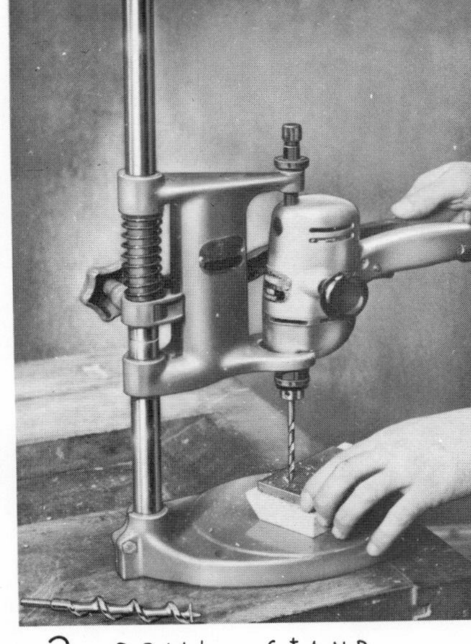

2 DRILL STAND

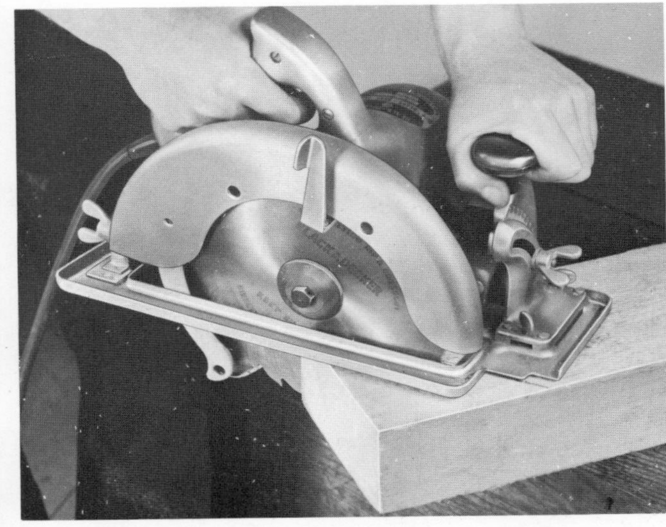

3 CIRCULAR SAW

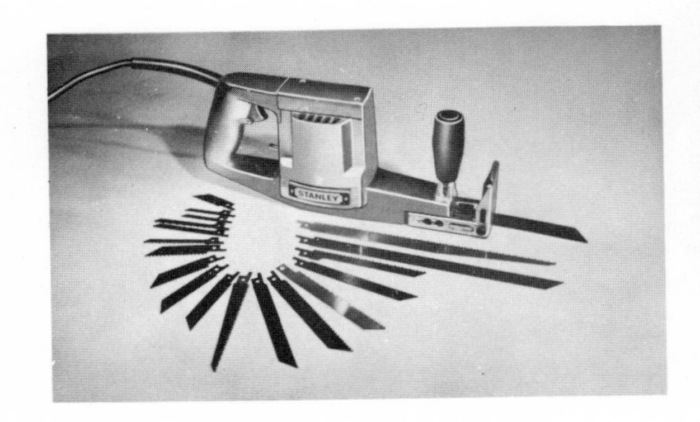

4 SABRE SAW

5 RECIPROCATING SAW

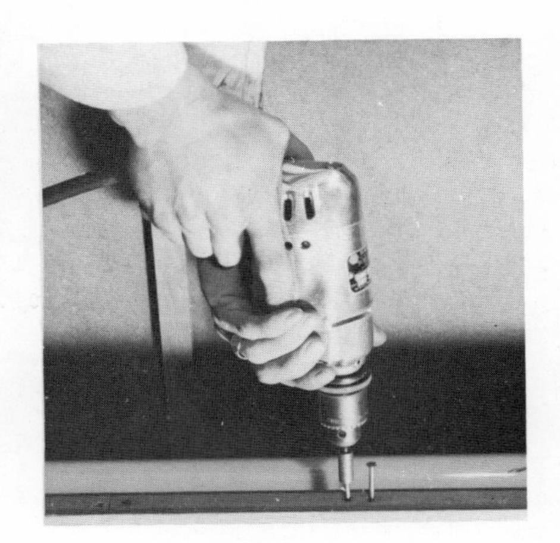

6 SCRUDRILL

PLATE I

PORTABLE ELECTRIC POWER TOOLS

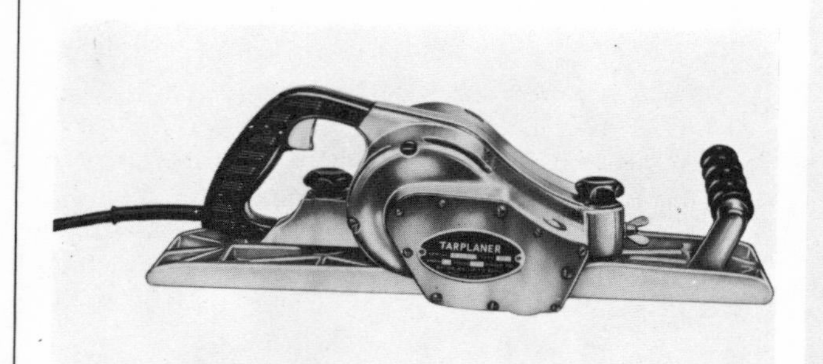

7 PLANER

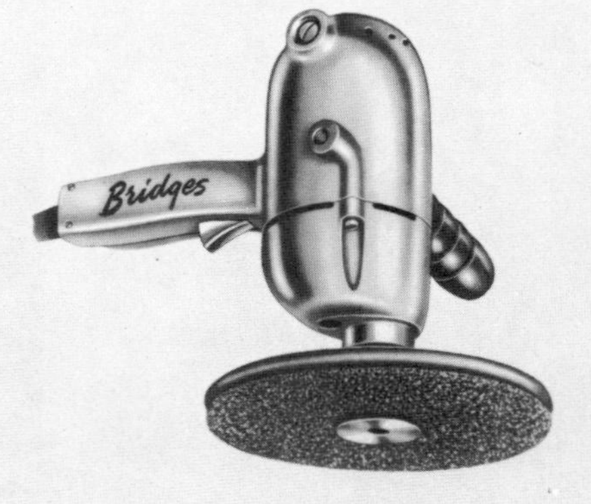

8 DISC SANDER

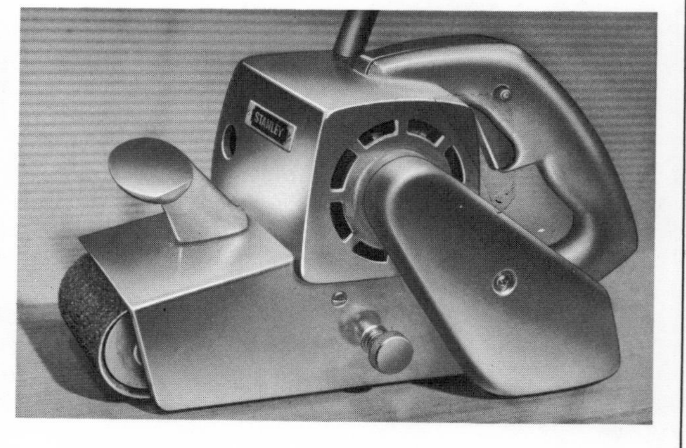

9 BELT SANDER

10 FINISHING SANDER

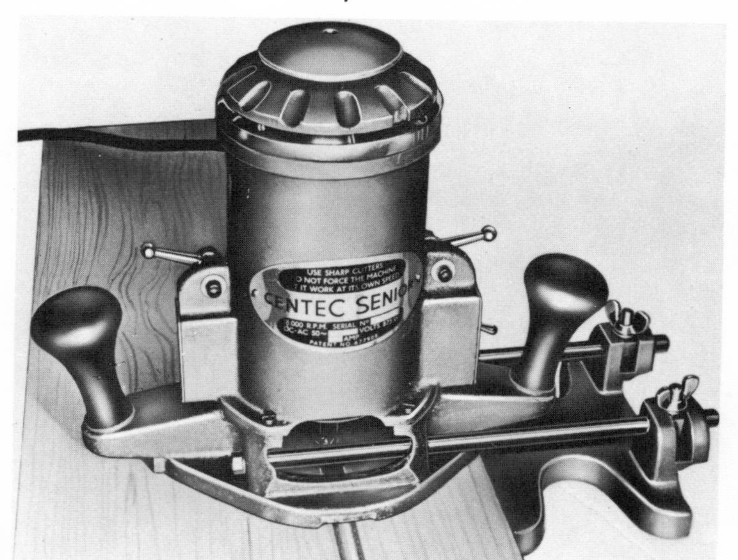

11 ROUTER

PLATE 2

tool is essential by reason of the speed with which the screws can be driven home. Pilot holes should always be made in the timber to take the shank of the screw. This avoids increasing the torque on the screwdriver and splitting of the timber.

Electric Planer

Plate 2.7. This machine combines in one handy portable tool the double advantages of machine-shop power and speed with the handiness of the jack plane.

The planer has ample power for hard woods, being fitted with 100 mm wide cutters. The depth of cut may be varied up to 3 mm by instant finger-screw adjustment and there is no limit to the width of surface it can plane. A hand grip with an automatic trigger switch ensures safety in operation. The high carbon chrome steel knives are easily removed for sharpening.

Whether planing with the grain, across the grain, or on end grain, the high speed cutters of the plane cut without tearing or break-away.

An adjustable fence is available for use when squaring edges of doors, window frames and similar work.

Portable Sanders

These are available in three different types, the disc sander, the belt sander and the orbital sander. Disc sanders range in size from 125 mm to 225 mm diameter, Plate 2.8. The head has a flexible rubber backing disc and the papers are fixed to this. The disc should not be used flat on the work, but at a slight angle with light pressure to prevent scoring of the surface of the work. Various accessories are available for use with this machine which include cup grinders, cup wire brushes and polishing pads.

Belt Sander

Plate 2.9. This machine is suitable for sanding large flat surfaces and consists of a continuous abrasive band fitted over rollers at each end of the machine. Sanding of framed joinery work with the grain can be accurately carried out and scratching at joints avoided. A dust collecting bag is fitted to the back end of the machine.

Orbital Sander

Plate 2.10. This machine is suitable for sanding flat or curved surfaces. The sanding paper is fitted to a flexible base and is ideal for finishing work as it reproduces at high speed the action of hand sanding in either the horizontal or vertical positions.

Electric Router

This machine is shown in Plate 2.11 and works on the same principle as the overhead-router. It is a most versatile machine and is used for moulding, rebating, grooving, etc. An adjustable fence is provided and the machine itself is adjustable for depth of cut.

Attachments, jigs and templets and a wide range of cutters are available for use with the router for shaped work, dovetailing and stair trenching. Plastic cutting and trimming using carbide-tipped blades can also be done.

Care and Use of Electric Power Tools

1. Check the nameplate voltage range of the tool with the voltage range of the supply line. Serious damage can result if supply line voltage does not fall within the ranges given.
2. An electric tool should always be earthed while in use. This protection will protect the operator against shock, should the tool develop an earth. Proper earthing is especially important where dampness is present or where abrasive dust is apt to create a short circuit.
3. Never drag the tool around by the cable and avoid any kinks or sharp bends.
4. Wear goggles in all abrasive tool operations.
5. Disconnect all electric tools from the supply point before making any adjustments.
6. Keep saws, blades, drills and cutters sharp and check each for tightness in the machine before starting.

Correct and adequate lubrication is the most important single factor in determining the life and service of any electric tool. All tools are properly lubricated at the factory and, under normal regular use, this lubrication will last for sixty days. Tools used constantly on production will need re-lubricating more often.

Bearings and gears contain sufficient grease for many months. Permanent lubricated bearings have sufficient lubrication packed in them at the factory to last the life of the bearings.

STEEL FORMWORK

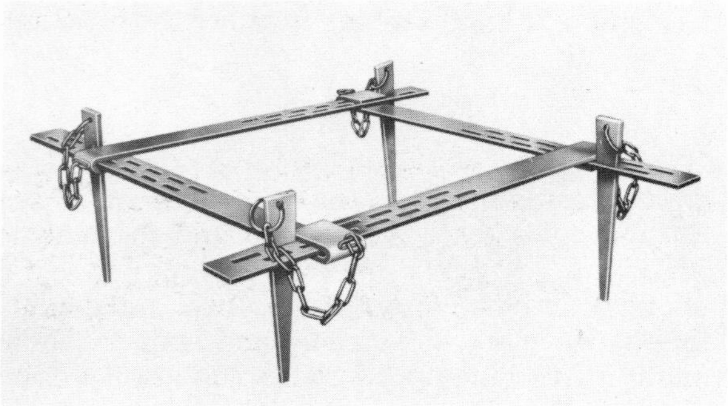

1 COLUMN CLAMPS

2

3 BEAM CLAMPS 4

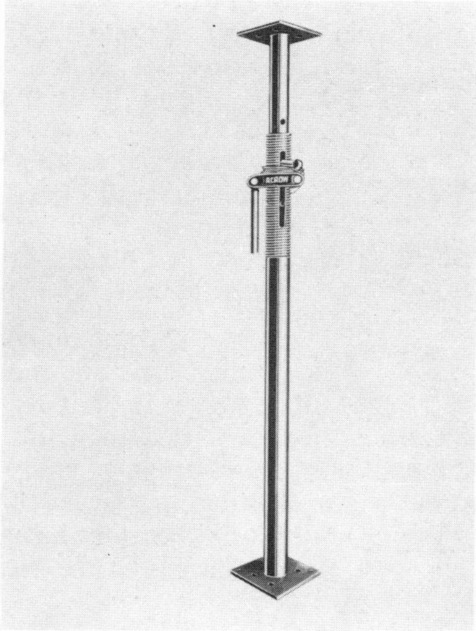

6 ADJUSTABLE PROP

5

PLATE 3

EXAMPLES OF GLUED LAMINATED WORK

1  2  3  4

LAMINATED PORTAL FRAMES

5

6 KAY BEAMS

7 DIAGONAL BOARDED PORTALS

8 BOWSTRING TRUSSES

PLATE 4

Index

INDEX

INDEX

INDEX